GRAPHICS PROCESSING UNIT-BASED HIGH PERFORMANCE COMPUTING IN RADIATION THERAPY

Series in Medical Physics and Biomedical Engineering

Series Editors: John G Webster, E Russell Ritenour, Slavik Tabakov, and Kwan-Hoong Ng

GRAPHICS PROCESSING UNIT-BASED HIGH PERFORMANCE COMPUTING IN RADIATION THERAPY

Edited by
Xun Jia
University of Texas Southwestern Medical Center
Dallas, USA

Steve B. Jiang
University of Texas Southwestern Medical Center
Dallas, USA

CRC Press
Taylor & Francis Group
Boca Raton London New York

CRC Press is an imprint of the
Taylor & Francis Group, an **informa** business

CRC Press
Taylor & Francis Group
6000 Broken Sound Parkway NW, Suite 300
Boca Raton, FL 33487-2742

First issued in paperback 2018

© 2016 by Taylor & Francis Group, LLC
CRC Press is an imprint of Taylor & Francis Group, an Informa business

No claim to original U.S. Government works

ISBN-13: 978-1-4822-4478-6 (hbk)
ISBN-13: 978-1-138-89432-7 (pbk)

Visit the Taylor & Francis Web site at
http://www.taylorandfrancis.com

and the CRC Press Web site at
http://www.crcpress.com

To all who made this book possible

Contents

Preface

G RAPHICS PROCESSING UNIT (GPU) is a specialized computer hardware originally designed to rapidly process graphics-related tasks, such as 3D view rendering in computer games. Due to their massive parallel processing capability, GPUs have been introduced into the scientific computing regime and are currently playing a critical role in the high-performance computing field. With their high processing power, cost-effectiveness, ease of deployment, access, and maintenance, GPUs have also attracted tremendous interest from the medical physics field in radiation therapy. Over the past several years, a number of exciting developments have been reported in a wide spectrum of problems, ranging from computed tomography reconstruction to Monte Carlo radiation transport simulation.

This book presents a collection of state-of-the-art research on GPU computing and its applications in medical physics problems in radiation therapy. After an overall introduction to GPU technology and its current applications in radiotherapy in Chapter 1, each of Chapters 2 through 20 was selected to present a specific application of GPU in a key radiotherapy problem. These chapters summarize the advancements achieved so far as well as present technical details and insightful discussions regarding the utilization of GPU in these problems. In addition, two chapters toward the end of the book present two real systems developed with GPU as a core component to achieve important clinical tasks in modern radiotherapy.

We hope this book will be of general interest to clinical and research physicists, graduate students, and other researchers who would like to learn about GPU computing for radiotherapy. We also hope it will inspire new and exciting developments that will be eventually translated to clinical practice to advance the radiotherapy technology. Finally, we sincerely thank all the outstanding contributing authors, without whom this work would not have been possible.

Editors

Dr. Xun Jia earned his MS in mathematics in 2007 and PhD in physics in 2009, both from the University of California, Los Angeles. After he received his postdoctoral training in medical physics from the Department of Radiation Medicine and Applied Sciences, University of California, San Diego, Dr. Jia joined the faculty team as an assistant professor in the same department in 2011. He then moved to Dallas, Texas, in 2013 and is currently an assistant professor and medical physicist at the Department of Radiation Oncology, University of Texas Southwestern Medical Center. Over the years, Dr. Jia has conducted productive research on developing numerical algorithms and implementations for low-dose cone-beam CT reconstruction and Monte Carlo radiation transport simulation on the GPU platform. His research has received federal and industrial funding support, as well as support from private funding institutions. Dr. Jia has published over 60 peer-reviewed research articles. He is currently serving as the section editor of the *Journal of Applied Clinical Medical Physics*.

Dr. Steve B. Jiang received his PhD in medical physics from the Medical College of Ohio in 1998. After completing his postdoctoral training at Stanford, he joined Massachusetts General Hospital, Harvard Medical School, in 2000 as an assistant professor of radiation oncology. In 2007, Dr. Jiang was recruited by the University of California, San Diego, as a tenured associate professor to build the Center for Advanced Radiotherapy Technologies, for which he was the founding and executive director. He was then promoted to full professor with tenure in 2011. In October 2013, Dr. Jiang joined the University of Texas Southwestern Medical Center as a tenured full professor, Barbara Crittenden Professor in cancer research, vice chair of the Radiation Oncology Department, and director of the Medical Physics and Engineering Division. He is a fellow of the Institute of Physics and the American Association of Physicists in Medicine. He currently serves on the editorial board for the *Physics in Medicine* *and Biology* journal and is an associate editor for the *Medical Physics* journal. Dr. Jiang's research in various areas of cancer radiotherapy has been funded by federal, charitable, and industrial grants for more than 10 million dollars, resulting in more than 130 peer-reviewed papers. He has supervised 24 postdoctoral fellows and 10 PhD students.

Contributors

Mingli Chen
Advanced Imaging Research
    Center
The University of Texas
    Southwestern Medical Center
Dallas, Texas

Rebecca Fahrig
Radiological Sciences Laboratory
Department of Radiology
Stanford University
Stanford, California

Michael M. Folkerts
Department of Physics
University of California, San Diego
La Jolla, California

Hao Gao
Department of Mathematics
School of Biomedical Engineering
Shanghai Jiao Tong University
Shanghai, People' Republic of
    China

Quentin Gautier
Department of Computer Science
University of California, San Diego
La Jolla, California

Yan Jiang Graves
Mobius Medical Systems, LP
Houston, Texas

Xuejun Gu
Department of Radiation
    Oncology
The University of Texas
    Southwestern Medical Center
Dallas, Texas

Minghao Guo
School of Biomedical Engineering
Shanghai Jiao Tong University
Shanghai, People's Republic of
    China

Sungsoo Ha
Visual Analytics and Imaging
    Laboratory
Department of Computer Science
Stony Brook University
Stony Brook, New York

and

Department of Computer Science
SUNY Korea
Songdo, South Korea

Sami Hissoiny
Department of Research
Elekta
Maryland Heights, Missouri

Robert A. Jacques (Deceased)
Department of Radiation Oncology
Johns Hopkins University
Baltimore, Maryland

Xun Jia
Department of Radiation
    Oncology
The University of Texas
    Southwestern Medical Center
Dallas, Texas

Steve B. Jiang
Department of Radiation
    Oncology
The University of Texas
    Southwestern Medical Center
Dallas, Texas

Soyoung Lee
J. Crayton Pruitt Family
    Department of Biomedical
    Engineering
University of Florida
Gainesville, Florida

Ruijiang Li
Department of Radiation Oncology
Stanford University
Stanford, California

Yifei Lou
Department of Mathematical
    Sciences
University of Texas at Dallas
Richardson, Texas

Weiguo Lu
Department of Radiation
    Oncology
The University of Texas
    Southwestern Medical Center
Dallas, Texas

Andreas Maier
Pattern Recognition Laboratory
Department of Computer Science
Friedrich-Alexander University
    Erlangen-Nuremberg
Erlangen, Germany

Todd R. McNutt
Department of Radiation
    Oncology
Johns Hopkins University
Baltimore, Maryland

Chunhua Men
Department of Research
Elekta
Maryland Heights, Missouri

Klaus Mueller
Visual Analytics and Imaging
    Laboratory
Department of Computer Science
Stony Brook University
Stony Brook, New York
and
Department of Computer Science
SUNY Korea
Songdo, South Korea

Wooseok Nam
Department of Medical Physics
Sunnybrook Health Sciences Centre
Toronto, Ontario, Canada

Eric Papenhausen
Visual Analytics and Imaging
    Laboratory
Department of Computer Science
Stony Brook University
Stony Brook, New York

Justin C. Park
Department of Radiation
    Oncology
University of Florida
Gainesville, Florida

Fei Peng
Department of Computer Science
Carnegie Mellon University
Pittsburgh, Pennsylvania

Jean-Marie Rocchisani
Unité de Formation et de
    Recherche de Santé, Médecine
    et Biologie Humaine
Sorbonne Paris Cité
Université Paris 13
and
Assistance Publique-Hôpitaux
    de Paris
Bobigny, France

H. Edwin Romeijn
H. Milton Stewart School of
    Industrial and Systems
    Engineering
Georgia Institute of Technology
Atlanta, Georgia

Sanjiv S. Samant
Department of Radiation
    Oncology
University of Florida
Gainesville, Florida

Sonja S.A. Samant
Duke Global Health Institute
Duke University
Durham, North Carolina

Bongyong Song
Department of Radiation Medicine
    and Applied Sciences
University of California, San Diego
La Jolla, California

William Y. Song
Department of Medical Physics
Sunnybrook Health Sciences
    Centre
and
Department of Radiation
    Oncology
University of Toronto
Toronto, Ontario, Canada

Allen Tannenbaum
Departments of Computer Science
    and Applied Mathematics/
    Statistics
Stony Brook University
Stony Brook, New York

Zhen Tian
Department of Radiation
    Oncology
The University of Texas
    Southwestern Medical Center
Dallas, Texas

Andres Uribe-Sanchez
Center for Advanced
    Radiotherapy Technologies
and
Department of Radiation
    Medicine and Applied Sciences
University of California, San
    Diego
La Jolla, California

Franck P. Vidal
School of Computer Science
and
United Kingdom and Research
    Institute of Visual Computing
Bangor University
Bangor, United Kingdom

Xiaoyu Wang
Center for Advanced
    Radiotherapy Technologies
and
Department of Radiation
    Medicine and Applied Sciences
University of California,
    San Diego
La Jolla, California

Lei Xing
Department of Radiation Oncology
Stanford University
Stanford, California

Wei Xu
Visual Analytics and Imaging
    Laboratory
Department of Computer Science
Stony Brook University
Stony Brook, New York
and
Computational
Science Center at Brookhaven
    National Lab
Upton, New York

Hao Yan
Department of Radiation
    Oncology
The University of Texas
    Southwestern Medical Center
Dallas, Texas

Masoud Zarepisheh
Department of Radiation Oncology
School of Medicine
Stanford University
Stanford, California

Xin Zhen
Department of Biomedical
    Engineering
Southern Medical University
Guangzhou, People's Republic of
    China

Ziyi Zheng
Visual Analytics and Imaging
    Laboratory
Department of Computer Science
Stony Brook University
Stony Brook, New York

CHAPTER 1

Introduction

Xun Jia and Steve B. Jiang

CONTENTS

1.1 INTRODUCTION

R ADIATION THERAPY is a form of cancer treatment that uses ionizing radiation to kill cancer cells. The success of modern radiation therapy heavily relies on computations. With the recent advancements in imaging and treatment technologies, it is faced with two conflicting requirements. They are the invention of novel algorithms for certain clinical purposes and the challenge to solve these problems arising from computations efficiently for clinical applications.

Advancing radiotherapy practice is the driving force for the continuous development of novel algorithms, which are usually associated with increasing complexity and size. For instance, driven by the desire to reduce imaging dose and improve image quality, cone-beam computed tomography (CT) [1,2] reconstruction has been evolving from the conventional analytical reconstruction method to the more advanced iterative one [3,4]. The complexity of each iteration step is comparable to that of the conventional

Feldkamp-Davis-Kress (FDK)-type reconstruction algorithm [5]. In recent times, the problem is further extended to the 4D imaging context [6,7] following the demands for visualizing and managing respiratory motion. The additional fourth dimension increases the problem size by a factor of ~10 compared with conventional 3D imaging techniques [8–11].

The other side of the story is the necessity of achieving sufficiently high efficiency, such that these advancements demonstrated in the labs can be translated to patient care in clinical practices where time is critical. In fact, while a number of encouraging developments have been constantly reported with great potentials to transform radiotherapy, the required high computational speed is often one of the main factors that impair their clinical applications. Hence, developing computational technologies to allow these novel algorithms in a routine environment is as important as developing the algorithms themselves.

During the last decade, the performance of computers has experienced a rapid growth resulting in advanced and a large number of processors. For years, the computer processor industry continuously delivered higher clock speed, which directly translated into higher processing power. This is a convenient way to achieve faster computations for the user as a code previously written for an old and slow central processing unit (CPU) can also run on a new and fast processor with no or little modifications. Yet due to the limitation of fabrication technology and the concern of power dissipation, multi-core or many-core chip designs have become popular. Although the processing power of one core on this new system is the same, or sometimes even lower than that of a conventional single-core processor, the cumulative power achieved in the system can be higher. One example is the graphics processing unit (GPU). A GPU integrates thousands of processing units on a single chip. It serves as a coprocessor on a computer system, which enables the CPU to off-load those computationally intensive tasks to the GPU and achieve a high processing efficiency.

Conventionally, problems ported to the GPU side were computer graphics related. For example, 3D scenes rendered in real time for computer game purposes in which highly realistic 2D images are generated based on 3D models according to light transport physics and human visual perception. The pursuit for more and more realistic images at a high frequency in computer games is actually the main driving force for advancements in the GPU technology. Over the past few years, this large consumer market has led to a steady increase in the processing power of GPUs, as clearly shown in Figure 1.1.

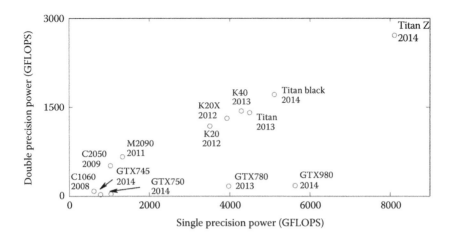

FIGURE 1.1 Double-precision peak performance in GFLOPS versus single-precision peak performance for several Nvidia GPU cards.

It did not take long for computer scientists to realize that GPU is not only a graphics processing accelerator but also a supercomputer that can potentially benefit scientific computation field as well. Soon after its birth, GPU was introduced to the scientific computing regime. Today, it has been widely utilized to attack challenging problems in a variety of different fields, such as physics, mathematics, chemistry, biology, and finance to name a few. Of particular interest is the fact that, as of November 2014, 53 supercomputers among the world's top 500 have used GPUs as coprocessors to boost their performance, indicating the solid role of GPU in the scientific computing regime [12].

In the field of radiation therapy, the utilization of GPU experienced a boom since 2006 [13,14]. This occurred as a combined consequence of the great demand of computations in radiotherapy, the vast development of GPU technology for scientific computation, and the advantages of GPU for radiotherapy clinical applications. Figure 1.2 presents the number of publications in two main journals in radiotherapy, *Medical Physics* and *Physics in Medicine and Biology*, which have GPU as a keyword. These publications reported a number of algorithms developed for GPU-based parallel computation, yielding dramatically accelerated processing speeds in a wide spectrum of problems. The trend of an increasing number of publications clearly reflects the researchers' interest in this novel technology. On the application side, GPU has been gradually adopted to the routine clinic by major vendors. Tomotherapy has replaced its traditional

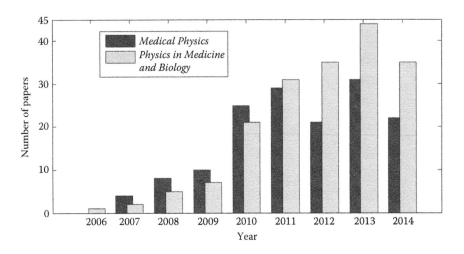

FIGURE 1.2 Number of GPU-related publications in *Medical Physics* and *Physics in Medicine and Biology* in recent years. Data obtained by searching the keywords "GPU" or "Graphics Processing Unit" in the two journals. Data for Year 2014 are for 11 months at the time of generating this figure.

56-core CPU cluster with GPUs to solve treatment planning problems. Both Elekta and RaySearch have utilized GPU in their products for dose calculations and inverse planning. Until now, using GPU has become a common approach to accelerate computation tasks in medical physics.

In light of the rapid growth of GPU-based high-performance computing technology and its utilizations in radiotherapy, this book is a comprehensive review of recent developments of the GPU technology in the field of radiotherapy. It is hoped that this book will offer a brief of the current utilization of GPU in many important clinical problems, and discuss not only its advantages but also its challenges. Knowing the limitations of this novel technology and potential solutions will be valuable for a better use of the GPU in radiotherapy practice. In addition, we have been developing computation algorithms and techniques primarily in the frame of serial processing over the years. However, GPU-based high-performance computing offers an easy way to realize parallel processing, which opens a new window for us to rethink about algorithm designs and implementations. Therefore, we also hope the book will inspire new developments in parallel computation regime, which will yield a substantial increase in computation speed to eventually bring those novel numerical techniques developed in research labs into radiotherapy clinical practice.

1.2 GPU HARDWARE STRUCTURE AND PROGRAMMING

1.2.1 GPU Hardware Structure

Figure 1.3a illustrates a typical hardware configuration of a GPU card. Being a coprocessor, GPU is typically plugged onto the motherboard of a desktop computer through its PCIe bus. Within a modern GPU card, there are hundreds or even thousands of processing units called stream processors. These processors are physically grouped into a set of single-instruction multiple data (SIMD) multiprocessors. SIMD refers to the mode that GPU executes codes in a parallel processing fashion, which will be presented later. Although the clock speed of each processor is relatively low compared to a modern CPU, the large number of processors available on a GPU card collaboratively leads to a much higher computational power.

Memory is another important component on a GPU card. Generally speaking, there are global memory, shared memory, and registers on a GPU card. In contrast to CPU-based programming in which one can obtain a descent performance without considering detailed memory structure, it is very important to use different memory spaces on GPU wisely according to their characteristics to achieve a high efficiency.

Global memory is the largest memory space on the card (up to several gigabytes nowadays) and is accessible to all the processors. It is used to store most of data in a GPU program. Global memory is also visible to the computer on which the GPU is installed. Hence, it is also used as the buffer space for CPU to transfer data from/to GPU. However, the memory

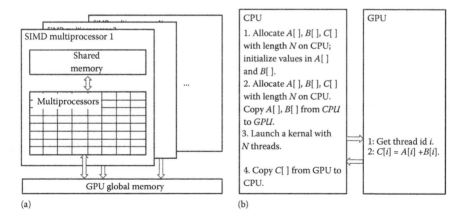

FIGURE 1.3 (a) Illustration of hardware structure of a GPU. (b) Execution path for a sample problem of vector addition.

bandwidth between the stream processor and the global memory is relatively low. Visits to the global memory should be minimized. Within the global memory space, one can allocate three different types of memory: linear memory, arrays, and constant memory. A linear memory space is the most commonly used one akin to the linear space in CPU programming. An array on the GPU is allocated and initialized from the CPU side. After that, it is bound to the so-called texture memory. During GPU code execution, the texture memory is read only by all stream processors. Its advantages include multidimensional spatial locality cache that effectively improves data access efficiency, as well as hardware-supported linear interpolations. Constant memory is also allocated and initialized from the CPU side. They are readable from each stream processor with cache supported.

On each SIMD multiprocessor, there is a relatively small memory space (e.g., 16 kB for most Nvidia GPU cards) called shared memory. It is accessible to all the processors inside the multiprocessor, but not to those on other multiprocessors. Because of the close proximity to the processors, data access to the shared memory is much faster than direct access to the global memory. Therefore, shared memory usually behaves as a user-managed cache space between each processor and the global memory to effectively improve data access speed.

Finally, each stream processor has a certain amount of registers. They are the space needed for the code execution in each particular GPU processor. Visiting the registers is also fast. Since the total amount of registers is limited, one has to carefully utilize them, for example, by reusing defined variables.

1.2.2 GPU Programming Mode

As mentioned previously, GPU executes a program in an SIMD fashion. Specifically, the code that is launched on the GPU side is called a kernel. When this part of the code is executed, GPU actually launches multiple copies of that with the number of copies defined by the programmer. Each of the copies is called a GPU thread. Furthermore, these threads are grouped into warps with 32 threads in a warp. The multiprocessors of a GPU then enumerate these warps to perform calculations. Within a warp, all the threads march along the same instruction path but possibly using different data sets. If the program leads to different instructions to those threads within a warp, for instance due to a condition statement on the thread index, a thread divergence problem occurs. In this situation, the warp runs through each branch path in a sequential order. Apparently, this thread divergence issue should be minimized to take full advantages of GPU's parallel processing capability.

We present in Figure 1.3b a typical code structure for GPU programming using a simple vector addition problem as an example. At the first step of this code, the CPU initiates the program and allocates spaces for three vectors, $A[]$, $B[]$ and the result $C[]$ in the CPU memory. It also initializes the values of A and B to be added. In Step 2, corresponding memory spaces for the three vectors are allocated on GPU, and the values of A and B are copied to the corresponding GPU counterparts. At this moment, a GPU kernel can be launched in Step 3 to actually perform the addition. In contrast to a CPU-based computation where only one or a few threads are utilized, the GPU kernel executes with a number of N threads, so each thread only performs calculation for one vector element. Specifically, the CPU turns control to GPU once the kernel starts. Each thread obtains its index, calculates $A[i] + B[i]$, and puts the result to $C[i]$ in the GPU's memory. On a modern GPU card, there are thousands of threads to perform this calculation concurrently. This step is fully agreed with the aforementioned SIMD programming mode, yielding a high parallel processing efficiency. After the kernel execution, GPU returns the control to CPU. The latter copies the resulting vector C from GPU to CPU memory.

1.2.3 GPU Programming Languages

Along with the vast advancements of GPU hardware, the associated programming languages or application programming interfaces (APIs) have also been experiencing a rapid development.

GPU was originally developed for rendering graphics, which is still the main focus of GPU nowadays. At the early stage of GPU programming for scientific computations, a scientific problem of interest was programmed through a graphics pipeline, where the algorithm was reformulated as a graphical rendering task. Program can be written in high-level language such as C for graphics (Cg) developed by Nvidia in collaboration with Microsoft and high-level shader language (HLSL) by Microsoft. Although pioneering studies achieved impressive accelerations, this approach encountered difficulties of mapping algorithms to graphical problems, tediousness of programming and poor program portability.

The great desire to maximally utilize the potential of GPUs led to modern GPU programming APIs with several nice features. (1) *Friendly interface.* Many of the APIs have offered interfaces with commonly used programming languages, for example, C, FORTRAN, MATLAB®, and Python.

This greatly reduces the learning curve for a new GPU programmer. It also facilitates the integration of GPU-based computations in other computation systems. (2) *Programmability.* Advanced APIs enable a user access to different capabilities of the GPU if needed, such as shared memory. One can, therefore, fully utilize GPU's power to achieve a high efficiency. (3) *Program portability.* Portability has been a major concern for GPU-based scientific computing. It is desirable to develop computer codes that run on different hardware structures, as opposed to only on one specific structure. Poor portability is actually a hurdle for clinical adoptions of GPU technology. Take medical physics field as an example, software vendors are sometimes reluctant to tie their product to only one specific GPU vendor.

Over the years, a set of modern APIs have been developed to facilitate GPU-based scientific computing. Compute unified device architecture (CUDA) developed by Nvidia to support the programming of its own GPUs is currently the most popular API. It provides developers direct access to the instruction set of GPUs through an interface in typical programming languages, such as C, and FORTRAN. The popularity of CUDA also comes from the available scientific libraries developed either by Nvidia or by the community, such as CURAND for random number generator, CULAPACK for linear algebra operations, and CUFFT for fast Fourier transform. Using these libraries substantially facilitates developing high-quality packages for scientific computations. Nvidia has also formed a community called CUDA zone to actively support the developments of CUDA and its applications in high-performance computing problems.

Another emerging API is open computing language (OpenCL) maintained by the nonprofit technology consortium Khronos Group. In contrast to CUDA, OpenCL enables cross-platform compatibility. It supports conventional CPUs, GPUs from different vendors, as well as heterogeneous platforms consisting of both CPUs and GPUs. The development in the OpenCL regime is not as advanced as in CUDA. For instance, the support for scientific computing such as libraries is relatively rare. Because of the focus on cross-platform portability, the efficiency of OpenCL may be compromised compared to CUDA.

1.3 ADVANTAGES OF GPU

While GPU has gathered a lot of research interests in a variety of scientific fields over the past several years, it is particularly attractive for medical physics problems in radiotherapy [13,14]. The apparent reason is that GPU offers an appropriately high level of computation power at a relatively low cost. As a matter of fact, most radiotherapy problems can be solved via a rapid parallel processing scheme. For instance, deformable registration calculations can be easily parallelized by using one GPU thread to compute the deformation vector at a voxel [15,16]. Superposition–convolution or pencil beam-type dose calculations can also be achieved at this voxel parallelization level [17–19]. These naturally existing parallelization schemes are compatible with GPU's SIMD processing structure, making GPU a favorable platform to accelerate these problems. On the cost side, with the continuous demands from computer game industry, GPU is extremely cost-effective: orders of magnitudes lower in price than a CPU cluster with a similar processing power.

Not only does GPU provide a high processing power, it actually offers the appropriate level of computation power that is commensurate with the problem size in radiotherapy. In fact, the size of a clinical problem in radiotherapy is usually at an intermediate scale, typically much smaller in size than those challenging problems in fundamental sciences, for example, in fluid dynamics. GPU provides an appropriate amount of power to overcome the computational burden using a centralized hardware, whereas large-scale distributed computations are probably an overkill, as internode communication would occupy a significant amount of the computation time, reducing the overall efficiency.

1.3.1 Compare with Other Parallel Processing Platforms

Parallel computation has been traditionally performed on CPU clusters, in which a number of CPUs are networked to collaboratively conquer a large problem. This setting, however, is less favorable for routine clinical applications. There are probably two models to use a cluster in a clinic. The first one is that each clinic set up a CPU cluster dedicated for its own use. The CPU cluster of this kind is probably of small scale to just meet the needs of computation power. Yet they are typically more costly than GPU with the same processing power. The additional burden of facility deployment and management is also an issue. While this setup has been employed in a few major clinical institutions, not every clinic has the budget and manpower to set up and maintain such a system. For instance, Tomotherapy originally developed a CPU cluster with 56 CPU cores to provide computations for treatment planning [20]. But the desire of cost reduction has resulted in the shift to a GPU-based system recently. The second mode of using a CPU cluster is to set it up in a centralized fashion, where different users remotely submit jobs to the large-scale cluster. In this scenario, the overhead due to internode communications may be a hurdle impacting the accelerations. Moreover, job scheduler is needed in this case, which cannot guarantee that a task is executed in a timely manner, hindering the workflow in some time-critical contexts. The concern of transferring protected patient data outside a clinical institution also exists and requires further investigations.

Another eye-catching platform for high-performance computing is cloud computing. With a large-scale CPU cluster like hardware at the backend, cloud computing uses modern virtualization technology to provide a user with processing power at the demand level in a "pay-as-you-go" mode. Yet the problems of cloud computing are similar to the large-scale CPU cluster. Overhead of data communication and concern about patient data security are potential issues hindering their applications. For instance, in a cloud-based treatment planning system [21], the speedup factor saturates to ~15 when more and more nodes were employed. This is essentially due to Amdahl's law [22]. The data communication overhead becomes nonnegligible, when the system becomes larger and larger, preventing a continuous increase in speed.

Recently, Intel Many Integrated Core (MIC) processor appeared as a strong competitor of GPU. It is particularly attractive to researchers because of its ×86 compatibility, which makes it possible to run existing CPU codes with minor modifications on the MIC processor. However, it was also found that substantial efforts were needed to achieve the optimal performance [23]. Running an existing code originally developed on the CPU platform may not achieve a high speedup factor, as parallel computing-related issues were not considered enough in the initial code developments, such as memory visit pattern and processor coordination.

1.4 CHALLENGES IN GPU-BASED PARALLEL PROCESSING

Despite the vast advantages, one cannot ignore some disadvantages and challenges when using GPU in radiotherapy.

First and probably most importantly, being a new platform that is quite different from the CPU platform we have been using for decades, GPU programming requires a lot of efforts to code from almost scratch. There are also much fewer libraries for scientific computing that one can use to help code developments, increasing the required effort for code development and maintenance. In addition, in contrast to CPU-based programming where compiler typically does a lot of code optimization, one probably needs to dig into a deeper level and considers the hardware structure, such as memory allocation and usage, in GPU-based programming to achieve a highly optimized code. Finally, some programming languages only support certain types of GPU, such as CUDA for Nvidia's cards. This means one loses code portability and potentially encounters limited usability of the developed codes.

Second, the hardware itself may have a certain limitations. For instance, the memory size of a GPU card is relatively small. Only a few GB memory is typically available on a modern card. Unlike in the CPU case in which one can easily expand memory size by plugging in more memory on the motherboard, the memory size on the GPU is fixed. Hence, in some problems with a large size, multiple GPU or frequent CPU–GPU data transfer is needed, which inevitably increase the computational burden. Support of double-precision calculations on GPU is also less than that of single precision. While Figure 1.1 undoubtedly demonstrates the increasing trend of GPU processing power, the peak power (GFLOPS) for the double-precision operation is much lower than the single-precision operation. The trend is further broken into two branches. The upper one consists of Nvidia's Tesla series and recent Titan cards, which are primarily used for scientific computing purposes. The lower branch contains consumer grade cards that are typically utilized for graphics rendering and double-precision operations are not critical for jobs as such.

Third, the hardware SIMD processing mode makes a GPU extremely capable of handling data parallel problems. One example is the aforementioned vector addition in which each GPU thread processes addition of two corresponding elements that are assigned to it. On the contrary, the so-called task parallel problems are not suitable for the GPU structure. One well-known example is Monte Carlo (MC) simulation in which different threads may run into completely different execution paths due to the randomness of the algorithm. Depending on the problems of interest, careful design of the algorithm considering the nature of GPU architecture is needed to achieve a high performance.

However, we have also witnessed the rapid developments of GPU computing in the past a few years. Many of these aforementioned issues have been constantly addressed. For instance, GPU programming languages have become more and more user-friendly and compatible with existing popular languages. OpenCL is increasingly used to solve the program portability problem. On the hardware side, specifications have experienced rapid and steady improvements, as illustrated by the processing power in Figure 1.1. Multi-GPU is also a practical solution. For instance, recent versions of CUDA support universal addressing over multiple cards to ease the programming process. The SIMD structure will probably not change on GPU and the inherent conflict with some problems, for example, Monte Carlo simulations, will remain. However, we have observed active research efforts toward developing GPU-friendly algorithms. It is just because of the conventional serial processing process that one has adopted the way of MC simulations. With the new platform and its potential to accelerate computations, it is worthy for us to develop new algorithms in order to be compatible with the hardware. We believe that the rapid evolution of GPU technology will greatly facilitate scientific computations and render them useful a greater number of clinical and research applications.

1.5 CONCLUSIONS

Over the past few years, GPU has emerged as a novel parallel processing platform for radiotherapy. We have witnessed the rapid growth of GPU-related researches and the gradual adoption of GPU technologies into radiotherapy clinic. The advantages of GPU place it at a unique position to help accelerating computations in important clinical problems. Its limitations, meanwhile, also require further investigations. We hope this book will serve as a review regarding the current utilizations of GPU in the field of radiotherapy, which will inspire new developments and help bringing this novel technology to the routine radiotherapy practice.

The rest of this book is organized as follows. Each of the remaining chapters will cover a specific application of GPU-based computations. These chapters can be generally divided into three categories. Chapters 2 through 12 present a number of problems related to medical imaging in radiotherapy. Chapters 13 through 20 are therapy-related problems such as dose calculation and treatment plan optimizations. Finally, Chapters 21 and 22 are two example systems that utilize GPU as a core component to achieve a certain clinical goal.

REFERENCES

1. Jaffray, D.A. and J.H. Siewerdsen, Cone-beam computed tomography with a flat-panel imager: Initial performance characterization. *Medical Physics*, 2000. **27**(6): 1311–1323.
2. Jaffray, D.A. et al., Flat-panel cone-beam computed tomography for image-guided radiation therapy. *International Journal of Radiation Oncology Biology Physics*, 2002. **53**(5): 1337–1349.

3. Sidky, E.Y., C.M. Kao, and X.H. Pan, Accurate image reconstruction from few-views and limited-angle data in divergent-beam CT. *Journal of X-Ray Science and Technology*, 2006. **14**(2): 119–139.

4. Jia, X. et al., GPU-based iterative cone beam CT reconstruction using tight frame regularization. *Physics in Medicine and Biology*, 2011. **56**: 3787.

5. Feldkamp, L.A., L.C. Davis, and J.W. Kress, Practical cone beam algorithm. *Journal of the Optical Society of America A: Optics Image Science and Vision*, 1984. **1**(6): 612–619.

6. Sonke, J.J. et al., Respiratory correlated cone beam CT. *Medical Physics*, 2005. **32**(4): 1176–1186.

7. Li, T.F. et al., Four-dimensional cone-beam computed tomography using an on-board imager. *Medical Physics*, 2006. **33**(10): 3825–3833.

8. Gao, H. et al., 4D cone beam CT via spatiotemporal tensor framelet. *Medical Physics*, 2012. **39**(11): 6943–6946.

9. Yan, H. et al., A hybrid reconstruction algorithm for fast and accurate 4D cone-beam CT imaginga. *Medical Physics*, 2014. **41**(7): 071903.

10. Tian, Z. et al., Low-dose 4DCT reconstruction via temporal nonlocal means. *Medical Physics*, 2011. **38**(3): 1359–1365.

11. Jia, X. et al., Four-dimensional cone beam CT reconstruction and enhancement using a temporal nonlocal means method. *Medical Physics*, 2012. **39**(9): 5592–5602.

12. Top 500. Top 500, The list. 2014; November 2004. Available from: http://www.top500.org/.

13. Pratx, G. and L. Xing, GPU computing in medical physics: A review. *Medical Physics*, 2011. **38**(5): 2685–2697.

14. Jia, X., P. Ziegenhein, and S.B. Jiang, GPU-based high-performance computing for radiation therapy. *Physics in Medicine and Biology*, 2014. **59**(4): R151.

15. Samant, S.S. et al., High performance computing for deformable image registration: Towards a new paradigm in adaptive radiotherapy. *Medical Physics*, 2008. **35**(8): 3546–3553.

16. Gu, X. et al., Implementation and evaluation of various demons deformable image registration algorithms on a GPU. *Physics in Medicine and Biology*, 2010. **55**(1): 207–219.

17. Gu, X. et al., GPU-based ultra fast dose calculation using a finite size pencil beam model *Physics in Medicine and Biology*, 2009. **54**(20): 6287–6297.

18. Hissoiny, S., B. Ozell, and P. Despres, Fast convolution-superposition dose calculation on graphics hardware. *Medical Physics*, 2009. **36**(6): 1998–2005.

19. Jacques, R. et al., Towards real-time radiation therapy: GPU accelerated superposition/convolution. *Computer Methods and Programs in Biomedicine*, 2010. **98**: 285–292.

20. Lu, W., A non-voxel-based broad-beam (NVBB) framework for IMRT treatment planning. *Physics in Medicine and Biology*, 2010. **55**(23): 7175–7210.

21. Na, Y.H. et al., Toward a web-based real-time radiation treatment planning system in a cloud computing environment. *Physics in Medicine and Biology*, 2013. **58**(18): 6525–6540.

22. Rodgers, D.P., Improvements in multiprocessor system design. *SIGARCH Computer Architecture News*, 1985. **13**(3): 225–231.
23. Mackay, D., Optimization and performance tuning for Intel® Xeon Phi™ coprocessors—Part 1: Optimization essentials. 2012; Available from: https://software.intel.com/en-us/articles/optimization-and-performance-tuning-for-intel-xeon-phi-coprocessors-part-1-optimization.

CHAPTER 2

Digitally Reconstructed Radiographs

Michael M. Folkerts

CONTENTS

2.1 INTRODUCTION

DIGITALLY RECONSTRUCTED RADIOGRAPHS (DRRs) have played an important role in radiotherapy, especially when it comes to patient alignment for image-guided radiotherapy (IGRT) before treatment delivery. DRR calculations have also received much attention with the advent of clinically viable statistical and iterative cone-beam computed tomography (CBCT) reconstruction techniques, where the computational complexity of the DRR calculation tends to be a bottleneck. The fundamental ray tracing component used in DRRs is also used in advanced Monte Carlo (MC) dose calculations and tends to be a performance bottleneck as well. In this chapter, we will explore the variety of DRR and ray tracing techniques, which have been implemented on GPU, with a focus on the imaging domain. We continue with an overview of relevant applications, then, in the following sections, set up the physical concept and integration problem, and review recently published GPU-based DRR techniques.

2.1.1 Applications and Motivation

With the advent of IGRT, highly conformal radiation fields have been used to directly target tumor volumes and spare critical structures based on CT imaging data. The issue of patient alignment with the treatment delivery device became a much more sensitive matter. The DRR is useful in this application because it provides a simulated x-ray projection image (radiograph) of the patient's CT data for comparison with real x-ray images acquired when the patient is lying on the treatment table. When these images match, it is an indication that the patient is aligned to the treatment device as expected.

Recently, CBCT imaging has become more commonly used for IGRT. Since traditional fractionated treatments may continue for several weeks, it is important to monitor the daily and weekly changes in the anatomy of a patient. However, there is growing concern within the community about excessive exposure to imaging radiation. Therefore, low-dose imaging protocols have become common, especially for younger patients. However, a low-dose protocol in conjunction with common imaging algorithms, like FDK [1], often results in poor image quality, typically amplifying common CBCT artifacts.

Fortunately, much research has gone into the development of advanced imaging algorithms specifically designed for low-dose imaging. In the literature, algorithms like these fall under the category of tomography: compressive sensing and iterative reconstruction. Each iteration of these algorithms involves two computationally intensive operations: a 2D projection operation (DRR) of the current 3D CT guess and a back projection of the (weighted) errors from the 2D imaging plane. Due to the CBCT geometry, hundreds of DRRs need to be generated at each iteration and can account for nearly one-third of the total CBCT reconstruction time [2]. Therefore, it is important to use high-performance computing hardware and advanced algorithms to generate the DRRs in a timely fashion to enable "real-time" iterative CBCT reconstruction.

It is desirable to study the effects of beam hardening and scatter contamination to evaluate and improve CBCT reconstruction techniques. By applying a CBCT reconstruction technique to energy-dependent (polyenergetic) DRRs it is possible to quickly model the effects of beam hardening. It has also been shown that the primary (unscattered) signal recorded at the detector plane is very well approximated by computing a polyenergetic DRR. By subtracting the DRR (primary) signal from a MC (primary and scatter) signal, and performing some denoising, it is possible to quickly recover the signal from only the scattered photons [3]. Having detailed information about each of these common artifacts enables researchers to study and model these effects as well as invent new ways of eliminating them in CBCT imaging.

MC simulations involve tracking particles along ray paths and computing line integrals. Using the Woodcock transportation method [4], one eliminates the need to compute the actual radiological path. However, for electron transport, ray tracing cannot be avoided due to the nature of the continuous scattering that the electron experiences. Therefore, it is desirable to have an efficient ray tracing algorithm to accelerate electron transport calculation times [5]. Finite pencil beam [6] and collapsed cone convolution [7] dose calculation algorithms depend on computing many radiological paths as well.

Now that we have a feel for the vast applications of DRRs in radiation therapy and the importance of speedy calculation times, in the following sections, we will present the general formulation of the DRR calculation and then explore various implementations on GPU.

2.2 RADIOLOGICAL PATH AND DIGITALLY RECONSTRUCTED RADIOGRAPHS

2.2.1 Beer–Lambert Law

For a given incident photon energy and material, it has been measured that x-ray intensity drops exponentially with the depth of penetration. Therefore, the x-ray signal intensity I measured by a detector placed behind a patient being illuminated with incident x-ray intensity and energy spectrum $I_0(E)$ is given by the generalized Beer–Lambert attenuation law:

$$I = \int dE I_0(E) \exp\left(-\int_L dl \mu(E, \vec{r})\right) \tag{2.1}$$

where

$\mu = \mu(E, \vec{r})$ is the energy- and position-dependent linear attenuation coefficient of the material through which the x-rays pass

L is the path from the x-ray source to a detector pixel.

The purpose of all DRR algorithms is to numerically evaluate the line integral in Equation 2.1 over the many ray paths that exist between the x-ray source and the detector plane.

2.2.2 Integration Problem

Equation 2.1 can be simplified and evaluated in many ways. First, the x-ray source can be modeled as monoenergetic by setting the spectrum to a constant intensity value $I_0(E) = I_0$ and by replacing energy-dependent attenuation coefficients with coefficients at an effective energy $\mu(E, \vec{r}) = \mu(E_{\text{eff}}, \vec{r}) = \mu(\vec{r})$. This is more consistent with CT imagers, which intrinsically reconstruct Hounsfield data at an effective energy. This results in a much simpler attenuation integral:

$$I = I_0 \exp\left(-\int_L dl \, \mu(\vec{r})\right) \tag{2.2}$$

To compute a simulated radiograph, it is necessary to discretize the integral in Equation 2.2 along the ray paths connecting the x-ray source to the detector plane (pixels). There are numerous techniques to numerically evaluate and approximate this integral [8].

The first choice to be made is how to represent the CT data. One choice is to use the voxelized representation of the CT dataset. For a DRR projection, each voxel n has an associated attenuation coefficient μ_n and an intersection length $l_{n,m}$ with each ray m traced from the source position to the center of a detector pixel. The attenuated ray path can be expressed as a sum over all n for each m:

$$\ln\left(\frac{I_0}{I_m}\right) = \sum_n \mu_n l_{n,m} \tag{2.3}$$

where I_m is the intensity measured by pixel m on the image plane. Since many of the $l_{n,m}$ values are zero, it is beneficial to use an efficient algorithm to compute the sum. We will see how this is done later on.

Another choice is to represent the CT attenuation data as a function of location $\mu_n(\vec{r})$ with known values at sample grid points. One could then rewrite the summation in Equation 2.3 by using a constant Δl_m and resampling, by various means, attenuation values at regular intervals along the ray paths:

$$\ln\left(\frac{I_0}{I_m}\right) = \sum_i \mu(\overrightarrow{r_{i,m}})\Delta l_m \tag{2.4}$$

where $\overrightarrow{r_{i,m}}$ are the collection of sample points within the CT volume uniformly spaced by Δl_m along the path from the x-ray source to the detector pixel. As we will see later, one may choose Δl_m to be different for each ray or the same for all rays ($\Delta l_m = \Delta l$). Consequently, the quality of the DRR becomes dependent on the resampling rate. In the following, we will discuss a variety of sampling techniques.

2.3 SIDDON'S ALGORITHM

The most widely accepted technique to compute DRRs in medical physics (via Equation 2.3) has been Siddon's algorithm [9]. It has been shown to be more accurate when compared to resampling techniques with similar sampling rates [10]. This algorithm models the CT data as voxels with uniform spacing along each axis. The voxels are considered as intersection volumes of orthogonal sets of equally spaced parallel planes (Figure 2.1). Siddon's algorithm utilizes the parameterization of the ray path from the x-ray source $\vec{r}_s$ to the center of a pixel on the detector $\vec{r}_d$:

$$\vec{r}(t) = \vec{r}_s + t(\vec{r}_d - \vec{r}_s) \tag{2.5}$$

This allows for a set of parametric values $\{t^u\}$ to be calculated for all voxel planes a ray intersects in each dimension $u = \{x, y, z\}$. Note that t values between 0 and 1 represent points on the ray path between the source and the detector, other values are ignored. These sets of intersection values are then merged together in ascending order with the difference between adjacent t values being the parametric difference Δt_m.

$$\{t\} = \text{merge}(\{t^x\}, \{t^y\}, \{t^z\}) \tag{2.6}$$

$$\Delta t_n = t_n - t_{n-1} \tag{2.7}$$

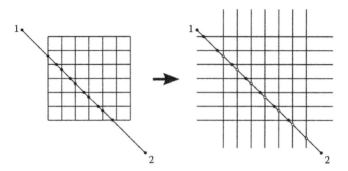

FIGURE 2.1 CT data being represented as intersection areas of orthogonal sets of parallel lines. The intersection points with horizontal lines are marked with solid dots on the right, whereas the intersection points with the vertical lines are marked with circles. The parameterization along the ray path and the intersection points play a key role in Siddon's algorithm.

The parametric difference can then be multiplied by the Euclidian distance from the source to the detector pixel to calculate the intersection length through the corresponding voxel. The voxel index $v(n) = v(i(n),$ $j(n), k(n))$ is calculated for each intersection length using a floor function, which returns only the integer part of a floating point number:

$$i(n) = 1 + \text{floor}\frac{\left(r_s^x + t_{\text{mid}}\left(r_d^x - r_s^x\right) - r_p^x\right)}{d^x}$$

$$j(n) = 1 + \text{floor}\frac{\left(r_s^y + t_{\text{mid}}\left(r_d^y - r_s^y\right) - r_p^y\right)}{d^y} \qquad (2.8)$$

$$k(n) = 1 + \text{floor}\frac{\left(r_s^z + t_{\text{mid}}\left(r_d^z - r_s^z\right) - r_p^z\right)}{d^z}$$

where

$t_{\text{mid}} = (t_n + t_{n-1})/2$

d^u is the voxel size along the u dimension

r_p^u is the point on the volume bounding plane, perpendicular to the u dimension, with the smallest t^u value.

The sum equivalent to Equation 2.3 is then computed for each pixel m in the detector:

$$\ln\left(\frac{I_0}{I_m}\right) = \sum_n \mu_{v(n)}\Delta t_n \qquad (2.9)$$

2.3.1 Modern Incremental Algorithm

Much of the computational time (41%) for the original version of Siddon's algorithm, described earlier, is spent computing voxel indexes in Equation 2.8. However, it turns out that it is only necessary to compute the indexes in Equation 2.8 once for the first voxel intersected by each ray (where $n = 1$). Then a dynamic ray-driven tracing or "stepping" algorithm can determine which voxel boundary will be crossed next by computing intersection locations t_{next}^u of the immediate down-range voxel planes:

$$t_f = \min\left(t_{next}^x, t_{next}^y, t_{next}^z\right) \qquad (2.10)$$

The difference between the parameter for each intersection $\Delta t_n = t_f - t_{n-1}$ is the parametric length of intersection for the $v(n)$th voxel and is multiplied by that voxel's attenuation value. Before incrementing n, the update $t_n = t_f$ is made, the t_{next} value is set to the next plane intersection down wind, and the voxel index $v(n)$ is incremented (not recomputed) by the appropriate stride based on the dimension of the voxel plane being crossed. Using the preceding calculations and updates for each intersection n, Equation 2.9 can be evaluated much faster than the original method. This modified approach can be referred to as Siddon's incremental (or stepping) algorithm because of the way the algorithm steps through the voxel space (Figure 2.2) but still uses Siddon's voxel plane and line intersection parameterization [11,12]. This algorithm appears to be an adaptation of a voxel traversal algorithm used for computer imaging [13]. To summarize, in the incremental algorithm, the voxel intersection length calculation is conducted on the fly, and voxel indexes are incremented rather than calculated for each intersection. In the next section, we will review a few implementations of Siddon's incremental algorithm on GPU.

2.3.2 Siddon's Incremental Algorithm on GPU

GPU implementations of Siddon's incremental algorithm for medical physics applications can be found in conference abstracts from 2010.

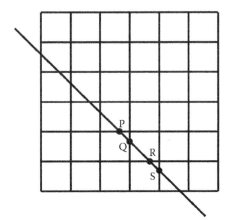

FIGURE 2.2   Illustration of the stepping algorithm. Starting at point P, the nearest of points Q and R is selected. The intersection length becomes Q–P and the algorithm repeats at point Q (and the nearest of points R and S is selected).

The late adoption of this algorithm in GPU computing is largely based on the limited programmability of GPUs before the advent of CUDA and the prevalence of other volume rendering methods more amicable to traditional graphics APIs. It is straightforward to parallelize the computation by assigning each GPU thread to compute just one ray integral. Using this technique researchers reported speedup factors, when comparing to single threaded CPU algorithms, of 47 on an Nvidia GeForce 8800 GTX [14] and 100 using an Nvidia Quadro FX 5800 [15]. One abstract reported the detailed result of a 512×384 projection image being computed from a $512 \times 512 \times 104$ phantom in 30 ms on an Nvidia Tesla C1060 [16].

2.4 FIXED STRIDE INTERPOLATION

Another common technique is also a ray-driven approach. In this case, the CT data are treated as piecewise linear and continuous rather than piecewise constant. Therefore, specific ray–voxel intersection lengths are not a concern, in fact, the concept of a voxel is abandoned in favor of interpolating between points of known attenuation values. In this case, Equation 2.4 is used with a constant sampling rate for all rays ($\Delta l_m = \Delta l$), and $\mu(\vec{r_{i,m}})$ being a trilinear interpolation function:

$$\ln\left(\frac{I_0}{I_m}\right) = \Delta l \sum_i \mu(\vec{r_{i,m}}) \qquad (2.11)$$

In general, it is more computationally expensive to compute the interpolated value in 3D. However, most GPUs contain special functional units that allow for fast hardware interpolation that is enabled via a specialized memory object known as a texture. Older CPU algorithms would use techniques like octtrees and opacity thresholding to speed up these types of calculations [17]. In addition to hardware interpolation, textures enable caching in older Nvidia GPU hardware where caching of regular global memory arrays is not available. Of course the performance of these types of algorithms is highly dependent on the sampling interval, and to avoid loss of information, a sampling interval equal to half of the smallest voxel dimension is typically used. All algorithms mentioned in the remainder of this section are implemented with Nvidia's CUDA [18].

2.4.1 Fixed Stride Interpolation on GPU

A straightforward implementation of the fixed stride interpolation algorithm uses one GPU thread per ray. A CUDA-based algorithm reported a speedup factor of 50 compared to an equivalent CPU implementation [19]. Another similar implementation achieved DRR rendering times of 35 ms from a $512 \times 512 \times 344$ CT volume [20]. An algebraic image reconstruction technique used 3D hardware trilinear interpolation sampling, but failed to report raw DRR calculations times [21]. A 2D–3D image registration algorithm likewise used 3D hardware interpolation, but reduces the computation time by randomly sampling rays within a subset of the detector area [22]. The full ROI timing reported for the registration algorithm represents about one-sixth of the imaging plane area, and therefore, one can extrapolate that full 750×750 DRR may be rendered from CT data with in-plane resolution of 512×512 in about 20 ms on an Nvidia GTX 580, very similar to the previously mentioned work. This trilinear ray sampling technique was also used in computing gradients for iterative 2D–3D registration algorithms, but no explicit DRR computation times were mentioned [23,24]. Another implementation compared OpenGL implementation to CUDA and reported speedup factors of 148 over CPU implementations; however, the voxel data size was never mentioned [25].

Some authors claim to have increased performance of this sampling technique by having multiple GPU threads contribute to the computation of one ray path [26]. The strategy is to improve caching behavior (reduce cache misses) by having threads predictably access localized regions of texture memory. The authors also claim to utilize fast, on-chip, shared memory to perform quick sum reduction tasks. Using this cache-aware approach, one can expect a projection of a $512 \times 512 \times 512$ CT dataset onto a 512×512 detector to take about 11 ms using an Nvidia Tesla C1060. However, it must be noted that in this work the number of samples along a given ray was limited to 128, meaning that at least one-fourth of the voxel data is not involved in computing a ray path (due to the local nature of the trilinear interpolation).

2.5 TEXTURE STACKING (SHEAR-WARP) AND TEXTURE SLICING

Before the advent of CUDA and OpenCL, many volume rendering algorithms were written with standard 2D–3D graphics libraries. Since low-level control over the GPU threads had not yet matured, researchers were left to deal with graphics primitives. The idea here was to utilize the GPUs' native texture mapping hardware (typically in-plane bilinear interpolation) to stack and sum the projections of individual 2D CT layers aligned perpendicular (or otherwise) to the projection axis. In cases where the CT layers or slabs are not parallel to the imaging plane, the image can be sampled and warped onto the imaging plane using a perspective transformation, which is equivalent to, and also known as the shear-warp factorization technique [27]. Typically, more than one copy of the CT data would be stored, each being a set of 2D texture planes aligned with each of the major axes of the CT data. In rare cases, a new set of CT slices is sampled from the 3D data at the proper perspective (perpendicular to the projection axis), and this resampling technique is referred to as texture slicing. Here, the general form of Equation 2.4 applies, with i looping over the set of 2D CT planes rather than sample points. The sampling rate is different for each ray based on its angle of incidence with each CT slab and the sampling of $\mu(\vec{r})$ is done in-plane

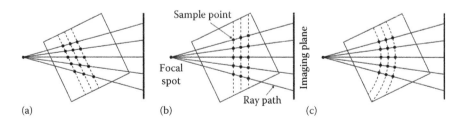

FIGURE 2.3   Illustration of CT aligned slices (a), perspective aligned slices (b), compared to the effective sampling scheme of the fixed stride interpolation algorithm (c).

using bilinear interpolation (Figure 2.3). Typically, the CT planes are sampled sequentially (with detector pixel-wise parallelism) and accumulated to a 2D buffer to render the final DRR image.

2.5.1 Texture Stacking (Shear-Warp) on GPU

Early work on this 2D texture sampling and stacking method was done using Silicon Graphics International (Milpitas, CA) workstations and a graphics library [28] and other similar graphics hardware [29]. Later, textures were used in volume rendering on an Nvidia GeForce 3 GPU that could render a 512×512 DRR in 71 ms from a CT volume of $256 \times 256 \times 256$ voxels [30]. More comprehensive DRR work also implemented and tested FDK, SART, and EM algorithms on GPU [31]. This work reports projection times ranging from 5 to 11 ms of $128 \times 128 \times 128$ voxel data onto a 128×128 imager using an Nvidia FX 5900 GPU. The 2D texture stacking technique was also used in an automatic patient positioning algorithm [32] but projection calculation times were not stated.

In some more recent work on 2D–3D image registration, an algorithm, adapted from the above methods, utilized 3D textures [33]. In this work, DRR calculation times of 17 ms were reported for ROIs of $300 \times 300 \times 48$ projected onto a 300×200 pixel imager using an Nvidia GeForce 7800 GTX. Other work on CPU attempts to improve the performance of this approach by removing empty voxels before rendering the 2D slices [34]. Other GPU implementations of this approach were explored where projections of $512 \times 512 \times 256$ voxel data to a 473×473 imager plane were reported to take 20–50 ms using an ATI X800 [35].

A GPU-optimized approach [36] was based on 2D image projections [37] but adapted to the 3D problem. High performance was achieved by storing subprojections in the GPU's on-chip shared memory. The algorithm was able to achieve 12.9 Giga updates per second (GUPS) on 64 Giga voxels with an Nvidia FX 5600 using CUDA (equiv to about 10 ms per projection) and 15.3 GUPS using OpenGL (8 ms).

2.5.2 Texture Slicing on GPU

A more computationally intensive technique known as texture slicing, used in volume rendering for computer graphics purposes, uses a similar approach; however, the 2D slabs are generated or resampled on the fly to be parallel to the imaging plane and are cached for use in graphics-related rendering processes [38]. A digital tomosynthesis paper also implemented this slicing technique where 2D planes, aligned perpendicular to the projection axis, were sampled using trilinear interpolation and accumulated to the frame buffer. It was claimed that a speedup factor of 67 was observed when compared to a single threaded CPU implementation [39]. Some DRR work explored both stacking and slicing on GPU where an Nvidia Quadro FX 3500 generated a 512×512 DRR of a 200 MB dataset in 78 ms [40]. Later, a variation of this slicing technique was implemented specifically for CBCT reconstruction purposes with a reported 21 ms average DRR projection time for $512 \times 512 \times 128$ data onto a 1024×768 imaging plane using an Nvidia GTX 580 [41].

2.5.3 Voxel Splatting on GPU

A variant of the 2D texture stacking approach, called Voxel Splatting, can be thought of as voxel projection stacking [42–44]. This splatting technique sums the 2D projection of all 3D cubic voxels to the imaging plane rather than using the, often off-axis, CT slices as a surrogate. After some preprocessing, namely, the removal of zero-valued or thresholded voxels, the projection of a translucent cube at the appropriate perspective and attenuation value is rendered directly to the imaging plane. For added realism, the x-ray source point can be randomly wobbled using a uniform Gaussian, which effectively simulates a finite focal spot. Using this wobbled splatting technique, DRR projections of 512×512 pixels can be rendered from clinically relevant datasets in 11–58 ms, based on various thresholding values and varying sizes of regions of interest, using an Nvidia Quadro FX 570M GPU [45].

2.6 POLYENERGETIC (ENERGY-DEPENDENT) DRR

All of the DRR implementations mentioned earlier used the effective energy assumption. In this section, we will step back and focus on the more general form of the DRR given by Equation 2.1. In reality, there is a characteristic x-ray spectrum that is generated by electron targets in CT imaging machines. To get a more realistic and accurate DRR, it is possible to simulate a discrete energy spectrum with corresponding weights w_i for a given energy E_i resulting in the discreet form of the polyenergetic equation for attenuated x-ray intensity:

$$I = \sum_i w_i \exp(-\Sigma'_j M(E_i, m_j) \rho_j l_j) \tag{2.12}$$

with radiological properties characterized by mass-attenuation coefficients $M(E,m) = \mu(E,m)/\rho_0$ having units of length squared per mass, where m_j represents the material (as an id number) for a given voxel, and ρ_j is its density. Here, we will assume the voxelized form of the CT data as in Equation 2.3, and the sum on j is restricted to voxels intersected by a given ray. This approach obviously requires some preprocessing of the CT data to compute material id and voxel density using a suitable segmentation algorithm and CT calibration curve.

The naive approach to generating the polyenergetic DRR involves a simple sum over the properly weighted DRRs from each energy bin of a discretized spectrum, which can be modeled with anywhere from 10 to 100 bins. The most recent work specifically addressing this topic claims to use the original implementation of Siddon's algorithm on GPU. Using a 16-bin Mohan (4 MV) energy spectrum and an OpenCL-based algorithm, which exploits data parallelism and task overlapping between CPU and GPU, it takes 382 ms to generate a polyenergetic DRR (~24 ms per energy bin on average) for CT data with in-plane resolution of 512×512, and a detector plane with 512×512 pixels [46]. This performance is very similar to a CUDA-based CBCT simulation tool, which reports calculation times, using an Nvidia GTX 580, ranging from 12 to 23 ms per energy bin for 512×384 pixel DRRs of digital phantoms with in-plane resolution of 256×256 and 512×512, respectively [3].

The standard approach to evaluating Equation 2.12 is very computationally intensive due to the fact that a DRR must be generated for each and every energy bin. However, through clever manipulation, it is possible to arrive at a form of Equation 2.12, which drastically reduces the polyenergetic DRR calculation time. The key is to factor out the energy dependence from the ray trace:

$$I = \sum_i w_i \prod_n \exp(-M(E_i,n) \Sigma'_{j,m_j=n} \rho_j l_j) \qquad (2.13)$$

Notice that the sum on j is restricted to voxels of material n only and represents a ray trace through the unconverted density data. The product over n accounts for material energy dependence and makes Equations 2.12 and 2.13 equivalent. In this approach, the DRR is computed by first computing a single modified ray trace through the density data for each material producing a set of n projections, one for each material m. The set of n projections is reused to evaluate the product over n for each energy E_i. Finally, the results are weighted and summed as indicated earlier in Equation 2.13.

This clever formulation was buried deep within statistical image reconstruction literature where computation speed is of no concern, but material composition and energy spectra are [47]. This formulation was independently rediscovered and presented orally at the annual conference of the American Association of Physicists in Medicine in 2013 where a speedup factor of 60 was realized for a 100-bin spectrum on CPU [48].

2.7 SUMMARY

DRRs have much utility within the field of medical physics, and with the advent of general purpose GPU programming, more and more high-speed algorithms have been developed. Although we have covered several ways to discretize Equation 2.2, they all become equivalent as the voxel and pixel sizes approach zero. It is difficult to pick out the best implementation since many were conducted on different and already outdated hardware. It is apparent that the algorithms that utilized on-chip shared memory tend to perform the best and are likely only limited by the GPU's internal bandwidth. Therefore it is up to researchers and practitioners to choose the best suited algorithm for their applications.

REFERENCES

1. L. A. Feldkamp, L. C. Davis, and J. W. Kress, *J. Opt. Soc. Am. A* **1**, 612 (1984).
2. H. Yan, X. Wang, F. Shi, T. Bai, M. Folkerts, L. Cervino, S. B. Jiang, and X. Jia, *Med. Phys.* **41**, 111912 (2014).
3. X. Jia, H. Yan, L. Cerviño, M. Folkerts, and S. B. Jiang, *Med. Phys.* **39**, 7368 (2012).
4. I. Lux and L. Koblinger, CERN Document Server (1991). http://cds.cern.ch/record/268101.
5. X. Jia, X. Gu, J. Sempau, D. Choi, A. Majumdar, and S. B. Jiang, *Phys. Med. Biol.* **55**, 3077 (2010).
6. X. Gu, D. Choi, C. Men, H. Pan, A. Majumdar, and S. B. Jiang, *Phys. Med. Biol.* **54**, 6287 (2009).
7. R. Jacques, R. Taylor, J. Wong, and T. McNutt, *Comput. Methods Programs Biomed.* **98**, 285 (2010).
8. M. H. K. Engel, ACM SIGGRAPH course notes, 1st edn. (A. K. Peters, 2004). http://www.real-time-volume-graphics.org/?page_id=12 (accessed on July 24, 2006).
9. R. L. Siddon, *Med. Phys.* **12**, 252 (1985).
10. F. Xu and K. Mueller, *Third IEEE International Symposium on Biomedical Imaging: From Nano to Macro 2006* (Arlington, Virginia, 2006), pp. 1252–1255.

11. G. Han, Z. Liang, and J. You, *1999 IEEE Nuclear Science Symposium Conference Record* (Seattle, WA, 1999), vol. 3, pp. 1515–1518.
12. F. Jacobs, E. Sundermann, B. De Sutter, M. Christiaens, and I. Lemahieu, *CIT J. Comput. Inform. Technol.* **6**, 89 (1998).
13. J. Amanatides and A. Woo, *Eurographics'87* (Amsterdam, the Netherlands, 1987), pp. 3–10. *Computer Graphics Forum* **6**(1), 66–68 (1987).
14. P. Després, F. Lacroix, and J. Carrier, *Med. Phys.* **35**, 2915 (2008).
15. J. Yuan, S. Chang, C. Tsang, W. Chen, and D. Jette, *Med. Phys.* **37**, 3141 (2010).
16. M. Folkerts, X. Jia, X. Gu, D. Choi, A. Majumdar, and S. Jiang, *Med. Phys.* **37**, 3367 (2010).
17. M. Levoy, *ACM Trans. Graph.* **9**, 245 (1990).
18. http://docs.nvidia.com/cuda/pdf/CUDA_C_Programming_Guide.pdf
19. S. Mori, M. Kobayashi, M. Kumagai, and S. Minohara, *Radiol. Phys. Technol.* **2**, 40 (2009).
20. O. M. Dorgham, S. D. Laycock, and M. H. Fisher, *IEEE Trans. Biomed. Eng.* **59**, 2594 (2012).
21. Y. Lu, W. Wang, S. Chen, Y. Xie, J. Qin, W.-M. Pang, and P.-A. Heng, *Sixth International Conference in Computer Graphics, Imaging and Visualization 2009 CGIV 09* (Tianjin, China, 2009), pp. 480–485.
22. G. J. Tornai, G. Cserey, and I. Pappas, *Med. Phys.* **39**, 4795 (2012).
23. W. Wein, B. Roeper, and N. Navab, *Proceedings of SPIE 5747, Medical Imaging 2005: Image Processing* (May 5, 2005), pp. 144–150. http://dx.doi.org/10.1117/12.595466.
24. M. Grabner, T. Pock, T. Gross, and B. Kainz, Automatic Differentiation for GPU-Accelerated 2D/3D Registration. In *Advances in Automatic Differentiation*, C. H. Bischof, H. M. Bücker, P. Hovland, U. Naumann, and J. Utke (eds.) (Springer, Berlin, Germany, 2008), pp. 259–269. http://link.springer.com/chapter/10.1007/978-3-540-68942-3_23).
25. A. Weinlich, B. Keck, H. Scherl, M. Kowarschik, and J. Hornegger, *High Performance and Hardware Aware Computing: Proceedings of the First International Workshop on New Frontiers in High-Performance and Hardware-Aware Computing (HipHaC '08)* (Lake Como, Italy, November 2008).
26. C.-Y. Chou, Y.-Y. Chuo, Y. Hung, and W. Wang, *Med. Phys.* **38**, 4052 (2011).
27. P. Lacroute and M. Levoy, *Proceedings of the 21st Annual Conference on Computer Graphics and Interactive Techniques* (Orlando, FL, July 24–29, 1994) (ACM, New York, 1994), pp. 451–458.
28. T. J. Cullip and U. Neumann, *Accelerating Volume Reconstruction with 3D Texture Hardware* (University of North Carolina at Chapel Hill, Chapel Hill, NC, 1993).
29. B. Cabral, N. Cam, and J. Foran, *Proceedings of the 1994 Symposium on Volume Visualization* (Tysons Corner, VA, October 17–18, 1994) (ACM, New York, 1994), pp. 91–98.
30. D. LaRose, *Iterative X-Ray/CT Registration Using Accelerated Volume Rendering* (Robotics Institute, Carnegie Mellon University, Pittsburgh, PA, 2001).

31. F. Xu and K. Mueller, *IEEE Trans. Nucl. Sci.* **52**, 654 (2005).
32. A. Khamene, P. Bloch, W. Wein, M. Svatos, and F. Sauer, *Med. Image Anal.* **10**, 96 (2006).
33. F. Ino, J. Gomita, Y. Kawasaki, and K. Hagihara, *Proceedings of the Fourth International Conference on Parallel and Distributed Computing, Applications and Technologies* (Springer-Verlag, Berlin, Germany, 2006), pp. 939–950. http://dx.doi.org/10.1007/11946441_84.
34. C. Bethune and A. J. Stewart, *J. Graph. GPU Game Tools* **10**, 55 (2005).
35. R. H. Gong, P. Abolmaesumi, and J. Stewart, *Annual International Conference of the IEEE Engineering in Medicine and Biology Society 2006 EMBS 06* (New York City, New York, 2006), pp. 1433–1436.
36. M. Knaup, S. Steckmann, and M. Kachelriess, *IEEE Nuclear Science Symposium Conference Record 2008 NSS 08* (Dresden, Germany, 2008), pp. 5153–5157.
37. P. M. Joseph, *IEEE Trans. Med. Imaging* **1**, 192 (1982).
38. J. Mensmann, T. Ropinski, and K. Hinrichs, *Fifth International Conference on Computer Graphics Theory and Applications (GRAPP 2010)* (Angers, France, May 17–21, 2010), pp. 190–198.
39. H. Yan, L. Ren, D. J. Godfrey, and F.-F. Yin, *Med. Phys.* **34**, 3768 (2007).
40. D. Ruijters, B. Ter Haar-Romeny, and P. Suetens, *Proceedings of the Sixth IASTED International Conference on Biomedical Engineering, BioMED* (2008), pp. 431–435.
41. M. Folkerts, X. Jia, D. Choi, X. Gu, A. Majumdar, and S. Jiang, *Med. Phys.* **38**, 3403 (2011).
42. W. Cai and G. Sakas, *2000 IEEE Nuclear Science Symposium Conference Record* (2000), vol. 3, pp. 19/12–19/17.
43. W. Birkfellner, R. Seemann, M. Figl, J. Hummel, C. Ede, P. Homolka, X. Yang, P. Niederer, and H. Bergmann, *Phys. Med. Biol.* **50**, N73 (2005).
44. Y. Long, J. Fessler, and J. M. Balter, *IEEE Trans. Med. Imaging* **29**, 1839 (2010).
45. J. Spoerk, C. Gendrin, C. Weber, M. Figl, S. A. Pawiro, H. Furtado, D. Fabri et al., *Z. Für Med. Phys.* **22**, 13 (2012).
46. L. Zhou, K. S. C. Chao, and J. Chang, *Med. Phys.* **39**, 6745 (2012).
47. I. A. Elbakri and J. A. Fessler, *IEEE Trans. Med. Imaging* **21**, 89 (2002).
48. M. Folkerts, X. Jia, and S. Jiang, *Med. Phys.* **40**, 426 (2013).

CHAPTER 3

Analytic Cone-Beam CT Reconstructions

Bongyong Song, Wooseok Nam,
Justin C. Park, and William Y. Song

CONTENTS

3.1 INTRODUCTION

THE 3D CONE BEAM computed tomography (CBCT) reconstructions can be done either analytically or iteratively. An analytic approach has an explicit formula for reconstructing the 3D volumetric images from a set of x-ray projection data. In particular, the well-known Feldkamp, Davis, and Kress (FDK) algorithm [1] offers a computationally efficient approximate formula that has an advantage of obtaining fair quality 3D images without requiring excess computations. Given the excessive number of voxels to be reconstructed for a CBCT image (often a few tens of millions of voxels), this low complexity formula has been the most commonly used CBCT reconstruction method in practice. It will be shown in

Section 3.4 that, by using a currently available off-the-shelf graphics processing unit (GPU) computer, near real-time 3D CBCT reconstruction is possible. This computational advantage came from the fact that the reconstruction algorithm is derived from a simple yet neat mathematical x-ray projection model that includes an infinitesimal focal spot of the x-ray source, pencil beams from the source without any scattering, no measurement noise at the detector, etc. For this reason, the analytic method is not too flexible to incorporate various nonideal factors in the real system or to leverage possible additional information that can be used for further improving the image quality.

On the other hand, an iterative CBCT reconstruction approach is capable of flexibly formulating a CBCT problem to model various nonidealities and/or incorporate prior information such that the solution of the problem well represents the desired CBCT image. These advantages are enabled by iteratively refining the reconstructed image to find a better and better solution to the CBCT problem. Due to the large problem size of CBCT, this iterative refinement is very computationally expensive. Fortunately, when the recent breakthrough in the iterative CBCT reconstruction algorithms is combined with the fast evolving parallel computing hardware, the concern of excess computational amount can be effectively addressed.

The issue of fast image reconstruction becomes most important when the CBCT is used for an online clinical purpose where near real-time image reconstruction is desired. It will be proved in Section 3.4 that more than 80 times speed acceleration is possible using an off-the-shelf GPU hardware by carefully implementing the FDK algorithm to fully utilize the massive parallel computing capability of the GPU.

3.2 FDK ALGORITHM FOR CBCT RECONSTRUCTION

The FDK algorithm is an *approximate* 3D CBCT reconstruction algorithm that resembles an exact 2D fan-beam CT reconstruction algorithm known as filtered back projection (FBP). For this reason, the 2D FBP principle is described first prior to describing the FDK algorithm for 3D CBCT reconstruction.

3.2.1 2D Filtered Back Projection

Consider a 2D parallel beam geometry illustrated in Figure 3.1. In an idealized projection model with monoenergetic, nondiffracting x-ray beam

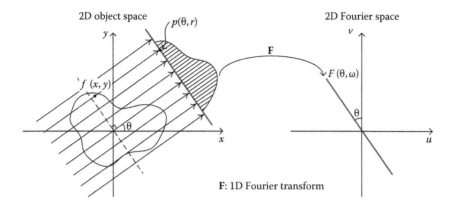

FIGURE 3.1   Illustration of 2D Fourier slice theorem.

from a point x-ray source, the attenuation of the intensity I_0 from the x-ray tube through the object along the line L can be characterized as:

$$\int_L f(x,y)\,dl = -\ln\frac{I}{I_0} \qquad (3.1)$$

where
  $f(x, y)$ denotes the x-ray attenuation coefficient of the pixel at (x, y)
  I and I_0 denote the incident and transmitted x-ray intensities.

For a given set of projection data over all angles, FBP algorithm can be derived from the following 2D Fourier slice theorem:

- *2D Fourier slice theorem*: Let $p(\theta, r)$ denote the parallel projection of a 2D object $f(x, y)$ onto a line with projection angle θ as shown in Figure 3.1. The 1D Fourier transform of $p(\theta, r)$ equals the radial line $F(\theta, \omega)$ in the 2D Fourier transform of $f(x, y)$ which is parallel to the projection line.

This implies that the 1D Fourier transform of a line integral (or parallel projection) of an object $f(x, y)$ obtained at angle θ represents the radial line in a 2D Fourier transform of $f(x, y)$ taken at the same angle as illustrated in Figure 3.1.

As a result, by taking parallel projections, that is, populating the 2D Radon space, one can essentially populate the 2D Fourier space of the

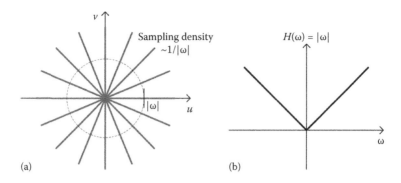

FIGURE 3.2 (a) 2D Fourier space filled with projection information and (b) a ramp filter in frequency domain.

object (after applying 1D Fourier transform for each projection). Because the 2D Fourier space is populated by radial lines, its density is inversely proportional to the distance ($\sim 1/|\omega|$) from the origin as illustrated in Figure 3.2a. The well-known FBP algorithm compensates for this nonuniformity by applying a ramp filtering ($H(\omega) = |\omega|$) shown in Figure 3.2b.

It can be shown that the filtered 2D Fourier space is transformed back to the object space by back projecting the filtered projection data. In summary, the FBP algorithm for parallel beams involves (1) obtaining projections from parallel x-ray source over uniformly spaced angles, (2) ramp filtering to the projection, and (3) back projection, which can be expressed as:

$$f(x,y) = \int_0^{2\pi} \tilde{p}(\theta, r(\theta, x, y)) d\theta \qquad (3.2)$$

where $\tilde{p}(\theta, \cdot)$ denotes the ramp-filtered parallel projection data given by:

$$\tilde{p}(\theta, r) = \int_{-\infty}^{\infty} p(\theta, \alpha) h(r - \alpha) d\alpha \qquad (3.3)$$

and the relationship between (θ, x, y) and r is illustrated in Figure 3.1.

In practice, the ramp filter $h(r)$ has such a simple representation in the Fourier domain (Figure 3.2b) and it is often found more computationally efficient to filter in the Fourier domain, that is, (1) take the fast Fourier transform (FFT) of the projection data, (2) multiply by the ramp filter

$(H(\omega) = |\omega|)$, and (3) take the inverse fast Fourier transform (IFFT) of the product. Also, it should be noted that the back projection can be taken from angles spanning from 0 to π by virtue of rotational symmetry.

Reconstruction for the 2D fan-beam geometry also utilizes the FBP principle because the equivalent parallel beam projections can be obtained by a set of fan-beam projections. This parallel beam conversion is illustrated in Figure 3.3 and is often called *rebinning*.

The FBP algorithm for 2D fan-beam geometry is given by [2]:

$$f(x,y) = \int_{0}^{2\pi} \frac{D^2}{U(\theta,x,y)^2} \hat{p}(\theta, r(\theta,x,y)) d\theta \tag{3.4}$$

where $\hat{p}(\theta, \cdot)$ denotes the ramp-filtered and fan-beam projection data given by:

$$\hat{p}(\theta, r, s) = \int_{-\infty}^{\infty} \cos(\beta) p(\theta, \alpha, s) h(r - \alpha) d\alpha. \tag{3.5}$$

Comparing Equation 3.5 with 3.3, it can be seen that a cosine weighting factor $(\cos(\beta))$ is applied to the projection data prior to ramp filtering. When the detector pixels are uniformly spaced, the density of the rays on the detector pixel with fan angle β decreases with $\cos(\beta)$. The projection data with less incident x-ray intensity are accordingly deemphasized by the "cosine weighting."

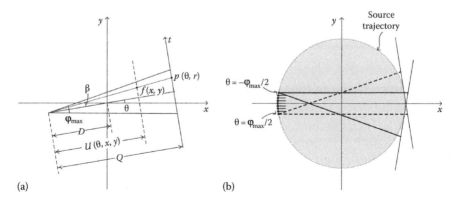

(a) (b)

FIGURE 3.3   (a) Illustration of fan-beam projection geometry and (b) conversion to parallel projections by rebinning.

Comparing Equations 3.4 and 3.2, the back projection process for fan-beam geometry involves a weighting factor $D^2/(U(\theta, x, y)^2)$ that is essentially inversely proportional to the square of the distance from the source to the pixel position (x, y). Since the source-to-pixel distance is a function of the source position or θ, different weights need to be computed for each source position even for the same pixel. As a result, serial computations of the weights for each pixel for every projection angle can noticeably increase the reconstruction time of the FBP algorithm. Fortunately, these weights for various pixels can be calculated in a parallel manner when reconstructed in the GPU-based parallel computing environment, greatly reducing the computational burden.

Unlike the parallel projections where short scan requires source trajectories only from 0 to π, fan-beam short scan requires additional source trajectories from 0 to $\pi + \varphi_{max}$ in order to complete the fan beam to parallel beam transformation via rebinning. Although this inherently introduces the data redundancy in the collected projection data, the circular symmetry in the circular fan-beam system enables effective removal of the redundancy by appropriate weighting applied to the projection data. The Parker's weighting [3] is the most commonly used weighting method for this purpose.

The overall procedure for 2D FBP method for the fan-beam geometry involves (1) projection, (2) cosine weighting, (3) Parker weighting (for short scan), (4) ramp filtering, and (5) back projection (with $1/U^2$ weighting).

3.2.2 3D Filtered Back Projection: FDK Algorithm

Similar to the 2D object reconstruction, an *exact* reconstruction of a 3D object relies on the 3D Fourier slice theorem that is built upon the 3D Radon transform. The 3D Radon transform is a generalization of 2D Radon transform where the 3D object is integrated over 2D planes. Specifically, a given point $R(x, y, z)$ in 3D Radon space is computed by integrating the 3D object over the unique plane that is passing the point and, at the same time, perpendicular to the line connecting the point (x, y, z) and the origin. As a 2D object reconstruction is developed based on 2D Fourier slice theorem, an exact reconstruction of 3D can be derived from 3D Fourier slice theorem:

- *3D Fourier slice theorem*: Let $R(v, r)$ denote the radial line along the unit vector v in the 3D Radon transform of a 3D object $f(x, y, z)$. The 1D Fourier transform of $R(v, r)$ equals $F(v, \alpha)$ where $F(\cdot)$ denotes the 3D Fourier transform of $f(x, y, z)$.

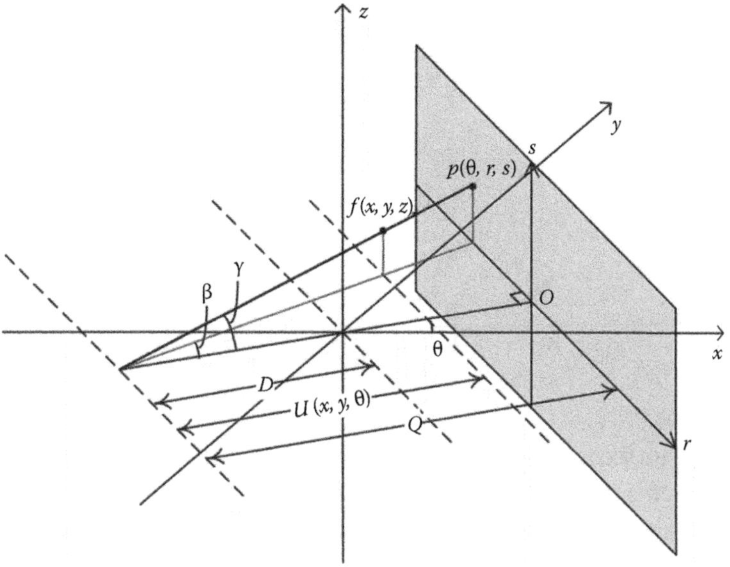

FIGURE 3.4   Illustration of the 3D circular cone-beam geometry.

This suggests that an exact reconstruction of a 3D object is possible as long as the 3D Radon transform space is fully populated. However, it can be easily seen that the circular cone-beam geometry (Figure 3.4) does not provide enough source trajectories to complete the 3D Radon transform. For example, any point on the z-axis in the Radon space except for the origin cannot be calculated because the source rotates on the x–y plane only.

The FDK algorithm [1] is therefore, an approximate, computationally efficient 3D reconstruction algorithm for circular cone-beam geometry. The FDK algorithm is very much similar to the FBP algorithm for the fan-beam geometry. In order to reconstruct a point (x, y, z), it looks at the effective fan-beam planes created by the cone-beam rays that pass through the x-ray source at different angles and the point (x, y, z). The 2D filtered back projection method is applied for each plane to compute the contribution of the angle θ. The contributions are integrated over all angles to reconstruct the point. Since these are coming from different planes, an exact 2D fan-beam reconstruction is not possible unless the point x is located in the central plane (x–y plane). However, for moderate cone-beam angles, this method reconstructs acceptable 3D images.

The FDK algorithm can be expressed as:

$$f(x,y,z) = \int_0^{2\pi} \frac{D^2}{U(\theta,x,y)^2} \breve{p}(\theta,r(\theta,x,y),s(\theta,x,y,z))\,d\theta \qquad (3.6)$$

where $\breve{p}(\theta,\cdot,\cdot)$ denotes the ramp-filtered fan-beam projection data given by:

$$\breve{p}(\theta,r,s) = \int_{-\infty}^{\infty} \cos(\gamma)p(\theta,\alpha,s)h(r-\alpha)\,d\alpha \qquad (3.7)$$

and the relationship between (θ, x, y, z) and (r, s) is illustrated in Figure 3.4. In computer implementations, the images and projections are discretized and the integral is replaced by a sum over projection angles. There are two approaches to conduct the sum:

- Voxel-driven approach—Each voxel reconstruction involves summing the corresponding detector pixel values at different projection angles. As a result, the back projection operation can be independently conducted for each voxel.

- Ray-driven approach—Detector pixel values are each distributed to all voxels that meet the line between the source and the detector pixel. Since a given voxel generally meets multiple rays, some form of serial computation is required to avoid "memory conflict."

A voxel-driven approach is commonly used for the back projection operations in the GPU computing environment [4] because reconstruction of different voxels can be conducted in a completely independent manner. Another advantage of the voxel-driven approach is simplicity of interpolation. When back projecting detector values to a voxel, the intersection point of the ray connecting the source and the voxel and the detector plane is generally not a center of a detector pixel. Therefore, the projection data for the intersection point are computed by interpolating nearby pixels according to their contributions. This involves 2D interpolation of the projection data where very efficient implementation is often offered by the GPU hardware itself which will be further discussed in the subsequent section. On the other hand, the ray-driven approach involves

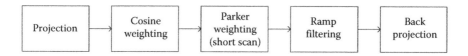

FIGURE 3.5 The overall 3D FDK algorithm workflow for a circular CBCT image reconstruction.

interpolation in the 3D image domain where, in this case, computationally more complex 3D interpolations need to be computed by the software program without leveraging the readily available hardware interpolation capability.

In practice, the CBCT reconstruction process can be classified depending on the dimensions of the imaging volume and the flat panel detector. When the size of the cone beam, defined by the flat panel detector, covering the imaging volume is adequately large, the CBCT mode is categorized as *full-fan* scans and a general 3D FDK procedure illustrated in Figure 3.5 can be applied for reconstruction. However, if the cone-beam projection is inadequate to cover the imaging volume, then the CBCT mode is categorized as *half-fan* scans. In this mode, the flat panel detector is shifted laterally depending upon the size of detector panel and the two laterally shifted half-fan projections with 180° apart can be combined to obtain one larger projection to reconstruct a larger field of view. As a result, the Parker weighting step in Figure 3.5 needs to be replaced by a different weighting scheme. The replaced weighting refers to half-fan weighting function for stitching two opposite projections into a single large one [5].

Finally, it should be noted that the steps in Figure 3.5 are required for each projection data. Therefore, these steps can be conducted either when projections from all angles become available or each time new projection data become available.

3.3 GPU IMPLEMENTATION

In this section, we describe an efficient implementation of the FDK algorithm using modern multi-core GPU platforms. The voxel-driven FDK implementation is GPU-friendly as the operations required for each data element, such as a detector pixel or a voxel, are highly disjointed to one another. Therefore, hazardous memory conflict, which is caused by multiple processes concurrently accessing the same memory address, can be avoided.

As mentioned earlier, all data for computer implementation are discretized in practice:

- Projection angle—θ

  The projection is taken at N_p discrete angles, that is, $\theta \in \Theta$, where $\Theta = \{\theta_0, \ldots, \theta_{N_p-1}\}$. It is further assumed that the angles are equally spaced, that is,

  $$\theta_k = \theta_0 + k\Delta\theta, \quad \text{where } \Delta\theta = \frac{2\pi}{N_p}.$$

- Projection data—$p(\theta, r, s)$

  Given a projection angle $\theta \in \Theta$, the projection data are captured by an $N_r \times N_s$ array of discrete detector elements. Therefore, the projection data are only measured on a collection of discrete points, each of which is identified by the center position of the detector element, $(r, s) \in \mathcal{R} \times \mathcal{S}$ where $\mathcal{R} = \{r_0, \ldots, r_{N_r-1}\}$ and $\mathcal{S} = \{s_0, \ldots, s_{N_s-1}\}$. Assuming a uniform detector array, we can let $r_k = r_0 + k\Delta r$ and $s_k = s_0 + k\Delta s$, where Δr and Δs are detector intervals along the horizontal and vertical directions, respectively.

- 3D object—$f(x, y, z)$

  The 3D object is defined on a set of voxels, positioned at $(x, y, z) \in \mathcal{X} \times \mathcal{Y} \times \mathcal{Z}$, where $\mathcal{X} = \{x_0, \ldots, x_{N_x-1}\}$, $\mathcal{Y} = \{y_0, \ldots, y_{N_y-1}\}$, and $\mathcal{Z} = \{z_0, \ldots, z_{N_z-1}\}$. The voxel position is determined in an absolute coordinate system, which is invariant to the source and detector geometries.

The integration over θ in Equation 3.6 is now replaced by a summation over Θ, which is a good approximation as long as the number of projection

TABLE 3.1   FDK Algorithm Outline

---

for each projection angle $\theta \in \Theta$,
    Cosine weighting – for each detector pixel, $(r, s) \in \mathcal{R} \times \mathcal{S}$
    Ramp filtering – for each detector pixel, $(r, s) \in \mathcal{R} \times \mathcal{S}$
    Back projection – for each voxel, $(x, y, z) \in \mathcal{X} \times \mathcal{Y} \times \mathcal{Z}$
end

N_p is large enough. For each projection angle $\theta \in \Theta$, the FDK algorithm is composed of three subsequent procedures—each procedure contains multiple threads running synchronously on the GPU cores. The FDK algorithm is outlined in Table 3.1 and each sub-procedure will be described in the sequel.

3.3.1 Cosine Weighting

Given a projection angle $\theta \in \Theta$, cosine weighting is applied to each detector pixel value as described in Table 3.2.

Note that the "**for**" statement in Table 3.1 refers to a "sequential-loop" operation as usually intended in standard computer programming. However, even though we use the same "**for**" statement, the one in Table 3.2 implies simultaneous creation of multiple synchronous threads for each $(r, s) \in \mathcal{R} \times \mathcal{S}$. Therefore, as long as the processor allows, up to $N_r N_s$ threads can simultaneously be created and processed. In the description given earlier, the cosine weight is readily derived that $\cos(\gamma(r,s)) = Q / \sqrt{Q^2 + r^2 + s^2}$. Note that, since the weights $\cos(\gamma(r, s))$ are common for all projections, they can be calculated in advance and stored in the memory to avoid computational redundancy.

3.3.2 Ramp Filtering

As explained in the previous section, a frequency-domain implementation of the ramp filtering is often found more computationally efficient than a spatial-domain implementation. Powerful FFT algorithms provide efficient implementation of discrete Fourier transform (DFT) to obtain frequency-domain representations of the projection data. However, for the frequency-domain implementation based on DFT, one should bear in mind that the multiplication of the frequency responses of the data and the filter appears as circular convolution in the spatial domain, rather than linear convolution as in Equation 3.7. To avoid any artifacts incurring from this difference, a sufficient number of zeros are padded at the end of the projection data before taking DFT. In other words, for the

TABLE 3.2   Cosine Weighting

for each $(r, s) \in \mathcal{R} \times \mathcal{S}$,
$\quad p'(\theta, r, s) = \cos(\gamma(r, s))p(\theta, r, s)$
end

TABLE 3.3   The Ramp Filtering

---

for each $s \in \mathcal{S}$,

$\quad P''(\theta,k,s) = \text{FFT}_{N_{FFT}}(p''(\theta,r,s))$

end

for each $s \in \mathcal{S}$ and $k = 0,\ldots, N_{FFT} - 1$,

$\quad \breve{P}(\theta,k,s) = P''(\theta,k,s)H(k)$

end

for each $s \in \mathcal{S}$,

$\quad \breve{p}(\theta,r,s) = \text{IFFT}_{N_{FFT}}(\breve{P}(\theta,k,s))$

end

---

cosine-weighted projection $p'(\theta,r,s)$, a zero-padded sequence $p''(\theta, r, s)$ is given by:

$$p'(\theta,r_0,s),\ p'(\theta,r_1,s),\ \ldots,\ p'(\theta,r_{N_r-1},s),\ \ \underbrace{0,\ldots,0}_{(N_{FFT}-N_r)\ zeros}\ ,\ \ s \in \mathcal{S},$$

where N_{FFT} is the FFT size. A rule of thumb is to pad $N_r - 1$ zeros. Also, assuming that we use a radix-2 FFT algorithm, the size of FFT should be an integer that is a power of 2. Therefore, we let N_{FFT} be the smallest power of 2 that is greater than or equal to $2N_r - 1$. The ramp filtering procedure is described in Table 3.3.

Note that, as in the cosine weighting procedure, the "**for**" statements in Table 3.3 imply creation of multiple synchronous threads. Also, in the earlier description, $\text{FFT}_{N_{FFT}}(\cdot)$ and $\text{IFFT}_{N_{FFT}}(\cdot)$ denote N_{FFT}-point FFT and IFFT operations, respectively, each of which may possibly consist of multiple threads on its own. Finally, $H(k)$ is the frequency response of the ramp filter, which is written as:

$$H(k) = \begin{cases} \dfrac{k}{N_{FFT}}, & 0 \le k < \dfrac{N_{FFT}}{2}, \\[3mm] 1 - \dfrac{k}{N_{FFT}}, & \dfrac{N_{FFT}}{2} \le k \le N_{FFT} - 1. \end{cases}$$

3.3.3 Back Projection

Before starting the iteration for the projection angle θ, the voxel values of the 3D object $f(x, y, z)$ are initialized to zeros. As the iteration goes on for the projection angle $\theta \in \Theta$, the back projection procedure calculates the data portion to be accumulated to each voxel value $f(x, y, z)$. In the voxel-driven approach, this portion corresponds to the ramp-filtered projection data $\bar{p}(\theta, r, s)$ at a point, which intersects with the ray connecting the source and the voxel on the detector plane. However, this intersection point does not necessarily coincide with the center of a detector element, at which the projection data are determined. Therefore, it is often approximated by suitable interpolation applied to the nearby detector pixels. Linear interpolation, for example, is widely used, by which, the projection data at a point (r, s), where $r_k \leq r < r_{k+1}$ and $s_l \leq s < s_{l+1}$, are approximated as:

$$\hat{p}(\theta, r, s) = (1-\lambda)(1-\rho)\breve{p}(\theta, r_k, s_l) + \lambda(1-\rho)\breve{p}(\theta, r_{k+1}, s_l)$$

$$+ (1-\lambda)\rho\breve{p}(\theta, r_k, s_{l+1}) + \lambda\rho\breve{p}(\theta, r_{k+1}, s_{l+1})$$

where
$$\lambda = (r - r_k)/(r_{k+1} - r_k)$$
$$\rho = (s - s_l)/(s_{l+1} - s_l).$$

It is worth remarking that some modern GPUs provide integrated hardware interpolators, which can perform interpolation given earlier in a very efficient manner [6]. In Table 3.4, the back projection procedure is delineated.

TABLE 3.4   Back Projection

---

for each voxel $(x, y, z) \in \mathcal{X} \times \mathcal{Y} \times \mathcal{Z}$

$\quad x' = x \cos\theta + y \sin\theta$ (axis rotation relative to the projection angle θ)

$\quad y' = -x \sin\theta + y \cos\theta$

$\quad U = D + x'$

$\quad r' = \dfrac{Qy'}{U}$ (intersection point on the detector plane)

$\quad s' = \dfrac{Qz}{U}$

$\quad$ **if** $r_0 \leq r' \leq r_{N_r-1}$ and $s_0 \leq s' \leq s_{N_s-1}$

$\quad\quad f(x, y, z) = f(x, y, z) + \dfrac{2\pi}{N_p} \dfrac{D^2}{U^2} \hat{p}(\theta, r', s')$ (summation for projection angle θ)

$\quad$ **end**

end

3.4 EXAMPLES AND PERFORMANCE

Figures 3.6 and 3.7 illustrate a few of FDK reconstructed slices of a phantom and patients, respectively, using an in-house written algorithm, accelerated with a single GPU card (NVIDIA GTX 295, Santa Clara, CA), implemented with the Compute Unified Device Architecture (CUDA) [6] environment complemented via C programming. An acceleration of

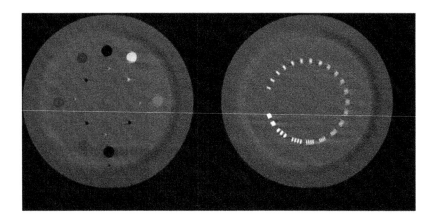

FIGURE 3.6 Contrast and spatial resolution slices of the reconstructed CatPhan 600 phantom using the FDK algorithm with 364 x-ray projections taken with a full-fan mode using the Varian TrueBeam's On-Board Imager. (Courtesy of OBI™, Varian Medical Systems, Palo Alto, CA.)

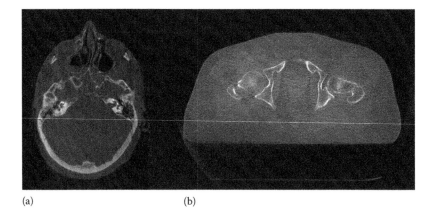

(a)   (b)

FIGURE 3.7 Example of (a) head-and-neck and (b) prostate patient slices reconstructed using the FDK algorithm with 657 x-ray projections taken with full-fan and half-fan modes, respectively, using the Varian TrueBeam's On-Board Imager. (Courtesy of OBI™, Varian Medical Systems, Palo Alto, CA.)

>80 times in computation speed is easily achieved compared with a CPU-only-based implementation using an Intel Core™ i7 CPU with 2.68 GHz clock speed, 12.0 GB DDR3 RAM, on a 64-bit Vista OS [4]. The reconstruction quality achieved by both CPU and GPU implementation is nearly identical with the CT number differences typically within one Hounsfield unit. This shows an excellent acceleration ratio compared with the CPU implementation, without losing numerical accuracy.

REFERENCES

1. Feldkamp LA, Davis LC, Kress JW. Practical cone-beam algorithm. *J Opt Soc Am A* 1984; A6:612–619.
2. Buzug TM. *Computed Tomography: From Photon Statistics to Modern Cone-Beam CT*, 1st edn. Springer-Verlag, Berlin, Germany, 2008.
3. Parker DL. Optimal short scan convolution reconstruction for fanbeam CT. *Med Phys* 1982; 9:254–257.
4. Park JC, Park SH, Kim JS, Han Y, Cho MK, Kim HK, Liu Z, Jiang SB, Song B, Song WY. Ultra-fast digital tomosynthesis reconstruction using general-purpose GPU programming for image-guided radiation therapy. *Technol Cancer Res Treat* 2011; 10(4):295–306.
5. Cho PS, Johnson RH, Griffin TW. Cone-beam CT for radiotherapy applications. *Phys Med Biol* 1995; 40:1863–1883.
6. NVIDIA. NVIDIA CUDA Programming Guide, Version 2.3.1, NVIDIA, Santa Clara, CA, August 26, 2009.

CHAPTER 4

Iterative Cone-Beam CT Reconstruction on GPUs

A Computational Perspective

Wei Xu, Ziyi Zheng, Eric Papenhausen,
Sungsoo Ha, and Klaus Mueller

CONTENTS

4.1 INTRODUCTION

ITERATIVE CONE-BEAM CT ALGORITHMS have become increasingly popular in recent years. They have been found useful when the projections are limited in number, irregularly spaced, or noisy. These conditions arise, for example, in low-dose CT [4,8,11,15], where one reduces the x-ray beam intensity or tube current per projection and/or cuts down on the total number of projections to lessen the radiation dose to the patient. Low-dose CT has become a mission of great importance in recent years due to reports that the x-ray energy imposed onto patients during a CT scan can cause cancer. But there are also other imaging scenarios that can lead to sparse x-ray data, such as lack of time for acquisition or reduced angular access. In any of these cases, analytical techniques, such as the Feldkamp cone-beam algorithm [5], tend to produce reconstructions with strong streak and noise artifacts, which make reading these images for diagnostics difficult. Iterative reconstruction methods, on the other hand, in particular when augmented with some form of regularization, such as total variation minimization (TVM) [10,11] or nonlocal means (NLM) filtering [1,7,17], can overcome these challenges. Based on numerical optimization, they produce reconstructions that best fit the data as well as some prior expectation of the object, formulated in the regularization function.

A drawback of iterative reconstruction methods, however, is that they suffer from high computational effort that cannot be met by traditional CPU-based platforms. Fortunately, the widespread availability of GPUs now enables these computationally challenging tasks to be performed inexpensively on the desktop. In the context of medical imaging, we demonstrated already in 2007 [13] that with just a single such board one could filter and back project cone-beam projections faster (at 50 projections/s) than they could be acquired by a modern flat panel gantry, enabling a new paradigm we called *streaming CT*. Since then, the performance of GPUs has increased nearly 10-fold, from about 500 GFlops (the NVIDIA GTX 8800 we used in 2007) to the 4.5 TFlops the NVIDIA GTX Titan (launched in 2014) delivers. And so, while our work in 2007 only allowed analytical Feldkamp-type cone-beam CT reconstructions to be performed efficiently, these newer boards can now also make iterative CT reconstructions computationally feasible [18].

Besides the high computational effort, another problem with iterative methods in general is that they require users to set many parameters, and if set incorrectly low image quality and slow reconstruction convergence are likely consequences. Furthermore, the intricate hardware of GPUs and the complex interplay between memory hierarchy and parallelism can also cause reductions in computational speed when reconstruction parameters are not set favorably (although they might work well on CPUs). This can lead to surprising relationships, as is demonstrated next.

4.1.1 Iterative Reconstruction Methods

Iterative methods can be broadly categorized into projection onto convex sets (POCS) algorithms (such as ART, SART, SIRT, and POCS) and statistical algorithms (such as EM, OS-EM, and MAP). For the purpose of demonstration, we select OS-SIRT (Ordered Subsets SIRT) [14]. OS-SIRT is a generalization of SART and SIRT, with SIRT having just one and SART having M subsets (with M being the number of projections). Its correction update is computed as:

$$v_j^{(k+1)} = v_j^{(k)} + \lambda \sum_{p_i \in OS_s} \frac{p_i - r_i}{\sum_{l=1}^{N} w_{ij}} \quad r_i = \sum_{l=1}^{N} w_{ij} \cdot v_j^{(k)} \tag{4.1}$$

Here, the weight factor w_{ij} determines the contribution of a voxel v_j to a ray r_i (starting from a projection pixel p_i in subset OS_s), and is typically given by the interpolation kernel. Hence, there are two parameters, the relaxation factor λ and the number of subsets S. Using these parameters, the effects of noise and sparse views on reconstruction quality can be controlled, but their choice can also affect the number of iterations needed to converge. Typically, noisier projections require larger subset sizes (smaller S) and/or smaller λ-settings. This is because the updates arising from the combined effect of larger subsets average out some of the noise while a smaller λ-setting dampens the impact of a noisy projection.

We use a 2D NLM filter [1] for regularization. The NLM filter is a nonlinear filter that replaces the pixel $P(x, y)$ with the mean of the pixels $P(x', y')$ whose Gaussian neighborhoods look similar to the neighborhood of $P(x, y)$. There are two windows: (1) the search window V, in which the algorithm looks for similar neighborhoods, and (2) the similarity window U which is essentially a Gaussian mask.

$$P_{NLM}(x,y) = \frac{\sum_{P(x',y') \in V(x,y)} w(x'-x, y'-y)P(x',y')}{\sum_{P(x',y') \in V(x,y)} w(x'-x, y'-y)}$$

$$w(x,y) = \exp\left(-\frac{\sum_{P(x',y') \in U(x,y)} G_a(x'-x, y'-y)\,|\,P(x',y') - P(x,y)\,|^2}{h^2}\right)$$

$$G_a(x,y) = \exp\left(-\frac{x^2 + y^2}{2a^2}\right) \tag{4.2}$$

Where

$V(x, y)$ is the 2D search window centered at $P(x, y)$

$U(x, y)$ is the 2D similarity window centered at $P(x, y)$ and $P(x', y')$, respectively

G_a is a 2D Gaussian kernel with standard deviation a, and the variable h acts as a parameter to control the smoothing. Thus, the NLM filter contains several parameters that can be tuned to achieve the best quality. With regard to performance, with NLM filtering being noniterative, the computational performance is stable, unlike TVM regularization. We have used the scheme described in [19] to make NLM filtering efficient on the GPU.

4.1.2 The Effect of GPU Parameters on the Speed of Iterative Reconstruction Methods

GPU-accelerated applications have a large number of parameters that can be tuned for optimal performance. Occupancy, thread granularity, and memory bandwidth are all examples of the types of parameters that can have a large impact on performance. Tuning one parameter too much can often lead to a sudden decrease in performance in some other aspect of the application. This is what has become known as *performance cliff*. A successful acceleration on GPUs must ensure that all processors are busy computing useful results and any processor idle time due to waiting for data to be read from memory is avoided. A key technology of GPUs in the regard is *hardware multi-threading*. It kicks in when a thread within a core ALU stalls due to a memory request (and others). In this case, another Single Instruction Multiple Threads (SIMT) resting at the same computational step is swapped in for execution. This hides the latency for the stalled thread. For example, the NVIDIA GTX 280 allows 128 more threads to be maintained than are SIMT executed. Hardware multi-threading, however, does not come without trade-offs. It requires memory since the contexts of all such threads must be maintained there. This typically limits the amount of threads that can be simultaneously maintained for latency hiding. Once these limits are reached, storage is deferred to memory units further away from the processor, which takes many more cycles to access.

These delicate mechanisms on the GPU and the need to ensure an ample degree of parallelism lead to interesting performance effects. While on CPUs the subset size of OS-SIRT does not influence the speed of computation—there is always the same number of volume updates—on GPUs; however, choosing a subset size has significant implications for the time required per iteration, as shown in Figure 4.1. We observe that an iteration step with SART is the slowest while for SIRT it is fastest. This is because the many projection–back projection context switches for SART disturb the parallelism and data flow. There are not enough threads to keep all processors busy and hide latencies.

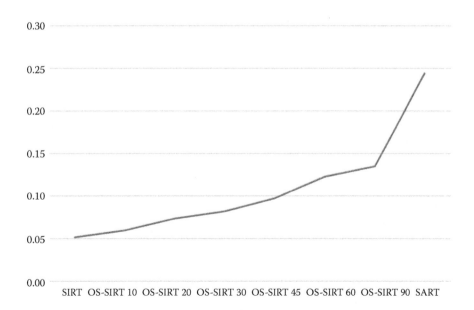

FIGURE 4.1   Performance of OS-SIRT for different numbers of subsets.

4.1.3 The Interplay of CT Reconstruction Convergence and GPU Execution Speed

So does this mean that SIRT is always the best subset scheme for GPUs? It is not, since SIRT takes many iteration steps (and therefore, time) to converge. So there is a natural trade-off, which needs to also be explored in the context of reconstruction quality. We have mentioned earlier that large subsets are required to achieve good reconstruction quality for noisy data, but determining the best size of the subsets is often left to guesswork or tedious hand tuning. The complexity of the GPU hardware makes this determination even harder since now the time required per iteration also varies per subset scheme, and not only by the number of iterations. To overcome these difficulties, we have devised a framework that optimizes the various parameters, given the quality of the projection data [16]. Here, we identified two (mildly) competing objectives—reconstruction speed and reconstruction quality. They are *mildly* competing since more computations can lead to better reconstructions, but there are always limits imposed by the data. In that regard, our framework can optimize the parameters according to two alternative objectives: (1) achieve the best reconstruction speed given a certain quality and (2) achieve the best reconstruction quality given a certain reconstruction speed.

Providing a visualization of this parameter space can help CT technologists and researchers trade-off these two or more objectives in an informed manner. This essentially comes down to visualizing the *Pareto frontier*—the set of parameter settings that all yield solutions that are optimal in the sense that improving one objective will not lead to a nonoptimal setting for another objective [6]. By assessing the Pareto frontier, an analyst can make trade-offs within this constrained set of parameters without having to consider the full ranges of parameters. We also exploit that fact that these optimized (learned) parameters can then be reused for novel data that were acquired under similar conditions or scenarios.

4.2 PARAMETER OPTIMIZATION FRAMEWORK

4.2.1 Assessing Reconstruction Quality Automatically—Image Quality Metrics

To properly optimize the reconstruction parameters, we require many observations of reconstruction results and an evaluation of their quality with respect to a gold standard. Most popular for the assessment of image quality in CT reconstruction have been statistical metrics such as the mean absolute error (MAE), root mean square (RMS), normalized RMS (NRMS), cross-correlation coefficient (CC), and R-factor. However, these metrics do not consider the fact that human vision is highly sensitive to structural information [12]. These properties are well captured in the gradient domain, for example, by ways of an edge-filtered image calculated via a Sobel Filter operator. We have labeled this group of metrics by prefix "E-." For example, the E-CC metric stands for the CC of two edge-filtered images. Another method to gauge structural information is structural similarity (SSIM) which combines luminance, contrast, and structure [12]. Given two signal images x and y, the SSIM index is defined as:

$$SSIM(x, y) = \left[l(x, y)\right]^{\alpha} \cdot \left[c(x, y)\right]^{\beta} \cdot \left[s(x, y)\right]^{\gamma} \qquad (4.3)$$

where
   α, β, and γ are parameters adjusting relative importance
   The terms $l(x, y)$, $c(x, y)$, and $s(s, y)$ are the luminance, contrast, and structure comparison functions, respectively.

These functions are computed from local image statistics. We have used these various quality metrics in our optimization framework.

4.2.2 Optimization Algorithm—The Ant Colony Simulation

We have used the ant colony algorithm to optimize our various parameters. The ant colony system optimization algorithm is a part of the family of swarm optimization algorithms. It is a modification of the ant system algorithm, which was designed to mimic the way ants find the shortest path from the ant nest to a food source. Initially, ants will choose paths randomly. Once an ant finds food, it will travel back to the nest and emit pheromones so other ants can follow that path to the food source. As other ants follow the pheromone trail, they emit pheromones as well which reinforces the trail. After some time, however, the pheromone trail will evaporate. Given multiple paths to a food source, the pheromones on the shortest path will have the least amount of time to evaporate before being reinforced by another ant. Over time, the ants will converge to the shortest path. The ant colony system was presented in [2] and was applied to the traveling salesman problem. It modifies the ant system algorithm [3] in several ways to lead to a faster convergence rate. After an ant crosses an edge, the pheromone value of that edge is decayed according to:

$$\tau_{ij} = (1-\varphi)\cdot\tau_{ij} + \varphi\cdot\tau_0 \qquad (4.4)$$

where τ_{ij} denotes the pheromone quantity on the edge from state i to state j. The pheromone decay coefficient, φ, determines how much pheromone is decayed after an ant chooses the edge from i to j. The initial pheromone value, τ_0, is the value every edge has at the beginning of the program. Equation 4.4 reduces the probability of multiple ants choosing the same path.

After all ants have chosen a path, the pheromone of each edge is updated as follows:

$$\tau_{ij} = (1-\rho)\cdot\tau_{ij} + \rho\cdot\Delta\tau_{ij}^{best} \qquad (4.5)$$

The variable τ_{ij} has the same meaning as Equation 4.1. The variable $\Delta\tau_{ij}^{best}$ evaluates to the inverse of the length of the best path if the edge from i to j was taken by the ant with the best path, otherwise it evaluates to zero. The variable ρ represents the evaporation rate. This leads to a pheromone increase on the edges taken by the ant that produced the best solution, while decaying the pheromones on all other edges. When transitioning from one state to another, the edge is selected probabilistically according to the following probability:

$$p_{ij}^k = \frac{\left(\tau_{ij}^\alpha\right)\left(\eta_{ij}^\beta\right)}{\sum\left(\tau_{ij}^\alpha\right)\left(\eta_{ij}^\beta\right)} \qquad (4.6)$$

where
  τ_{ij} determines the amount of pheromone on the edge from i to j
  η_{ij} defines some predetermined desirability of that edge (e.g., the inverse of the edge weight).

The variables α and β are weighting factors for τ_{ij} and η_{ij}, respectively. The variable p_{ij}^k is the probability that an ant will select an edge that goes from state i to state j during the kth iteration.

4.2.3 Multi-Objective Optimization—Setting Fixed Objectives

For our first experiment, we chose the optimization objective "best reconstruction speed, given a certain quality." We first simulated, from a baby head CT scan (size 256^2), 180 projections at uniform angular spacing of ($-90°$, $+90°$) in a parallel projection viewing geometry. We then added different levels of Gaussian noise to the projection data to obtain SNRs of 15, 10, 5, and 1.

Using the simulated projection data, we computed a representative set of reconstructions, sampling the parameter space in a comprehensive manner. We then evaluated these reconstructions with the E-CC perceptual quality metric. Adaptive sampling was used to drive the data collection into more "interesting" parameter regions (those that produce more diverse reconstruction results in terms of the quality metrics). The observations with the higher marks, according to some grouping, subsequently received higher weights in determining the reconstruction algorithm parameters. We used a fast-decaying Gaussian function to produce this weighting.

Figure 4.2 summarizes the various parameters obtained for the various data scenarios mentioned earlier. The "Best Subset" and "Best Lambda"

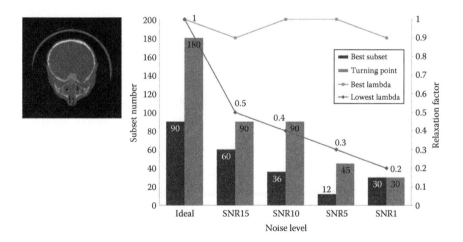

FIGURE 4.2 Optimal parameter settings for the baby head dataset (shown on the left): subset number and relaxation factor as a function of imaging condition (SNR) and the turning point and lowest lambda for each SNR level. (From Zheng, Z. et al., *Phys. Med. Biol.*, 58(21), 7857, 2013. With permission.)

values denote the parameter settings that promise to give the best results, in terms of the quality metric (and objective). The "Lowest Lambda" and "Turning Point" values describe the shape of the λ-curve as a function of the number of subsets. The λ-factor is always close to 1 for small subsets and then linearly (as an approximation) falls off at the "Turning Point" to value "Lowest Lambda" when each subset only consists of one projection (which is SART). This summary plot helps practitioners to pick the best-performing number of subsets and the associated λ for a given expected SNR level. For example, we observe that low SNR requires a low number of subsets, while less noisy data can use a higher number of subsets. This trend is well confirmed by prior studies and field experience and thus, validates the correctness of our general approach. Similar curves can also be derived for the "best reconstruction quality, given a certain speed" objective.

4.2.4 Multi-Objective Optimization—Assessing the Entire Pareto Frontier

While the previous application evaluated all parameter combinations exhaustively, more elaborate parameter combinations and schedules require a more strategic approach. For this, we took advantage of the ant colony optimization framework described in Section 4.2.2. We also changed the number of objectives to three—quality, speed, and radiation dose (defined by the mA setting of the tube and the number of projections acquired).

Figure 4.3 gives a block diagram of our optimization scheme [18]. The input are a set of low-dose projections obtained at some dose D, and the gold standard is the equivalent regular dose scan. We perform iterative

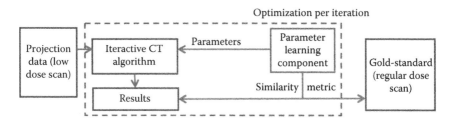

FIGURE 4.3 Our overall parameter optimization framework. (From Zheng, Z. et al., *Phys. Med. Biol.*, 58(21), 7857, 2013. With permission.)

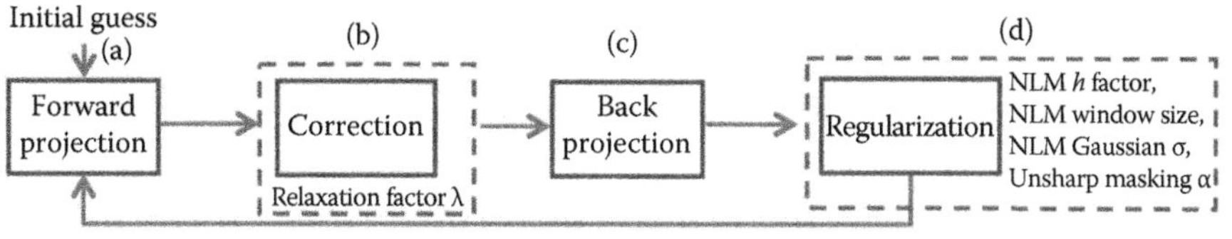

FIGURE 4.4 Iterative reconstruction and parameters. (a) Simulate x-ray projections from present reconstruction, (b) compute difference between real projection and estimated projections multiplied by λ, (c) add differences to present reconstruction, (d) regularize updated reconstruction. (From Zheng, Z. et al., *Phys. Med. Biol.*, 58(21), 7857, 2013. With permission.)

reconstruction with NLM regularization in the low-dose acquisitions. The optimization is done per iteration in the iterative reconstruction. Figure 4.4 summarizes our iterative reconstruction pipeline and the parameters we optimize.

For the optimization, we model the iterative CT reconstruction process as a path searching problem. In this graph of nodes and edges, each node represents a unique state of the image that is being reconstructed, while each edge represents the computation of a correction pass and a regularization pass at a given setting of parameters. The edge weight represents the time cost, which in our case can be uniformly set to 1 since all passes have a fixed constant cost. Each node is tagged with a score that encodes a quality metric. The overall goal is to find the node that has the highest score at a given cost. Since all edges have the same weight, the problem is reduced to find the best score after a fixed number of steps.

The ant colony optimization scheme can be employed to search for good paths in discrete graphs. It launches a large number of artificial ants searching for the best score. In our case, each artificial ant independently moves through the regularized iterative CT reconstruction graph and receives a score based on image quality and reconstruction time. We optimize six parameters including the relaxation factor λ, the h factor, Gaussian blur factor, window and mask size for the NLM filter, and an unsharp masking parameter α. These parameters can be different per iteration, resulting in an astronomical search space. For example, assuming we allow 100 discretized values for each parameter and run the pipeline for 10 iterations, the search space will be 10,120. This search space is so huge that simple

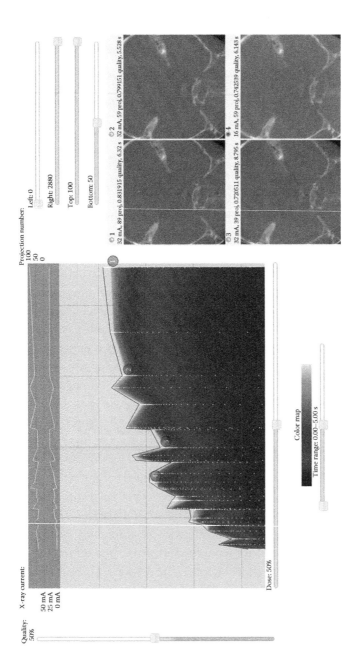

FIGURE 4.5 (**See color insert.**) The quality *vs.* dose plot. The numbers inserted into the plot refer to the image matrix (the light box) on the right. Image 1 is the image of the highest quality but requires a high dose. Images 3 and 4 require much less dose but have streak artifacts or blurred features, respectively. Image 2 seems to be a good compromise—lower dose than image 1 but still offering good quality. It could serve as a starting point for further exploration of the plot, where the user would mouse-click at desirable plot locations and insert the corresponding images into the light box. (From Zheng, Z. et al., *Phys. Med. Biol.*, 58(21), 7857, 2013. With permission.)

exhaustive search algorithms will fail to find the optimal solution in a reasonable amount of time, hence the need for a metaheuristic algorithm such as the ant colony algorithm.

Once all solutions have been computed, they can be visualized in the visual interface we called DQS Advisor and depicted in Figure 4.5. The main window of the interface shows the 2D scatter plot of dose vs. quality with color mapped to computation time (or dose vs. time with color mapped to quality). The two graphs on the top of the main window visualize dose composition—mAs and number of projections. Finally, at the bottom right there is what we call the "Light Box"—four image slots by which users can place noteworthy reconstruction results for comparative studies. The locations/origins of these images in the main plot are indicated by circled numbers.

By means of this interface, users can easily recognize and balance trade-offs in dose, quality, and computation time and so are able to make well-informed choices for these parameters. Once they have decided on these meta-parameter settings, they can click on the plot to find the setting for the six run-time parameters, such as the relaxation factor λ, the h factor for NLM filtering, and so on.

4.2.5 Learning from Experience—Applying the Optimized Parameters for Novel Reconstructions

The results of the parameter optimization can be reapplied to correctly set the parameters for any new CT scan taken at similar scenarios, such as CT scanner, imaging task, patient size, and the like. In this case, the data stored in the interface are regarded knowledge and the parameter optimization as knowledge acquisition. Enabling this knowledge transfer is the ultimate use of our system.

We did some first tests if the learnt parameter settings can be used to reconstruct data from a different scan and anatomy. For this purpose, we optimized the parameters for the central slice of a head scan and then shifted the slice to another autonomy region in the head phantom. Figure 4.6 shows the image quality of this new dataset with (a) 50 iterations of OS-SIRT with a constant $\lambda = 1$, (b) 360 projection Feldkamp, (c) 50 iterations of ASD-POCS [11], and (d) 50 iterations with optimized parameters. We observe that even in this new scan, the parameters can guide the reconstruction to achieve better detail reproduction than ASD-POCS. Some streaks remain for both ASD-POCS and our approach.

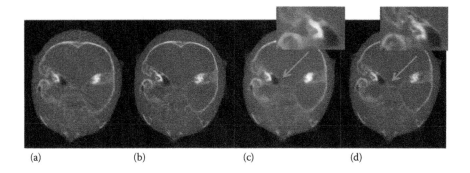

(a)   (b)   (c)   (d)

FIGURE 4.6   A central slice of a reconstruction for a new dataset. (a) Low-dose OS-SIRT, (b) gold-standard FDK by 360 projections, (c) low-dose ASD-POCS, and (d) optimized low dose. (a), (c), and (d) use 72 projections and 50 iterations. (From Zheng, Z. et al., *Phys. Med. Biol.*, 58(21), 7857, 2013. With permission.)

4.3 CONCLUSIONS

In this chapter, we studied iterative CT reconstruction algorithms from the perspective of GPU acceleration. We used OS-SIRT as an example, but similar observations also apply to other predictor–corrector schemes of this nature. A basic requirement for a successful GPU implementation is to keep the parallel pipelines full and this makes schemes that switch between these two modes frequently less amenable for acceleration. There is also a complex interaction between the need for parallelism and the required levels of the CT reconstruction parameters to achieve fast convergence and high quality in the presence of noise, and at the same time keep the radiation dose to the patient low. Achieving optimal configurations here is too tedious to achieve by simple manual tuning. We described a visual multi-objective optimization framework called the DQS Advisor [18] to help in this task and also transfer these optimized settings to novel CT scans of similar nature.

Optimization can go beyond the identification of favorable settings of CT parameters. It can also work on the actual GPU program code level. We have described such a framework in [9]. It uses ant colony optimization to create a path through a collection of code snippets to construct the most efficient program given the reconstruction task at hand. This *Coding Ants* framework, as we call it, alleviates CT researchers from the tedious task of hand-optimizing GPU code, which would require a significant level of GPU expertise otherwise.

REFERENCES

1. A. Buades, B. Coll, J. Morel. A non-local algorithm for image denoising. *Proceedings of the Computer Vision and Pattern Recognition*, 2: 60–65, 2005.
2. M. Dorigo, L. Gambardella. Ant colony system: A cooperative learning approach to the traveling salesman problem. *IEEE Transactions on Evolutionary Computation*, 1(1): 53–66, 1997.
3. M. Dorigo, V. Maniezzo, A. Colorni. Ant system: Optimization by a colony of cooperating agents. *IEEE Transactions on Systems, Man, and Cybernetics—Part B*, 26(1): 29–41, 1996.
4. J. Evans, D. Politte, B. Whiting, J. O'Sullivan, J. Williamson. Noise-resolution tradeoffs in x-ray CT imaging: A comparison of penalized alternating minimization and filtered backprojection algorithms. *Medical Physics*, 38: 1444–1458, 2011.
5. L. Feldkamp, L. Davis, J. Kress. Practical cone-beam algorithm. *The Journal of the Optical Society of America A*, 1: 612–619, 1984.
6. D. Fudenberg, J. Tirole. Nash equilibrium: Multiple Nash equilibria, focal points, and Pareto optimality. In: D. Fudenberg and J. Tirole (eds.) *Game Theory*. MIT Press, Cambridge, MA, pp. 18–23, 1991.
7. J. Huang, J. Ma, N. Liu, H. Zhang, Z. Bian, Y. Feng, Q. Feng, W. Chen. Sparse angular CT reconstruction using non-local means based iterative-correction POCS. *Computers in Biology and Medicine*, 41(4): 195–205, 2011.
8. X. Jia, B. Dong, Y. Lou, S. Jiang. GPU-based iterative cone-beam CT reconstruction using tightframe regularization. *Physics in Medicine and Biology*, 56: 3787–3807, 2011.
9. E. Papenhausen, Z. Zheng, K. Mueller. Creating optimal code for GPU-accelerated CT reconstruction using ant colony optimization. *Medical Physics*, 3(40): 031110, 2013.
10. L. Rudin, S. Osher, E. Fatemi. Nonlinear total variation based noise removal algorithms. *Physica D*, 60: 259–268, 1992.
11. E. Sidky, X. Pan. Image reconstruction in circular cone-beam computed tomography by constrained, total-variation minimization. *Physics in Medicine and Biology*, 53: 4777–4807, 2008.
12. Z. Wang, A. Bovik, H. Sheikh, E. Simoncelli. Image quality assessment: From error visibility to structural similarity. *IEEE Transactions on Image Processing*, 13(4): 600–612, 2004.
13. F. Xu, K. Mueller. Real-time 3D computed tomographic reconstruction using commodity graphics hardware. *Physics in Medicine and Biology*, 52: 3405–3419, 2007.
14. F. Xu, W. Xu, M. Jones, B. Keszthelyi, J. Sedat, D. Agard, K. Mueller. On the efficiency of iterative ordered subset reconstruction algorithms for acceleration on GPUs. *Computer Methods and Programs in Biomedicine*, 98(3): 261–270, 2010.
15. W. Xu, K. Mueller. Efficient low-dose CT artifact mitigation using an artifact-matched prior scan. *Medical Physics*, 39: 4748–4760, 2012.

16. W. Xu, K. Mueller. Learning effective parameter settings for iterative CT reconstruction algorithms. In: *10th International Meeting on Fully 3D Image Reconstruction in Radiology and Nuclear Medicine*, Beijing, China, pp. 20–23, September 2009.

17. W. Xu, K. Mueller. Evaluating popular non-linear image processing filters for their use in regularized iterative CT. In: *IEEE Medical Imaging Conference*, Knoxville, TN, pp. 2864–2865, October 2010.

18. Z. Zheng, E. Papenhausen, K. Mueller. DQS advisor: A visual interface and knowledge-based system to balance dose, quality, and reconstruction speed in iterative CT reconstruction with application to NLM-regularization. *Physics in Medicine and Biology*, 58(21): 7857–7873, 2013.

19. Z. Zheng, W. Xu, K. Mueller. Performance tuning for CUDA-accelerated neighborhood denoising filters. In: *Workshop on High Performance Image Reconstruction (Fully 3D Image Reconstruction in Radiology and Nuclear Medicine)*, Potsdam, Germany, pp. 52–55, July 2011.

CHAPTER 5

4DCT and 4D Cone-Beam CT Reconstruction Using Temporal Regularizations

Hao Gao, Minghao Guo, Ruijiang Li, and Lei Xing

CONTENTS

5.1 INTRODUCTION

IN 4DCT [1,2] OR 4DCBCT [3,4], a temporal sequence of CT images are reconstructed instead of a single static image, for which a respiratory signal is often recorded to assist the binning of projection data to their belonging temporal frames, and then the images can be reconstructed either independently frame by frame (so-called 3D method) or altogether using temporal regularization methods (so-called 4D method).

In this chapter, we will first introduce a total variation (TV)-based iterative image reconstruction example to illustrate the improvement from 3D method to 4D method, and then review various 4D regularization methods. It is important to note that some of these emerging 4D methods have been studied in the imaging fields other than 4DCT/CBCT, and the systematic evaluation of these 4D regularizations for 4DCT/CBCT is beyond the scope of this chapter.

5.2 FROM 3D METHOD TO 4D METHOD

The regularized iterative reconstruction method we consider is as follow:

$$\min_{X} R(X), \text{ subject to } AX = Y, \tag{5.1}$$

where

X represents a dynamic sequence of images to be reconstructed

Y is the collected projection data

$R(X)$ is the regularization term to be specified in the next section

A is the 4DCT/CBCT system matrix that is specified by a particular scanning geometry and often modeled by the x-ray transform

AX represents the forward model that computes the projections through a dynamic changing image X per the projection binning from the respiratory signals.

In this chapter, a new parallel algorithm with the O(1) cost per thread is implemented on a GPU computer for the efficient computation of A and its adjoint [5]. It is worthy of mentioning here that although $AX = Y$ formally in Equation 5.1, in practice the iterative method for computing Equation 5.1 should stop when $\|AX - Y\| < \varepsilon$ is reached. The reason is that the x-ray transform model may not completely describe the actual x-ray photon transport in tissue, for example, the scattering effect, the unknown geometry mismatch, or noise pattern [6]. Therefore, if $AX = Y$

were required, the reconstructed X would contain the artifact due to the forward model mismatch.

5.2.1 TV Regularization

L1 sparsity regularized iterative reconstruction has been extensively studied over the past decade. Since the early works [7,8], TV transform [9] has been commonly used to sparsify the image when the image itself may not be sparse. For a dynamic sequence of 3D images $X = \{X_n, n \leq N_t\}$, with each 3D image $X_n = \{X_{ijkn}, i \leq N_x, j \leq N_y, k \leq N_z\}$, mimicking the first-order gradient, the spatial and temporal TV transform can be defined respectively as:

$$DX\big|_{ijk} = \begin{bmatrix} D_x X \\ D_y X \\ D_t X \end{bmatrix}_{ijk} = \begin{bmatrix} X_{i+1,j,k} - X_{ijk} \\ X_{i,j+1,k} - X_{ijk} \\ X_{i,j,k+1} - X_{ijk} \end{bmatrix} \quad \text{and} \quad D_t X\big|_{ijkn} = X_{ijk,n+1} - X_{ijkn}. \quad (5.2)$$

In 3D method, only the spatial regularization is considered, that is, the anisotropic TV norm

$$\| W_{3D,A} X \|_1 = \sum_n \sum_{i,j,k} (|D_x X| + |D_y X| + |D_z X|) \quad (5.3)$$

or the isotropic TV norm

$$\| W_{3D} X \|_1 = \sum_n \sum_{i,j,k} \sqrt{|D_x X|^2 + |D_y X|^2 + |D_z X|^2}. \quad (5.4)$$

The latter Equation 5.4 is often preferred, since it promotes the smooth gradient isotropically while Equation 5.3 tends to promote the smooth gradient along x, y, z coordinate directions only.

In contrast, the temporal regularization is also considered in 4D method. Using isotropic TV norm for spatial regularization, again there are at least two ways to define 4DTV regularization, that is, the anisotropic 4DTV norm

$$\| W_{4D,A} X \|_1 = \| W_{3D} X \|_1 + w_t \sum_{i,j,k,n} |D_t X| \quad (5.5)$$

or the isotropic 4DTV norm

$$\| W_{4D} X \|_1 = \sum_{i,j,k,n} \sqrt{ | D_x X |^2 + | D_y X |^2 + | D_z X |^2 + | D_t X |^2 }. \qquad (5.6)$$

Equation 5.5 is commonly used when a prior image X_0 is available, that is, $|D_t X| = |X - X_0|$ [8]. (For clarity, it is not necessary to supply an additional "prior" image here; rather a common strategy for X_0 is a statically reconstructed X using all projection views that has no temporal information, but improved signal-to-noise ratio with averaged spatial resolution.)

Here we compare 3D method and 4D method with isotropic TV norm, that is,

$$\min_X \| W_{3D} X \|_1, \quad \text{subject to } AX = Y; \qquad (5.7)$$

$$\min_X \| W_{4D} X \|_1, \quad \text{subject to } AX = Y. \qquad (5.8)$$

Please note that Equation 5.7 can be solved frame by frame, while Equation 5.8 needs to be solved altogether. Next we briefly describe a commonly used Alternating Direction Method of Multipliers (ADMM) algorithm [10] (or so-called Split Bregman method [11]) for solving the L1-type optimization problems discussed earlier.

5.2.2 Solution Algorithm via ADMM

In this section, we first consider ADMM for solving

$$\min_X \| WX \|_1, \quad \text{subject to } AX = Y. \qquad (5.9)$$

A key for solving L1-type problem (Equation 5.9) with nondifferentiable L1 norm is to first split out the sparsity term by introducing a dummy variable $Z = WX$, that is,

$$\min_{(X,Z)} \| Z \|_1, \quad \text{subject to } AX = Y, WX = Z. \qquad (5.10)$$

Then the augmented Lagrangian of Equation 5.10 is

$$L(X,Z) = \frac{\mu_Y}{2} \| AX - Y + u_Y \|_2^2 + \frac{\mu_Z}{2} \| WX - Z + u_Z \|_2^2 + \| Z \|_1. \qquad (5.11)$$

Following the parameter convention, Equation 5.11 is transformed into

$$L(X,Z) = \frac{\|\, AX - Y + u_Y \,\|_2^2}{2} + \frac{\mu}{2}\left(\|\, WX - Z + u_Z \,\|_2^2 + \lambda\,\|\, Z \,\|_1 \right). \qquad (5.12)$$

Then according to ADMM, Equation 5.10 can be solved by the following iteration:

$$\begin{cases} X^{k+1} = \min_X L(X, Z^k) \\ Z^{k+1} = \min_Z L(X^{k+1}, Z) \\ u_Y^{k+1} = u_Y^k + AX^{k+1} - Y \\ u_Z^{k+1} = u_Z^k + WX^{k+1} - Z^{k+1} \end{cases}. \qquad (5.13)$$

Here, the first step is a L2 problem, and from its optimal condition it is equivalent to solve the following linear system:

$$(A^T A + \mu W^T W)X^{k+1} = A^T \left(Y - u_Y^k\right) + \mu W^T \left(Z^k - u_Z^k\right) \qquad (5.14)$$

that can be solved efficiently using conjugate gradient method together with fast computation of AX and $A^t X$ [5].

The second step has an explicit solution, that is,

$$Z^{k+1} = S\left(WX^{k+1} + u_Z^k, \lambda\right) \qquad (5.15)$$

with S, so-called soft-shrinkage formula, defined as

$$S(x,\lambda) = \text{sgn}(x)\cdot\max(|\,x\,| - \lambda, 0). \qquad (5.16)$$

Note that Equation 5.15 is a component-wise operation for the input vector that performs Equation 5.16 for each scalar component. Moreover, the fact that the second step of Equation 5.13, that is, the reduced scalar version of the L1-norm problem, has the analytical solution (Equation 5.16) is exactly the motivation to introduce the dummy variable $Z = WX$ in Equation 5.9.

Next, for isotropic TV norm, similarly we introduce the dummy variable $Z = [Z_x \ Z_y \ Z_z \ Z_t]$, that is, $Z_x = D_x X$, $Z_y = D_y X$, $Z_z = D_z X$, and $Z_t = D_t X$. Then, corresponding to Equation 5.10,

$$\min_{(X,Z)} \left\| \sqrt{Z_x^2 + Z_y^2 + Z_z^2 + Z_t^2} \right\|_1, \quad \text{subject to } AX = Y, DX = Z. \quad (5.17)$$

As a result, we still arrive at Equation 5.13, in which the only difference is for the second step Equation 5.16, which now is replaced by the so-called isotropic soft-shrinkage formula, that is,

$$\left[Z_x^{k+1} \quad Z_y^{k+1} \quad Z_z^{k+1} \quad Z_t^{k+1} \right]$$

$$= \frac{[x_x \quad x_y \quad x_z \quad x_t]}{\sqrt{x_x^2 + x_y^2 + x_z^2 + x_t^2}} \cdot \max\left(\sqrt{x_x^2 + x_y^2 + x_z^2 + x_t^2} - \lambda, 0 \right), \quad (5.18)$$

where $x = DX^{k+1} + u_Z^k$.

Therefore, both the 3D method (Equation 5.7) and the 4D method (Equation 5.8) can be efficiently solved by the simple ADMM iterative scheme (Equation 5.13) with the isotropic soft-shrinkage formula. As mentioned earlier, it is important to keep in mind of the forward model mismatch in practice, and therefore, the iteration loop (Equation 5.13) should stop when $\|AX - Y\|$ is sufficiently small. On the other hand, for fair comparison, it is important to note that the optimal parameters (μ, λ) may not be the same for the 3D method and the 4D method, and therefore, need to be optimized individually.

5.2.3 Parallelization via ADMM

Due to the iterative nature, the reconstruction speed is an important question to be addressed in order for the clinical use of iterative methods. It happens that ADMM has the desirable feature in terms of parallelizable optimization, or so-called distributed optimization [10]. Unlike the parallelization of A and A^t, the ADMM algorithm itself is parallelizable which will be briefly presented here.

To simplify the discussion and illustrate the key idea, we consider

$$\min_X \frac{1}{2} \|AX - Y\|_2^2 + \lambda \| X \|_1. \quad (5.19)$$

First of all, CT is a separable problem in terms of forward model, that is,

$$\|AX - Y\|_2^2 = \sum_i \|A_i X - Y_i\|_2^2, \tag{5.20}$$

where i indexes the projection views with N projections. Note that the temporal dependence of X is ignored here, which however, could be considered as a double index of i.

For parallelized computation of the large problem with X into many smaller problems with x_i only, we introduce the dummy variable that $x_i = X$, so that Equation 5.19 becomes

$$\min_{(X, x_i)} \frac{1}{2} \sum_i \|A_i x_i - Y_i\|_2^2 + \lambda \|X\|_1, \quad \text{subject to } x_i = X. \tag{5.21}$$

Then, its augmented Lagrangian is

$$L(X, x) = \frac{1}{2} \sum_i \left(\|A_i x_i - Y_i\|_2^2 + \mu \|x_i - X + u_i\|_2^2 \right) + \lambda \|X\|_1. \tag{5.22}$$

And therefore, the ADMM iterative scheme consists of the following:

$$\begin{cases} x_i^{k+1} = \arg\min_{x_i} \frac{1}{2} \|A_i x_i - Y_i\|_2^2 + \frac{\mu}{2} \|x_i - X^k + u_i^k\|_2^2, i \leq N \\ X^{k+1} = \arg\min_X \frac{\lambda}{\mu} \|X\|_1 + \frac{1}{2} \sum_i \|x_i^{k+1} - X + u_i^k\|_2^2 \\ u_i^{k+1} = u_i^k + \left(x_i^{k+1} - X^{k+1} \right) \end{cases} \tag{5.23}$$

in which the first step simply updates each x_i as a much smaller L2 problem that is completely parallelizable.

Moreover, the second shrinkage step can be transformed to

$$X^{k+1} = \arg\min_X \frac{\lambda}{\mu N} \|X\|_1 + \frac{1}{2} \left\| \frac{\sum_i \left(x_i^{k+1} + u_i^k \right)}{N - X} \right\|_2^2. \tag{5.24}$$

Then Equation 5.23 can be rewritten as

$$
\begin{cases}
x_i^{k+1} = \arg\min_{x_i} L(X^k, x), \quad i \leq N. \\[2mm]
\bar{X}^{k+1} = \dfrac{\sum_i \left(x_i^{k+1} + u_i^k \right)}{N}. \\[2mm]
X^{k+1} = S\left(\bar{X}^{k+1}, \dfrac{\lambda}{\mu N} \right). \\[2mm]
u_i^{k+1} = u_i^k + \left(x_i^{k+1} - X^{k+1} \right).
\end{cases}
\tag{5.25}
$$

A simple interpretation of the iterative scheme Equation 5.25 is that x_i is first computed based on its projection data, then averaged together and thresholded to update X, and then the corresponding contribution is distributed to x_i through the update of u_i. The scheme is completely parallelizable. It is important to note that (1) the L2 problem of x_i could be solved much faster due to a much smaller system matrix than X, which could be formed explicitly, so that the L2 problems can be solved by one-time Cholesky factorization and fast triangular matrix inversions; (2) it may be a better option to consider multiple projections for each x_i instead of a single projection, for example, according to the temporal binning in 4DCT/CBCT, so that x_i can be reconstructed with better accuracy during iterations.

5.2.4 Numerical Results

Here we compare TV-based 3D method (i.e., 3DTV) and 4D method (i.e., 4DTV) using the experimental data from the Siemens helical 4DCT system: (1) the normal-pitch normal-dose scan (Figure 5.1), (2) the normal-pitch low-dose scan (Figure 5.2), (3) the high-pitch normal-dose scan (Figure 5.3), and (4) the high-pitch low-dose scan (Figure 5.4). Here the normal pitch $p = 0.09$ with the normal dose rate $d = 296.2$ mAs and the low dose rate $d = 159.4$ mAs, while the high pitch $p = 0.36$ with the normal dose rate $d = 259.3$ mAs and the low dose rate $d = 129.1$ mAs. And the 2D slice images presented in figures are 512×512 with 0.98 mm in-plane resolution, and axial 1.32 mm axial resolution.

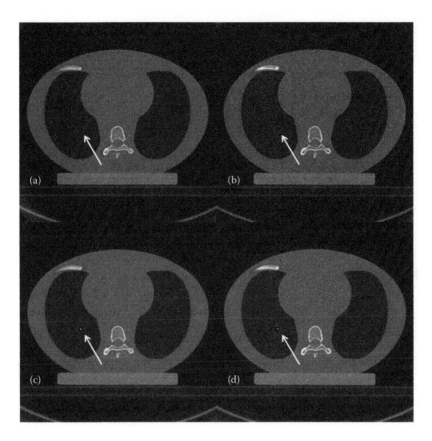

FIGURE 5.1 The normal-pitch normal-dose results. (a) 3DTV with 100% data; (b) 3DTV with 25% data; (c) 4DTV with 100% data; (d) 4DTV with 25% data.

We can draw from the visual assessments of Figures 5.1 through 5.4: (1) 4DTV has better image quality than 3DTV in all situations, and it is important to notice that 4DTV is able to reconstruct the details that 3DTV fails to, for example, the point mark in the low-dose cases (Figures 5.2 and 5.4) and (2) 4DTV has more consistent performance than 3DTV for high-pitch or low-dose or undersampled data (i.e., 25% data).

5.3 ADVANCED TEMPORAL REGULARIZATION METHODS

5.3.1 High-Order Local Methods

Inspired by improved image quality via TV regularization which is a local sparsity transform method, various high-order local regularization

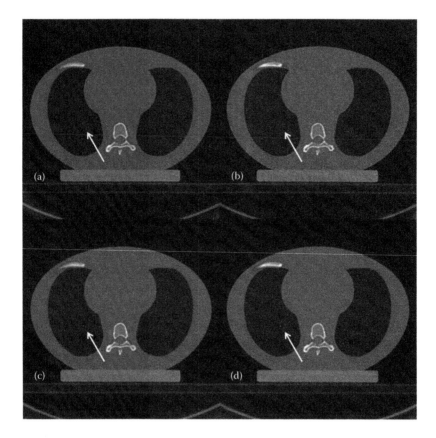

FIGURE 5.2 The normal-pitch low-dose results. (a) 3DTV with 100% data; (b) 3DTV with 25% data; (c) 4DTV with 100% data; (d) 4DTV with 25% data.

methods have been proposed. Here, we introduce a tight frame approach that naturally generalizes TV transform to high order and also the iso-tropic TV norm, so-called tensor framelet (TF) [12,13]. In addition, its tight-frame mathematical structure allows the fast computation of TF transform and its adjoint.

Here we consider a simple TF example, that is, one-level piecewise-linear TF transform,

$$WX = \frac{1}{\sqrt{3}} \begin{bmatrix} D_0 X \\ D_1 X \\ D_2 X \end{bmatrix}, \qquad (5.26)$$

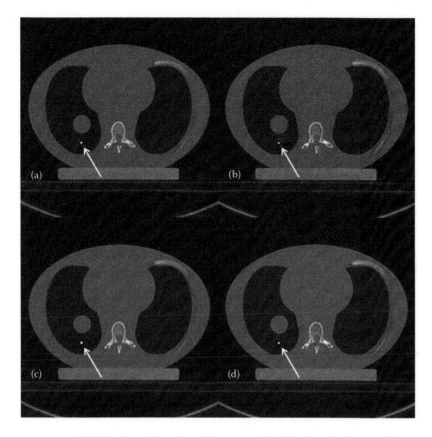

FIGURE 5.3 The high-pitch normal-dose results. (a) 3DTV with 100% data; (b) 3DTV with 25% data; (c) 4DTV with 100% data; (d) 4DTV with 25% data.

in which the averaging operator, the first-order derivative operator, and the second-order derivative operator are defined, respectively, as

$$D_0X\big|_{ijkn} = \begin{bmatrix} D_{0x}X \\ D_{0y}X \\ D_{0z}X \\ D_{0t}X \end{bmatrix}_{ijkn} = \frac{1}{4}\begin{bmatrix} X_{i+1,j,k,n}+2X_{ijk}+X_{i-1,j,k,n} \\ X_{i,j+1,k,n}+2X_{ijk}+X_{i,j-1,k,n} \\ X_{i,j,k+1,n}+2X_{ijk}+X_{i,j,k-1,n} \\ X_{i,j,k,n+1}+2X_{ijk}+X_{i,j,k,n-1} \end{bmatrix}, \qquad (5.27)$$

$$D_1X\big|_{ijkn} = \begin{bmatrix} D_{1x}X \\ D_{1y}X \\ D_{1z}X \\ D_{1t}X \end{bmatrix}_{ijkn} = \frac{\sqrt{2}}{4}\begin{bmatrix} X_{i+1,j,k,n}-X_{i-1,j,k,n} \\ X_{i,j+1,k,n}-X_{i,j-1,k,n} \\ X_{i,j,k+1,n}-X_{i,j,k-1,n} \\ X_{i,j,k,n+1}-X_{i,j,k,n-1} \end{bmatrix}, \qquad (5.28)$$

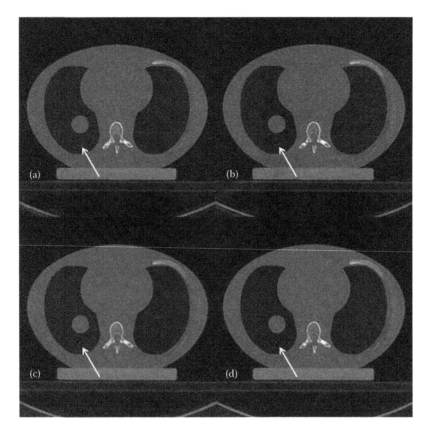

FIGURE 5.4 The high-pitch low-dose results. (a) 3DTV with 100% data; (b) 3DTV with 25% data; (c) 4DTV with 100% data; (d) 4DTV with 25% data.

$$
D_2 X\big|_{ijkn} = \begin{bmatrix} D_{2x}X \\ D_{2y}X \\ D_{2z}X \\ D_{2t}X \end{bmatrix}_{ijkn} = \frac{1}{4}\begin{bmatrix} -X_{i+1,j,k,n}+2X_{ijk}-X_{i-1,j,k,n} \\ -X_{i,j+1,k,n}+2X_{ijk}-X_{i,j-1,k,n} \\ -X_{i,j,k+1,n}+2X_{ijk}-X_{i,j,k-1,n} \\ -X_{i,j,k,n+1}+2X_{ijk}-X_{i,j,k,n-1} \end{bmatrix}. \tag{5.29}
$$

Here the coefficients are chosen so that $W^T W = I$, that is, the TF transform is conservative in the sense of the left inverse.

Then similar to isotropic TV norm (Equation 5.6), the TF norm is defined as

$$
\| WX \|_1 = \lambda_0 \| D_0 X \|_1 + \lambda_1 \| D_1 X \|_1 + \lambda_2 \| D_2 X \|_1 \tag{5.30}
$$

with

$$\| D_m X \|_1 = \sum_{i,j,k,n} \sqrt{| D_{mx} X |^2 + | D_{my} X |^2 + | D_{mz} X |^2 + | D_{mt} X |^2}, \quad m = 0,1,2.$$

(5.31)

It is apparent Equation 5.30 naturally generalizes the concept of isotropic TV norm with high-order gradients on multilevel. Similarly, there exist fast thresholding formulas for TF corresponding to Equations 5.16 and 5.18 so that TF-regularized iterative image reconstruction (Equation 5.9) can be conveniently solved by the ADMM algorithm given earlier.

It is worthy to note that TF is more efficient than the standard tight frame method for high-dimensional problems. Take one-level piecewise-linear case in 4DCT/CBCT for example. In terms of the memory, the standard tight frame requires ~$3^4 N$ memory, while TF requires ~$3 \times 4N$ memory, where $N = N_x \cdot N_y \cdot N_z \cdot N_t$. In terms of the computational cost, the standard tight frame needs ~$3^{2\times4} N$ operations, while TF needs ~$3 \times 4^2 N$ operations. The interested readers may refer to [12,13] for more details on TF and its applications to 4DCBCT [12], tomotherapy MVCT [13], and breast CT [14,15].

5.3.2 Nonlocal Methods

In contrast to the local regularization methods discussed earlier, we consider in this section the nonlocal methods, for example, nonlocal mean [16], nonlocal TV [17], and BM3D [18]. Different from the local methods that usually regularize the images locally by convolving images with various filters of certain embedded priors, in nonlocal methods the pixel values are often compared with its neighboring pixels to form certain weighted nonlocal transform matrix that reflects the similarity of patches and is applied to other regions or images for image regularization. Here we only review the use of temporal nonlocal mean (TNLM) method that was developed for 4DCT [19], that is,

$$R(X) = \sum_{n=1}^{N_t-1} J(X_n, X_{n+1}).$$

(5.32)

Equation 5.32 enforces the joint similarity for every neighboring temporal pair of images, and the similarity metric J consists of the similarity measure of each pixel in one image with respect to every pixel in the other image, that is,

$$J(X_m, X_n) = \sum_{ijk} \sum_{i'j'k'} w_{ijkm,i'j'k'n} \mid X_{ijkm} - X_{i'j'k'n} \mid^2,$$ (5.33)

where the weighting factor is the following, subject to a normalization factor,

$$w_{ijkm,i'j'k'n} \propto e^{-\|p_m(i,j,k) - p_n(i',j',k')\|_2^2 / h^2}.$$ (5.34)

In Equation 5.34, $p_m(i, j, k)$ refers to a collection of pixels within a patch centered at the pixel (i, j, k) of the mth image with h as a controlling parameter.

TNLM (Equation 5.32) regularization is essentially a nonlinear method due to the weight dependence on the images in Equation 5.34. A standard technique to handle this is to deal with TNLM iteratively, that is, to use the previous iterate X^k to compute w^k. As a result, TNLM with w^k is a linear transform on X^{k+1}, and therefore, TNLM can also be conveniently solved by the ADMM algorithm.

5.3.3 Low-Rank Models

The low-rank models [20] are another type of global regularization methods, where the sparsity is enforced through the principal value thresholding for the image matrix. Let us first consider the primitive low-rank model for which we rewrite the dynamic images in the matrix form

$$X = \begin{bmatrix} X_1 & \cdots & X_n & \cdots & X_{N_t} \end{bmatrix},$$ (5.35)

Then the low-rank regularized image reconstruction is

$$\min_X \| X \|_*, \text{ subject to } AX = Y,$$ (5.36)

where $\|\cdot\|_*$ is the nuclear norm as the convex relaxation for minimizing the rank of X, in the same spirit as the L1 norm as the convex relaxation of minimizing the number of nonzeros in X. Mathematically, the nuclear norm of X is equal to the summation of singular values of X in the matrix form Equation 5.35. The motivation for rank minimization is that the matrix of similar images tends to have a low rank, for example, the rank of a static sequence of images is one.

The low-rank model Equation 5.36 can be also efficiently solved by ADMM, in which what is new is the nuclear-norm subproblem

$$X^{k+1} = \arg\min_X \frac{1}{2} \| X - X^k \|_2^2 + \lambda \| X \|_*, \tag{5.37}$$

whose solution is explicitly given by the so-called singular value thresholding formula [21], that is,

$$X^{k+1} = T(X^k, \lambda) := U \cdot \operatorname{diag}(\max(\sigma - \lambda, 0)) \cdot V^T, \quad \text{with } X^k = U \cdot \operatorname{diag}(\sigma) \cdot V^T. \tag{5.38}$$

As the variants of Equation 5.36, one may also consider (1) rank in the transform domain, for example, rank of similar patches [22] or rank of sparsity transform of images [23], and (2) tensor versions of low-rank models [24,25].

In addition, an alternative way to regularize the rank is through the matrix factorization approach [26], that is,

$$(L,R) = \arg\min_{(L,R)} \| WL \|_1 + \lambda \| R \|, \quad \text{subject to } AX = Y, X = LR. \tag{5.39}$$

Here the rank regularization on X is through the explicit decomposition model that each column of X is a linear combination of the basic components L with its rank to be at most the number of columns of L, while each row of R represents the linear combination coefficients of a column of X in L.

Moreover, it is possible to develop the iterative scheme that uses both the sparsity and the rank, at least in two ways: (1) the simultaneous rank-sparsity regularization [27,28], that is,

$$X = \arg\min_X \| X \|_* + \lambda \| WX \|_1, \quad \text{subject to } AX = Y \tag{5.40}$$

and (2) the rank-sparsity decomposition regularization [23,29,30], that is,

$$(X_L, X_S) = \arg\min_{(X_L, X_S)} \| X_L \|_* + \lambda \| WX_S \|_1, \quad \text{subject to } AX = Y, X = X_L + X_S. \tag{5.41}$$

While the motivation for Equation 5.40 is clear, the rank-sparsity decomposition is motivated by the observation that the slowly varying background X_L and sparse contrast changes X_S or motion can be characterized, respectively, by a nuclear norm and a L1 sparsity norm.

5.3.4 Learning-Based Algorithms

When the prior images similar to the current images to be reconstructed are available, the learning-based algorithms are attractive in the sense that they potentially improve the imaging quality by constructively learning from the existing library. For example, the prior image X_0 can be conveniently utilized in PICCS [8] to reduce the streaking artifacts via the local regularization

$$R(X) = \| WX \|_1 + \lambda \| W_t(X - X_0) \|_1 . \tag{5.42}$$

Similarly, in the prior-image constrained low-rank (PCLR) model [31], the prior image X_0 can also be conveniently utilized to globally regularize the image, that is,

$$R(X) = \| [X \quad X_0] \|_* . \tag{5.43}$$

Similarly the variants of Equation 5.43 may be considered, for example, the patch- or transform-based prior-image rank constraints.

Moreover, one may also consider other advanced learning techniques, such as dictionary learning [32,33] and adaptive tight frame (ATF) method [34,35]. In dictionary learning, that is,

$$D = \underset{(D,\alpha)}{\arg\min} \| PX_0 - D\alpha \|_2^2 + \lambda \| \alpha \|_1, \tag{5.44}$$

a dictionary D is sought such that the prior image X_0 is sparsely represented in dictionary D with the sparse coefficients α. It is assumed that the current image X to be reconstructed can also be sparsely represented by D, since X is similar to X_0. And therefore, the dictionary learning–based image reconstruction is formulated as

$$X = \underset{(X,\alpha)}{\arg\min} \| AX - Y \|_2^2 + \mu \| PX - D\alpha \|_2^2 + \lambda \| \alpha \|_1 . \tag{5.45}$$

Note that the dictionary learning has no structure information. In contrast, in ATF method the new tight frame filters H are constructed so that the prior image X_0 is sparsely represented under the new tight frame transform $W(H)$, for example, the TF-based ATF scheme [35]

$$H = \underset{(H,\alpha)}{\arg\min} \| W(H)X_0 - \alpha \|_2^2 + \lambda \| \alpha \|_1, \quad \text{subject to } W^T W = I, \qquad (5.46)$$

where $W^T W = I$ is strictly constrained so that $W(H)$ is a tight frame transform. And then this learned TF that is adaptive to X_0 can be used in the conventional way, that is,

$$\underset{X}{\min} \| W(H)X \|_1, \quad \text{subject to } AX = Y. \qquad (5.47)$$

Comparing with dictionary learning, ATF has the apparent advantage in speed [34].

5.3.5 Others

Other temporal regularization methods that are also potentially interesting 4DCT/CBCT include, but not limit to: (1) the temporal Fourier transform method, that is, the L1-norm of Fourier transform with respect to the temporal dimension, which has been commonly used for dynamic MRI [36]; (2) Lp sparsity with $0 \le p < 1$, as a closer approximation of L0 sparsity than L1 sparsity [37,38]; (3) group sparsity that may be considered for temporal regularization where each temporal sequence of a spatial pixel or, pixels within a decomposed region per image intensity or the prior image, forms a group [39]; and (4) Kalman filter techniques [40] that have been extensively studied in other imaging fields.

5.4 CONCLUSION

Based on TV regularization, we have demonstrated the improved image quality from 3D method to 4D method using 4D helical CT Siemens data. ADMM, as a powerful solution algorithm for iterative reconstruction methods, has been reviewed, with the emphasis on its parallelizable nature that should be useful in dealing with the large-scale computing problems in 4DCT/CBCT. In addition, various temporal regularization techniques that may be suitable for 4DCT/CBCT have been reviewed, and the interested readers may refer to their references to develop and further optimize their own 4D methods that are pertinent to 4DCT/CBCT.

REFERENCES

1. Low, D., M. Nystrom, E. Kalinin, P. Parikh, J. Dempsey, J. Bradley, S. Mutic et al., A method for the reconstruction of four-dimensional synchronized CT scans acquired during free breathing, *Medical Physics* 30, 1254–1263 (2003).

2. Keall, P. J., G. Starkschall, H. Shukla, K. M. Forster, V. Ortiz, C. W. Stevens, S. S. Vedam, R. George, T. Guerrero, and R Mohan, Acquiring 4D thoracic CT scans using a multislice helical method, *Physics in Medicine & Biology* 49, 2053–2067 (2004).

3. Sonke, J.-J., L. Zijp, P. Remeijer, and M. van Herk, Respiratory correlated cone beam CT, *Medical Physics* 32, 1176–1186 (2005).

4. Li, T., L. Xing, P. Munro, C. McGuinness, M. Chao, Y. Yang, B. Loo, and A. Koong, Four-dimensional cone-beam computed tomography using an on-board imager, *Medical Physics* 33(10), 3825–3833 (2006).

5. Gao, H., Fast parallel algorithms for the x-ray transform and its adjoint, *Medical Physics* 39, 7110–7120 (2012).

6. Buzug, T. M., *Computed Tomography: From Photon Statistics to Modern Cone-Beam CT.* Springer, Berlin, Germany (2008).

7. Sidky, E. Y., C.-M. Kao, and X. Pan, Accurate image reconstruction from few-views and limited-angle data in divergent-beam CT, *Journal of X-Ray Science and Technology* 14, 119–139 (2006).

8. Chen, G. H., J. Tang, and S. Leng, Prior image constrained compressed sensing (PICCS): A method to accurately reconstruct dynamic CT images from highly undersampled projection data sets, *Medical Physics* 35, 660–663 (2008).

9. Rudin, L., S. Osher, and E. Fatemi, Nonlinear total variation based noise removal algorithms, *Journal of Physics D* 60, 259–268 (1992).

10. Boyd, S., N. Parikh, E. Chu, B. Peleato, and J. Eckstein, Distributed optimization and statistical learning via the alternating direction method of multipliers, *Foundations and Trends in Machine Learning* 3(1), 1–122 (2011).

11. Goldstein, T. and S. Osher, The split Bregman algorithm for l1 regularized problems, *SIAM Journal on Imaging Sciences* 2, 323–343 (2009).

12. Gao, H., R. Li, Y. Lin, and L. Xing, 4D cone beam CT via spatiotemporal tensor framelet, *Medical Physics* 39, 6943–6946 (2012).

13. Gao, H., X. Sharon Qi, Y. Gao, and D. A. Low, Megavoltage CT imaging quality improvement on TomoTherapy via tensor framelet, *Medical Physics* 40(8), 081919 (2013).

14. Zhao, B., H. Gao, H. Ding, and S. Molloi, Tight-frame based iterative image reconstruction for spectral breast CT, *Medical Physics* 40(3), 031905 (2013).

15. Ding, H., H. Gao, B. Zhao, H.-M. Cho, and S. Molloi, A high-resolution photon-counting breast CT system with tensor-framelet based iterative image reconstruction for radiation dose reduction, *Physics in Medicine & Biology* 59(20), 6005 (2014).

16. Buades, A., B. Coll, and J.-M. Morel, A review of image denoising algorithms, with a new one, *Multiscale Modeling & Simulation* 4(2), 490–530 (2005).

17. Gilboa, G. and S. Osher, Nonlocal operators with applications to image processing, *Multiscale Modeling & Simulation* 7(3), 1005–1028 (2008).

18. Dabov, K., A. Foi, V. Katkovnik, and K. Egiazarian, Image denoising by sparse 3-D transform-domain collaborative filtering, *IEEE Transactions on Image Processing* 16(8), 2080–2095 (2007).

19. Jia, X., Y. Lou, B. Dong, Z. Tian, and S. Jiang, 4D computed tomography reconstruction from few-projection data via temporal non-local regularization. In: *Medical Image Computing and Computer-Assisted Intervention— MICCAI 2010*, pp. 143–150. Springer, Berlin, Germany, 2010.

20. Recht, B., M. Fazel, and P. A. Parrilo, Guaranteed minimum-rank solutions of linear matrix equations via nuclear norm minimization, *SIAM Review* 52(3), 471–501 (2010).

21. Cai, J. F., E. J. Candès, and Z. Shen, A singular value thresholding algorithm for matrix completion, *Journal on Optimization* 20, 1956–1982 (2010).

22. Schaeffer, H. and S. Osher, A low patch-rank interpretation of texture, *SIAM Journal on Imaging Sciences* 6(1), 226–262 (2013).

23. Gao, H., H. Yu, S. Osher, and G. Wang, Multi-energy CT based on a prior rank, intensity and sparsity model (PRISM), *Inverse Problems* 27(11), 115012 (2011).

24. Li, L., Z. Chen, G. Wang, J. Chu, and H. Gao, A tensor PRISM algorithm for multi-energy CT reconstruction and comparative studies, *Journal of X-Ray Science and Technology* 22(2), 147–163 (2014).

25. Golub, G. H. and C. F. Van Loan, *Matrix Computations*, vol. 3. JHU Press, Baltimore, MD, 2012.

26. Cai, J.-F., X. Jia, H. Gao, S. B. Jiang, Z. Shen, and H. Zhao, Cine cone beam CT reconstruction using low-rank matrix factorization: Algorithm and a proof-of-principle study, *IEEE Transactions on Medical Imaging* 33, 1581–1591 (2014).

27. Lingala, S. G., Y. Hu, E. DiBella, and M. Jacob, Accelerated dynamic MRI exploiting sparsity and low-rank structure: kt SLR, *IEEE Transactions on Medical Imaging* 30(5), 1042–1054 (2011).

28. Gao, H., Y. Lin, C. B. Ahn, and O. Nalcioglu, PRISM: A divide-and-conquer low-rank and sparse decomposition model for dynamic MRI, *CAM Report* 11-26, 2011.

29. Gao, H., J. F. Cai, Z. Shen, and H. Zhao, Robust principle component analysis based four-dimensional computed tomography, *Physics in Medicine & Biology* 56, 3181–3198 (2011).

30. Otazo, R., E. Candès, and D. K. Sodickson, Low-rank plus sparse matrix decomposition for accelerated dynamic MRI with separation of background and dynamic components, *Magnetic Resonance in Medicine* 73, 1125–1136, doi: 10.1002/mrm.25240 (2014).

31. Gao, H., L. Li, and X. Hu, *Compressive Diffusion MRI—Part 3: Prior-Image Constrained Low-Rank Model (PCLR)*. ISMRM, Salt Lake City, UT, 2013.

32. Elad, M. and M. Aharon, Image denoising via sparse and redundant representations over learned dictionaries, *IEEE Transactions on Image Processing* 15(12), 3736–3745 (2006).

33. Xu, Q., H. Yu, X. Mou, L. Zhang, J. Hsieh, and G. Wang, Low-dose X-ray CT reconstruction via dictionary learning, *IEEE Transactions on Medical Imaging* 31(9), 1682–1697 (2012).

34. Cai, J.-F., H. Ji, Z. Shen, and G.-B. Ye, Data-driven tight frame construction and image denoising, *Applied and Computational Harmonic Analysis* 37(1), 89–105 (2014).

35. Zhou, W., J.-F. Cai, and H. Gao, Adaptive tight frame based medical image reconstruction: A proof-of-concept study for computed tomography, *Inverse Problems* 29(12), 125006 (2013).

36. Lustig, M., J. M. Santos, D. L. Donoho, and J. M. Pauly, kt SPARSE: High frame rate dynamic MRI exploiting spatio-temporal sparsity. In: *Proceedings of the 13th Annual Meeting of ISMRM*, vol. 2420, Seattle, WA, 2006.

37. Candes, E. J., M. B. Wakin, and S. P. Boyd, Enhancing sparsity by reweighted ℓ1 minimization, *Journal of Fourier Analysis and Applications* 14(5–6), 877–905 (2008).

38. Sidky, E. Y., R. Chartrand, and X. Pan, Image reconstruction from few views by non-convex optimization. In: *Nuclear Science Symposium Conference Record, 2007* (*NSS'07*), vol. 5, pp. 3526–3530. IEEE, Honolulu, HI (2007).

39. Huang, J., X. Huang, and D. Metaxas, Learning with dynamic group sparsity. In: *IEEE 12th International Conference on Computer Vision, 2009*, pp. 64–71. IEEE, Kyoto, Japan, 2009.

40. Kalman, R. E., A new approach to linear filtering and prediction problems, *Journal of Fluids Engineering* 82(1), 35–45 (1960).

CHAPTER 6

Multi-GPU Cone-Beam CT Reconstruction

Hao Yan and Xiaoyu Wang

CONTENTS

6.1 INTRODUCTION

Cone-beam CT (CBCT) [1,2] is widely used in image-guided radiation therapy (IGRT) for pretreatment patient setup and reconstruction plays a vital role in CBCT. Over the years, conventional Feldkamp-Davis-Kress (FDK)-type reconstruction algorithms [3] have been used in commercial CBCT systems. The reason is that FDK is based on an analytical formula, making it robust in different clinical contexts and simple enough to yield acceptable computation efficiency. Even though, an iterative reconstruction (IR) approach is still desired. One particular reason is that compared to FDK, IR is capable of reconstructing CBCT images with better image quality from noisy and under-sampled x-ray projections [4–24]. This allows delivering lower CBCT imaging dose to the patient, addressing the clinical concern on the excessive cumulated imaging dose during the whole treatment course of the CBCT-based IGRT [25–27]. However, computational inefficiency becomes a great obstacle, preventing IR being applied in the clinic. It is mainly because of the large problem size and the iterative nature of the algorithm. Specifically, an IR algorithm usually reconstructs a CBCT image by solving an optimization problem using an iterative numerical algorithm. Inside each iteration step, a forward projection and a backward projection are typically computed, both of which have complexities similar to those of the FDK-type reconstruction algorithm. Since a number of iterations are required to yield a clinically acceptable image quality, the overall computation time is much longer than that of a typical FDK algorithm. In addition, FDK algorithm performs the back projection sequentially, making it feasible to conduct reconstruction immediately after data acquisition starts. In contrast, an IR method requires all the projections at the initializing stage, prohibiting the concurrent execution of data acquisition and reconstruction. While graphics processing units (GPUs) have been employed recently to accelerate the IR process [14,19,28–33], it is still necessary to further boost the efficiency for the time-critical IGRT environment.

This chapter is on a multi-GPU-based CBCT IR system development. While using multiple GPUs is a straightforward idea for further efficiency boost, inter-GPU parallelization is not a trivial problem. Specifically, communications among GPUs should be handled with care to achieve satisfactory efficiency, as different GPUs only hold their own memory. From a parallel computing point of view, conventional memory organization in a parallel processing task is either distributed memory, where each unit holds its own memory space and inter-unit data communication is conducted in, for example, a CPU cluster [34], or shared memory, where all processing units share a common memory space in, for example, a GPU. Yet, CBCT reconstruction on a multi-GPU platform attains a hybrid structure of distributed and shared memory. It is a careful design of the data allocation and communication among GPUs that maximally exploit the potential of all GPUs, as will be shown in this chapter.

6.2 GENERAL CONSIDERATIONS OF INTER-GPU PARALLELIZATION

6.2.1 Typical Structure of an IR Algorithm

Essentially, IR solves a linear equation $Pf = g$, where f is the unknown image and g is the measurement of f, that is, the projections measured at certain angles, and P is a projection operator corresponding to those angles. Mathematically, image reconstruction from few projections is an ill-posed problem, as there are infinitely many solutions satisfying Pf = g. For such kind of problems, regularization based on assumptions about the solution f has to be performed in order to discard those undesirable solutions. For more detailed description and reviews of IR algorithms, readers are referred to the literature [35–40]. In general, the following two steps are iteratively conducted in a typical IR algorithm. First, the forward projections of the reconstructed image f should match the measurement g. In our IR system, this fidelity condition is enforced by solving the least-square minimization problem $\min_f E[f] = \|Pf - g\|_2^2$ using a conjugate gradient least square (CGLS) algorithm [14,19,41]. Second, a regularization step is preformed to regularize the reconstructed CBCT image f. Example regularization methods include quadratic constrain [42,43], total variation (TV) minimization [4,6,14], which assumes that the solution is piecewise constant, and tight frame (TF) reconstruction [19], which assumes that the image has a sparse representation under the TF basis [44–47], an over-complete wavelet basis. An additional regularization step is to enforce the positivity of the solution by truncating the negative values in f to zeros, because the reconstructed CBCT image f physically represents x-ray attenuation coefficients, which are supposed to be nonnegative. In the rest of this chapter, we will use the TF-based IR approach as an example method for demonstration.

6.2.2 Multi-GPU System Setup and Overall Structure

We have developed a multi-GPU IR system to improve the computational efficiency. The system is built on a desktop workstation with two Nvidia GTX590 GPU cards. Each of the two cards contains two identical GPUs, so that there are four GPUs available. These GPUs are labeled as GPU 1–4 in the rest of this chapter. For each GPU, there are 512 thread processors, each of which attains a clock speed of 1.2 GHz. All processors share 1.5 GB GDRR5 global memory at a 164 GB/s memory bandwidth. Among GPUs on different cards, data transfer is through the computer motherboard via PCIe-16 bus. Between GPUs on the same card, data transfer directly through a PCI switch on the card, instead of through the motherboard. The system is written in CUDA 4.0 [48], a C language extension that allows for the programming of each individual GPU, as well as inter-GPU communications.

Typical CS-based CBCT reconstruction algorithms share a set of key operations in common. These operations are implemented in the

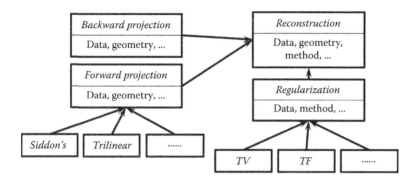

FIGURE 6.1 Illustration of the overall structure of the multi-GPU cone-beam CT reconstruction system. (Reprinted from Yan, H. et al., *Med. Phys.*, 41(11), 111912, 2014.)

system as individual modules, so that each specific algorithm can be used as building block. The overall structure is schematically illustrated in Figure 6.1. At the beginning of the reconstruction process, a user is asked to select an algorithm to perform reconstruction. After that, the reconstruction process launches and the corresponding modules are invoked. For instance, forward x-ray projection is a key module in all IR algorithms, where the x-ray projections of the currently reconstructed images are computed. Siddon's algorithm [49] and trilinear interpolation algorithm [50] are currently available in our system for this purpose (with others to be supported in the future). Another example is a regularization module. Minimization of a TV term is conducted in a TV-based reconstruction algorithm [4,14] to remove noise and any undesired streak artifacts, while preserving image edges. Shrinkage of tight frame (TF) coefficients is performed in TF-based approaches for the same purpose [19]. These two modules are currently supported in our system, and other module, such as dictionary learning–based regularization [51], will be incorporated in future.

6.2.3 Dataset Management

Generally speaking, two data sets are involved in two different domains in the CBCT reconstruction: one is in the image domain and the other in the projection domain. When it comes to multi-GPU, at least one of the two, if not both, needs to be partitioned and each GPU stores one portion. Hence, inter-GPU communications are needed. Practically, there are an infinite

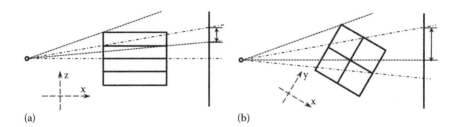

(a) (b)

FIGURE 6.2 Illustrations of projection overlap issue when partitioning the CBCT volume. (a) Side view of a partition with planes parallel to the xoy plane. (b) Top view of a partition with planes containing the rotational axis. Dash lines and dash-dot lines are projection regions of two sub-volumes and arrows indicate the overlap regions on the detector. (Reprinted from Yan, H. et al., *Med. Phys.*, 41(11), 111912, 2014.)

number of ways to divide the data. It is for practical considerations that one is preferred over the others.

One possible partitioning is to divide the CBCT volume into sub-volumes. Each GPU holds one sub-volume and all the projection data. Figure 6.2 illustrates two different ways of partitioning. Let the rotation axis be the z axis. The partition using planes that are parallel to the xoy plane is shown in Figure 6.2a and the one using planes that contain the rotational axis is in Figure 6.2b. Each GPU performs forward and backward projection for its dedicated sub-volume. Due to the prospective projection geometry in CBCT, the projections of different sub-volumes overlap with each other on the detector. The overlapped part from different GPUs has to be added every time after the forward projection operation is computed, causing cumbersome computational burdens. In contrast, the backward projection is straightforward.

Another way to partition is to divide the data in the projection domain. Each GPU stores a subset of projections at certain projection angles, plus the entire CBCT volume data. While each GPU can perform forward projection to its assigned angles, the backward projections to the image domain from different angles are to be accumulated from all GPUs, causing extra cost of data communication.

For a typical clinical case with an image resolution of $512 \times 512 \times 70$ voxels and a projection resolution of 512×384 pixels with 120 projection angles, the amount of data necessary to be communicated between GPUs in these two domains are similar. However, the second approach, the partition in the projection domain, avoids cumbersome treatments of the projection overlapping, and is therefore, chosen in our multi-GPU parallelization.

6.3 MULTI-GPU IMPLEMENTATION

6.3.1 Forward and Backward Projections

In our IR system, the CGLS step solves the least square problem $\min_f E[f] = \|Pf - g\|_2^2$. It includes three types of operations, namely, the computation of forward projection Pf, backward projection $P^T g$, and other vector–vector or scalar–vector operations. Under the partitioning in the projection domain, the backward projection operation requires special attentions. This operation computes $f = P^T g = \sum_{i=1}^{4} P_i^T g_i$, where g is a vector in the projection domain and is divided into different parts g_i, for $i = 1, \ldots, 4$. Each of them corresponds to a set of projection angles assigned to a GPU. The back projection operator can also be split into $P_i^T, i = 1, \ldots, 4$, and each sub-matrix represents the back projection in the corresponding angles. Note that each GPU keeps different g_i but the same CBCT volume data f. First of all, an intermediate variable $f_i = P_i^T g_i$ is computed at each GPU by employing the back projection formula derived in Jia et al. [19], which efficiently calculates the back projection results corresponding to projection angles at each GPU without any GPU memory writing conflict problem. After that, a parallel reduction among GPUs is conducted to compute the summation over all these intermediate variables. In this step, GPU 2 passes f_2 to GPU 1 and the summation $f_{12} = f_1 + f_2$ is calculated at GPU 1. The same operation is performed simultaneously at GPUs 3 and 4, leading to $f_{34} = f_3 + f_4$ at GPU 3. The reduction is then conducted one more time between GPUs 1 and 3 to get the final back projection result. The updated back projection result f is immediately broadcasted to all GPUs along the reverse path as in the parallel reduction, resulting in the same updated copy of f at all GPUs for later use. This process is illustrated in Figure 6.3a.

The forward projection operation, that is, the computation of $g = Pf$, is straightforward. Each GPU computes the projections of the CBCT volume data according to its designated projection angles, namely, $g_i = P_i f$, for $i = 1, \ldots, 4$. This task now becomes a standard forward x-ray digitally reconstructed radiograph calculation inside each GPU [52,53].

All the remaining operations are simply vector addition or scalar–vector multiplications, either in the CBCT image domain or in the CBCT

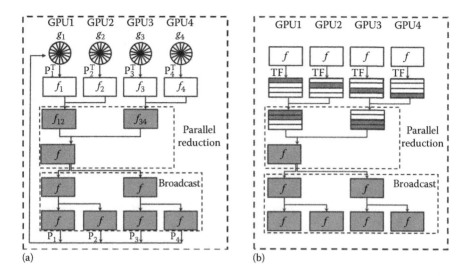

(a)　　　　　　　　　　　　　(b)

FIGURE 6.3　Workflow for the steps of (a) backward projection and (b) regularization. (Reprinted from Yan, H. et al., *Med. Phys.*, 41(11), 111912, 2014.)

projection domain. Take the calculation of the differences between the measurement and the forward projections of the current CBCT volume data as an example, such tasks are highly parallelizable. For operations in the projection domain, each GPU computes a subset of the projections and within a GPU, threads can independently process different pixels. For operations in the CBCT image domain, each GPU processes the whole CBCT volume data. Since there are apparently redundant computations, some alternative implementation may be available, for example, having a GPU only process a sub-volume. However, it may also involves additional inter-GPU communications, for example, for updating the sub-volume data, which leads to a relatively large overhead and may not be worthwhile for this simple job.

6.3.2　Regularizations and Multiple Resolution

There are three operations contained in the image regularization step in the TF reconstruction algorithm [19], namely decomposing the current CBCT image into the TF space, performing a shrinkage operation on these coefficients, and reconstructing the CBCT image from the updated coefficients. These operations at different voxels are independent of each other. In the multi-GPU implementation, each GPU performs regularization on a sub-volume data and within a GPU, each thread is responsible for the computations at a voxel. Note that no inter-GPU data transfer of the boundary layer between adjacent sub-volumes is needed before the regularization step, since each GPU already holds the same CBCT volume. After each GPU processes the designated sub-volume, the reduction of the results to the first GPU, as well as the posterior broadcasting of the entire volume to all GPUs, is conducted in a way similar to the backward projection step, as shown in Figure 6.3b. We have manually tuned regularization coefficients for each tested case to ensure a balance between removing streaking/noise artifacts and maintaining small fine structures, but it is still an open problem regarding appropriately selecting regularization coefficients adaptively under the IR framework.

Another feature that can be employed in iterative CBCT reconstructions is the multi-resolution [14,19]. To allow the freedom of handling reconstructions at different image resolutions, each of our modules takes relevant quantities as inputs, for example, voxel size and voxel numbers. At the beginning of each resolution level, the program sets these quantities and feeds them into the modules. When switching from a low-resolution level to a high-resolution level, it is necessary to up-sample the reconstructed CBCT image to get a proper initial value for the next resolution level. Because of the fact that the up-sampling is a relatively simple job and is not frequently performed, it is handled on only one GPU with no parallelization and the result is then broadcasted to other GPUs.

6.4 EXPERIMENTAL RESULTS

The computational time of TF-based CBCT reconstruction on our system is reported in Tables 6.1 and 6.2 for the CGLS step and the regularization step, respectively. We report here the computational time with a variety of resolutions separately, as the computational time depends on the number of projections and image resolution. Only real patient cases, namely,

TABLE 6.1   Computation Time (s) per Iteration for the CGLS Step in Single- and Multi-GPU Reconstructions

Protocol	Resolution	Single GPU	Multi-GPU	Speedup
Full-fan (121 projections)	$512 \times 512 \times 70$	9.15	2.67	3.43
	$256 \times 256 \times 70$	3.15	0.87	3.62
	$128 \times 128 \times 70$	1.26	0.36	3.50
Half-fan (164 projections)	$512 \times 512 \times 70$	10.20	3.09	3.30
	$256 \times 256 \times 70$	3.42	0.96	3.56
	$128 \times 128 \times 70$	1.32	0.36	3.67

TABLE 6.2 Computation Time (s) per Iteration for the Regularization Step in Single- and Multi-GPU Reconstructions

Protocol	Resolution	Single GPU	Multi-GPU	Speedup
Full-fan (121 projections)	$512 \times 512 \times 70$	1.58	0.80	1.98
	$256 \times 256 \times 70$	0.40	0.20	2.00
	$128 \times 128 \times 70$	0.10	0.06	1.67
Half-fan (164 projections)	$512 \times 512 \times 70$	1.59	0.80	1.99
	$256 \times 256 \times 70$	0.40	0.21	1.90
	$128 \times 128 \times 70$	0.10	0.06	1.67

the head-neck patient in the full-fan mode and the thorax/pelvis patients in the half-fan mode, which are of clinical interest, are included in this table. We report the time per iteration in these tables, because the total reconstruction time linearly increases with the number of iterations, and depending on the requirement on the final image quality, the number of iterations may vary significantly. In addition, we run our reconstruction code on a single GPU in our system, to quantitatively demonstrate the efficiency gain provided by the multi-GPU system. While the CGLS at each iteration step is also iterative by itself, we have used fixed number of iterations in the CGLS step as 3 in all the cases, which are found to be sufficient to ensure the projection condition.

The dependence of the computational time on the image resolution is plotted in Figure 6.4, which allows us to analyze the overhead of parallel processing. The calculation time in the CGLS step is plotted in a log–log scale in Figure 6.4a, where the data points form straight lines in both the single-GPU and the multi-GPU cases, indicating that computation time scales with image resolution following a power law. A linear fit in this plot leads to a function of the computational t and the transverse plane resolution x as $t_{CGLS} \sim x^{1.46}$ for the single-GPU case and $t_{CGLS} \sim x^{1.50}$ for the multi-GPU case. Note that the number of reconstructed transverse slices is fixed in all cases. The CGLS algorithm consists of mainly forward and backward projection operations. These operations scale linearly with the resolution x, as they are proportional to the number of voxels a ray line traverses. Yet, the observed scaling power ~1.5 is mainly due to the overheads in multi-GPU parallelization. The small amount of vector–vector operations in the CGLS algorithm also contributes to an increase of the scaling power. Moreover, it is also observed from Table 6.1 that the computation time is longer for the half-fan case, which is attributed to the larger number of projections. As for the acceleration

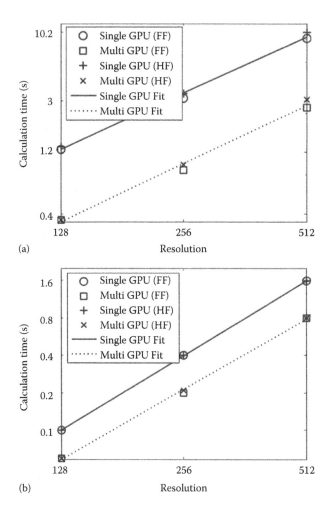

FIGURE 6.4 Computation time in (a) CGLS and (b) the regularization step, plotted as a function of CBCT image resolution for the single- and multi-GPU cases. Both full-fan (FF) and half-fan (HF) cases are plotted, and the lines correspond to the best fit. (Reprinted from Yan, H. et al., *Med. Phys.*, 41(11), 111912, 2014.)

ratio, it is found that a speedup of over three times is achieved in all cases using four GPUs.

The same analysis is performed for the regularization term, and the results are shown in Figure 6.4b. Again a power-law scaling of computation time is observed, yielding a best fit of $t_{reg} \sim x^{1.99}$ for the single-GPU implementation and $t_{reg} \sim x^{1.87}$ for multi-GPU. For regularization, the computation time should be proportional to the total number of voxels in the CBCT image, which is confirmed by the exponent of ~2 (note the resolution along the third dimension is kept constant in all the different cases). Because this step is a pure CBCT image domain processing, the computation time is independent of the number of projections, as shown in Table 6.2. Yet, the acceleration ratio under the four GPUs is only up to 2. This is because processing in the image domain is a relatively small size problem. The multi-GPU overhead under these circumstances is relatively large compared to the processing time, significantly impacting the speedup factor. This overhead effect is particularly severe for the case with the lowest CBCT image resolution, as indicated by the low speedup factors.

Table 6.3 summarizes the total computation time per iteration. The CGLS step dominates the overall computation due to its much more complicated operations. Hence, when combining the two steps, we still observe acceleration factors of 3.03–3.38, under the multi-GPU implementation. These acceleration factors are quite satisfactory considering the amount of data communication among the GPUs, from a parallel computing point of view. Again, the computation time for the half-fan case is slightly longer than that of the full-fan case, due to the use of more projections. The total computation time that is more clinically important is listed in the second half of Table 6.3. In particular, the reconstruction time is controlled to be ~25 s for the two cases, and the acceleration factors of ~3.1 have been achieved using the quad-GPU system.

6.5 CONCLUSION AND DISCUSSIONS

This chapter introduces the development of a multi-GPU system to achieve efficiency boost to IR algorithm. To avoid cumbersome implementations and minimize communication overhead, inter-GPU parallelization is carefully designed. Detailed analyses of computation time in each step, their relation to image resolution, and the acceleration factors, have been conducted. In summary, a boost in computational efficiency can be achieved by multi-GPU implementation, which will facilitate the use of low-dose IR in IGRT clinical practice. As an example, a total speedup factor of ~3.1 is achieved using four GPUs.

While all the results regarding the computational efficiency are generated under the TF reconstruction algorithm, the conclusions are expected to hold for most other IR algorithms, as long as the structure of those algorithms can be organized in a way similar to that of the TF algorithm [14].

TABLE 6.3 Computation Time (s) per Step, Total Time with and without Multiresolution Techniques in Single- and Multi-GPU Reconstructions

Protocol	Resolution	Per Iteration			Total (without MR)			Total (with MR)		
		Single	Multi	Factor	Single	Multi	Factor	Single	Multi	Factor
Full-fan (121 projections)	512×512×70	10.73	3.47	3.09	107.3	34.7	3.09	78.2	24.8	3.15
	256×256×70	3.55	1.07	3.32						
	128×128×70	1.36	0.42	3.24						
Half-fan (164 projections)	512×512×70	11.79	3.89	3.03	117.9	38.9	3.03	85.2	24.7	3.11
	256×256×70	3.82	1.17	3.26						
	128×128×70	1.42	0.42	3.38						

The total reconstruction time corresponds to the highest resolution case.

For instance, TV is similar to TF regarding how the regularization is imposed, as it is also completely performed in the CBCT image domain and can be parallelized among GPUs in the same fashion as described in this chapter. However, for some regularization based on advanced imaging processing techniques, such as dictionary learning [51], the regularization part may occupy significant longer time due to its algorithmic complexity.

It is noticed that some single-GPU-based studies had reported comparably short time to that realized in our multi-GPU system. This can be mainly attributed to the following two reasons. (1) A small number of projections, for example, 40 projections, were used in those studies [19,31], which reduces the reconstruction time. However, such an extremely few-projection protocol may not be clinical feasible, according to the comprehensive studies under CS-based iterative CBCT reconstruction [23] regarding the number/exposure of the projections versus image quality. (2) A small number of iterations were used. In a typical IR process, the first few iteration steps outline the main CBCT image content, while the posterior steps gradually improve image quality. Since it is the fine structures that are important for many clinical applications, a certain minimum number of iteration steps are indeed necessary to ensure an acceptable level of image quality.

REFERENCES

1. Jaffray, D.A. and J.H. Siewerdsen, Cone-beam computed tomography with a flat-panel imager: Initial performance characterization. *Medical Physics*, 2000. **27**(6): 1311–1323.
2. Jaffray, D.A. et al., Flat-panel cone-beam computed tomography for image-guided radiation therapy. *International Journal of Radiation Oncology Biology Physics*, 2002. **53**(5): 1337–1349.
3. Feldkamp, L.A., L.C. Davis, and J.W. Kress, Practical cone beam algorithm. *Journal of the Optical Society of America A: Optics Image Science and Vision*, 1984. **1**(6): 612–619.
4. Sidky, E.Y., C.M. Kao, and X. Pan, Accurate image reconstruction from few-views and limited-angle data in divergent-beam CT. *Journal of X-Ray Science and Technology*, 2006. **14**(2): 119–139.
5. Song, J. et al., Sparseness prior based iterative image reconstruction for retrospectively gated cardiac micro-CT. *Medical Physics*, 2007. **34**: 4476.
6. Sidky, E.Y. and X.C. Pan, Image reconstruction in circular cone-beam computed tomography by constrained, total-variation minimization. *Physics in Medicine and Biology*, 2008. **53**(17): 4777–4807.
7. Chen, G.H., J. Tang, and S.H. Leng, Prior image constrained compressed sensing (PICCS): A method to accurately reconstruct dynamic CT images from highly undersampled projection data sets. *Medical Physics*, 2008. **35**(2): 660–663.

8. Leng, S. et al., High temporal resolution and streak-free four-dimensional cone-beam computed tomography. *Physics in Medicine & Biology*, 2008. **53**(20): 5653–5673.

9. Wang, J. et al., Dose reduction for kilovotage cone-beam computed tomography in radiation therapy. *Physics in Medicine & Biology*, 2008. **53**(11): 2897.

10. Wang, J., T. Li, and L. Xing, Iterative image reconstruction for CBCT using edge-preserving prior. *Medical Physics*, 2009. **36**: 252.

11. Yu, H. and G. Wang, Compressed sensing based interior tomography. *Physics in Medicine & Biology*, 2009. **54**(9): 2791.

12. Yu, H. and G. Wang, A soft-threshold filtering approach for reconstruction from a limited number of projections. *Physics in Medicine & Biology*, 2010. **55**(13): 3905–3916.

13. Choi, K. et al., Compressed sensing based cone-beam computed tomography reconstruction with a first-order method. *Medical Physics*, 2010. **37**: 5113.

14. Jia, X. et al., GPU-based fast cone beam CT reconstruction from undersampled and noisy projection data via total variation. *Medical Physics*, 2010. **37**(4): 1757–1760.

15. Yu, H. and G. Wang, SART-type image reconstruction from a limited number of projections with the sparsity constraint. *Journal of Biomedical Imaging*, 2010. **2010**: 3.

16. Defrise, M., C. Vanhove, and X. Liu, An algorithm for total variation regularization in high-dimensional linear problems. *Inverse Problems*, 2011. **27**: 065002.

17. Ritschl, L. et al., Improved total variation-based CT image reconstruction applied to clinical data. *Physics in Medicine & Biology*, 2011. **56**: 1545.

18. Tian, Z. et al., Low-dose CT reconstruction via edge-preserving total variation regularization. *Physics in Medicine & Biology*, 2011. **56**: 5949.

19. Jia, X. et al., GPU-based iterative cone-beam CT reconstruction using tight frame regularization. *Physics in Medicine & Biology*, 2011. **56**: 3787.

20. Fessler, J.A., Assessment of image quality for the new CT: Statistical reconstruction methods. In: *54th AAPM Annual Meeting*, 2012. http://web.eecs.umich.edu/~fessler/papers/files/talk/12/aapm-fessler-iq.pdf.

21. Xu, Q. et al., Low-dose x-ray CT reconstruction via dictionary learning. *IEEE Transactions on Medical Imaging*, 2012. **31**(9): 1682–1697.

22. Lee, H. et al., Improved compressed sensing-based cone-beam CT reconstruction using adaptive prior image constraints. *Physics in Medicine & Biology*, 2012. **57**(8): 2287.

23. Yan, H. et al., A comprehensive study on the relationship between the image quality and imaging dose in low-dose cone beam CT. *Physics in Medicine & Biology*, 2012. **57**(7): 2063–2080.

24. Niu, T. et al., Accelerated barrier optimization compressed sensing (ABOCS) for CT reconstruction with improved convergence. *Physics in Medicine & Biology*, 2014. **59**(7): 1801.

25. Brenner, D.J. and C.D. Elliston, Estimated radiation risks potentially associated with full-body CT screening. *Radiology*, 2004. **232**(3): 735–738.
26. Hall, E.J. and D.J. Brenner, Cancer risks from diagnostic radiology. *British Journal of Radiology*, 2008. **81**(965): 362–378.
27. Brix, G. et al., Radiation exposures of cancer patients from medical X-rays: How relevant are they for individual patients and population exposure? *European Journal of Radiology*, 2009. **72**(2): 342–347.
28. Pan, Y. and R. Whitaker, Iterative helical cone-beam CT reconstruction using graphics hardware: A simulation study. In: *Proceedings of SPIE 7961, Medical Imaging 2011: Physics of Medical Imaging*, Vol. 79612N, Lake Buena Vista, FL, 2011.
29. Yan, M. et al., EM + TV based reconstruction for cone-beam CT with reduced radiation. In: *Proceedings of the Seventh International Conference on Advances in Visual Computing*, Las Vegas, NV, 2011, pp. 1–10.
30. Stsepankou, D. et al., Evaluation of robustness of maximum likelihood cone-beam CT reconstruction with total variation regularization. *Physics in Medicine & Biology*, 2012. **57**(19): 5955.
31. Park, J.C. et al., Fast compressed sensing-based CBCT reconstruction using Barzilai-Borwein formulation for application to on-line IGRT. *Medical Physics*, 2012. **39**: 1207.
32. Cui, J. et al., Distributed MLEM: An iterative tomographic image reconstruction algorithm for distributed memory architectures. *IEEE Transactions on Medical Imaging*, 2013. **32**: 957–967.
33. Zheng, Z., E. Papenhausen, and K. Mueller, DQS advisor: A visual interface and knowledge-based system to balance dose, quality, and reconstruction speed in iterative CT reconstruction with application to NLM-regularization. *Physics in Medicine & Biology*, 2013. **58**(21): 7857.
34. Hennessy, J.L. and D.A. Patterson, *Computer Architecture: A Quantitative Approach*, 4th edn., 2006. Morgan Kaufmann, Amsterdam, the Netherlands.
35. Herman, G.T. and A. Lent, Iterative reconstruction algorithms. *Computers in Biology and Medicine*, 1976. **6**(4): 273–294.
36. Eggermont, P.P.B., G.T. Herman, and A. Lent, Iterative algorithms for large partitioned linear systems, with applications to image reconstruction. *Linear Algebra and its Applications*, 1981. **40**: 37–67.
37. Qi, J. and R.M. Leahy, Iterative reconstruction techniques in emission computed tomography. *Physics in Medicine & Biology*, 2006. **51**(15): R541.
38. Tang, J., B.E. Nett, and G.H. Chen, Performance comparison between total variation (TV)-based compressed sensing and statistical iterative reconstruction algorithms. *Physics in Medicine & Biology*, 2009. **54**(19): 5781–5804.
39. Beister, M., D. Kolditz, and W.A. Kalender, Iterative reconstruction methods in X-ray CT. *Physica Medica*, 2012. **28**(2): 94–108.
40. Nuyts, J. et al., Modelling the physics in the iterative reconstruction for transmission computed tomography. *Physics in Medicine & Biology*, 2013. **58**(12): R63.

41. Hestenes, M.R. and E. Stiefel, Methods of conjugate gradients for solving linear systems. *Journal of Research of the National Bureau of Standards*, 1952. **49**(6): 409–436.
42. Shi, H. and J.A. Fessler, Quadratic regularization design for iterative reconstruction in 3D multi-slice axial CT. In: *Nuclear Science Symposium Conference Record, IEEE*, San Diego, CA, 2006. pp. 2834–2836. IEEE, New York.
43. Shi, H.R. and J.A. Fessler, Quadratic regularization design for 2-D CT. *IEEE Transactions on Medical Imaging*, 2009. **28**(5): 645–656.
44. Cai, J.F., R.H. Chan, and Z.W. Shen, A framelet-based image inpainting algorithm. *Applied and Computational Harmonic Analysis*, 2008. **24**(2): 131–149.
45. Cai, J.F., S. Osher, and Z.W. Shen, Split Bregman methods and frame based image restoration. *Multiscale Modeling & Simulation*, 2009. **8**(2): 337–369.
46. Cai, J.F., S. Osher, and Z.W. Shen, Linearized Bregman iteration for frame based image deblurring. *SIAM Journal on Imaging Sciences*, 2009. **2**(1): 226–252.
47. Cai, J.F. and Z.W. Shen, Framelet based deconvolution. *Journal of Computational Mathematics*, 2010. **28**(3): 289–308.
48. NVIDIA, *NVIDIA CUDA Compute Unified Device Architecture, Programming Guide*, Version 4.0, 2011. NVIDIA, Santa Clara, CA.
49. Siddon, R.L., Fast calculation of the exact radiological path for a three-dimensional CT array. *Medical Physics*, 1985. **12**(2): 252.
50. Watt, A. and M. Watt, *Advanced Animation and Rendering Techniques: Theory and Practice*, 1992. Addison-Wesley, Reading, MA.
51. Bai, T. et al., 3D dictionary learning based iterative cone beam CT reconstruction. *International Journal of Cancer Therapy and Oncology*, 2014. **2**(2): 020240.
52. Folkerts, M. et al., SU-EI-35: A GPU optimized DRR algorithm. *Medical Physics*, 2011. **38**: 3403.
53. Jia, X. et al., GDRR: A GPU tool for cone-beam CT projection simulations. *Medical Physics*, 2012. **39**(6): 3890.
54. Yan, H. et al., Towards the clinical implementation of iterative low-dose cone-beam CT reconstruction in image-guided radiation therapy: Cone/ring artifact correction and multiple GPU implementation. *Medical Physics*, 2014. **41**(11): 111912.

CHAPTER 7

Tumor Tracking and Real-Time Volumetric Imaging via One Cone-Beam CT Projection

Ruijiang Li and Steve B. Jiang

CONTENTS

7.1 INTRODUCTION

RESPIRATORY-INDUCED TUMOR MOTION IS a major issue in radiotherapy and requires careful management [1]. To account for the uncertainty caused by breathing motion, the conventional approach is to expand the target volume by adding a safety margin around the tumor [2]. This inevitably leads to a large volume of normal tissues being irradiated and increased toxicity, which is the main limiting factor for dose escalation and potential gain in local control. Therefore, it is important to reduce the breathing motion–induced safety margin. This is crucial for stereotactic body radiation therapy (SBRT), which is ablative not only to the tumor, but to normal tissues as well [3,4].

Margin reduction can be realized by either beam tracking [5–7], where the radiation beam dynamically conforms to the moving tumor during respiration, or beam gating [8–14], where the radiation beam is enabled only where the tumor moves inside the beam. An important prerequisite for these delivery techniques is a precise knowledge of tumor location at any given time during dose delivery. This calls for a method to directly image and localize the tumor accurately in real time.

The advent of an on-board x-ray imaging device mounted on the medical linear accelerator (LINAC) provides a useful tool to obtain valuable anatomic information of the patient [15–22]. One unique characteristic of the on-board imaging device is that the gantry rotation is very slow (~1 min per rotation), due to mechanical constraints. Compared with a conventional CT scanner, where a full gantry rotation can be completed in a very short time (<0.5 s), the slow gantry rotation of the on-board imaging device makes it extremely difficult to obtain volumetric information of patient's anatomy in real time.

An intermediate approach is to obtain the time-resolved patient anatomy during respiratory motion. The basic idea is to acquire x-ray projections over many breathing cycles for several complete gantry rotations and then group them into several independent datasets, according to the corresponding breathing phase for each projection image [23–30]. One volumetric image is reconstructed for each breathing phase, resulting in a 4D cone-beam CT (CBCT) dataset, which consists of several volumetric images at discrete breathing phases. A fundamental assumption of this technique is that the projections in the same breathing phase correspond to exactly the same patient anatomy. This requires that the patient breathe in a perfectly regular and reproducible manner during the entire scan, which lasts for several minutes due to the slow gantry rotation. In practice, however, this is rarely true, especially for lung cancer patients whose lung functions have already been compromised. It is frequently observed that during radiotherapy, lung cancer patients exhibit irregular breathing patterns with breath-to-breath changes in amplitude, period, baseline, and shape [31–33]. As a result, different patient anatomies at different breathing cycles are mapped into the same phase, which inevitably leads to motion artifacts. Moreover, the 4D CBCT approach is retrospective, requiring all projection data to be collected before the images can be reconstructed. Thus, it cannot provide the volumetric anatomy in real time.

We have recently made a breakthrough in image-guided radiotherapy by reconstructing volumetric images and localizing lung tumors in near real time using a single x-ray projection image [34,35]. The method is based on an accurate and efficient patient-specific lung motion model and uses CT images acquired during patient simulation prior to treatment as the reference anatomy. Our method is able to build a snapshot of the patient anatomy from limited information acquired in an instant and thus, is robust to patient breathing irregularities. In the following, sections, we will describe this technique in more detail.

7.2 PCA-BASED LUNG MODEL

We have proposed an efficient patient-specific lung motion model based on principal component analysis (PCA) [36]. In the PCA lung motion model [36,37], the deformation vector field (DVF) relative to a reference image as a function of space and time is approximated by a linear combination of the sample mean vector and a few eigenvectors corresponding to the largest eigenvalues, that is,

$$x(t) \approx \bar{x} + \sum_{k=1}^{K} u_k w_k(t). \qquad (7.1)$$

where u_k are the eigenvectors obtained from PCA and are functions of space only. The scalars $w_k(t)$ are PCA coefficients and are functions of time only. It is worth mentioning that the huge-dimensional eigenvectors are fixed once the PCA motion model is constructed. It is the temporal evolution of a few scalar PCA coefficients that drives the dynamic lung motion over time. This is critical in order to derive the entire lung motion from very limited information.

There are primarily three reasons why the patient-specific PCA lung motion model is suitable for this work [36]. First, PCA provides the best linear representation of the lung DVFs in the least mean-square-error sense. Second, the PCA motion model imposes inherent regularization on its representation of lung motion: if any two voxels move similarly, their motion represented by PCA will also be similar. The combined effect is that a few scalar variables, that is, PCA coefficients, are sufficient to accurately derive the dynamic lung motion. Third, the PCA motion model is capable of representing tumor motion that is beyond that of the training set (4DCT scan). Because the PCA coefficients $w_k(t)$ are free parameters, given three eigenvectors, the PCA motion model will allow the tumor to move freely in the 3D space and thus, tumor motion is not constrained to that during the 4DCT scan. This distinct feature enables the PCA motion model to capture a wide range of tumor motion trajectories.

We have done comprehensive evaluations of the PCA lung motion model using clinical 4DCT scans for eight lung patients [36]. Figure 7.1

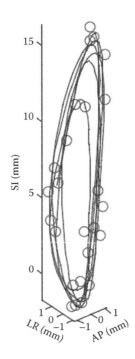

FIGURE 7.1 The 3D tumor positions in a lung cancer patient during an 18-s cine CT scan (circles) and the 3D trajectory obtained from the PCA motion model using the first two coefficients and eigenvectors.

shows the 3D tumor positions in a lung cancer patient during an 18-s cine CT scan and the 3D trajectory obtained from the PCA motion model using the first two coefficients and eigenvectors. As can be seen, the PCA model is able to capture this complex motion trajectory. The average 3D motion modeling error throughout the entire lung was found to be <1 mm for all patients, which is sufficient for most clinical applications.

7.3 OPTIMIZATION MODEL FOR REAL-TIME CBCT

7.3.1 Formulation of the Real-Time CBCT Problem

Once we have obtained a parameterized PCA lung motion model, we seek a set of optimal PCA coefficients such that the projection of the reconstructed volumetric image corresponding to the new DVF matches with the measured x-ray projection. Denote f_0 as the reference image, f as the reconstructed image, y as the measured projection image, and P as the projection matrix which computes the projection image of f. Note that in fixed-angle geometry, for example, fluoroscopy, P is a constant, while for rotational geometry, P will be a function of time. The cost function is

$$min. J(w,a,b) = \left\| P \cdot f(x,f_0) - a \cdot y - b \cdot 1 \right\|_2^2$$
$$s.t. \; x = \bar{x} + U \cdot w \tag{7.2}$$

where
  U is a matrix whose columns are the PCA eigenvectors
  w is a vector comprised of the PCA coefficients to be optimized
  x is the parameterized DVF
  $\|\cdot\|_p$ denotes standard vector l_p-norm.

Because the computed and measured projection images may have different intensity levels, we assumed there exists a linear relationship between them and introduced two auxiliary parameters a, b to account for the differences in the image intensities. For clarity of notation, we have suppressed the time index under P, f, x, y, w, a, b.

7.3.2 Optimization Using an Efficient Alternating Algorithm

To find the optimal values for w, a, b, the optimization algorithm alternates between the following two steps:

Step 1:

$$(a_{n+1}, b_{n+1})^T = (Y^T Y)^{-1} Y^T P \cdot f_n \tag{7.3}$$

Step 2:

$$w_{n+1} = w_n - \mu_n \cdot \frac{\partial J_n}{\partial w_n}, \tag{7.4}$$

where
$Y = [y, 1]$
$\partial J / \partial w = (\partial x / \partial w) \cdot (\partial f / \partial x) \cdot (\partial J / \partial f)$
$\qquad = 2 \cdot U^T \cdot (\partial f / \partial x) \cdot P^T \cdot (P \cdot f - a \cdot y - b \cdot 1).$

At each iteration, given the updated PCA coefficients in Equation 7.4, the DVF is updated according to Equation 7.2 and the reconstructed image f_{n+1} is obtained through trilinear interpolation. It turns out that $\partial f / \partial x$ is a linear combination of the spatial gradients of the reference image evaluated at the neighboring eight grid points, weighted by the appropriate fractional part of the DVF.

It is easy to see that the alternating algorithm discussed earlier is guaranteed to converge. In step 1, the update for a, b is the unique minimizer of the cost function with fixed w. Step 2 is a gradient descent method with variable w and fixed a, b, where the step size μ_n is found by Armijo's rule [38] for line search. Therefore, the cost function always decreases at each step. Note that the cost function is lower bounded by zero. The alternating algorithm discussed earlier is guaranteed to converge for all practical purposes.

At this point, we have reconstructed the volumetric image given the reference image, the parameterized DVF, and the measured projection image. Because we seek to minimize the mean square error in Equation 7.2, it is the best estimate of the current patient geometry in the least mean square error sense.

7.3.3 Tumor Localization via Deformation Inversion Based on an Efficient Fixed-Point Algorithm

In order to get the current tumor position, it is important to distinguish between two different kinds of DVFs: push-forward DVF and pull-back DVF. The DVF found by Equation 7.2 is a pull-back DVF, which is defined on the coordinate of the new image and tells how each voxel in the new image moves. It cannot be used directly to calculate the tumor position in the new image unless the DVF is rigid everywhere. To get the correct tumor position, we need its inverse, that is, the push-forward DVF, which is defined on the coordinate of the reference image and tells how each voxel in the reference image moves. We adopted an efficient fixed-point algorithm for deformation inversion, which has been shown to be about 10 times faster and yet 10 times more accurate than conventional gradient-based iterative algorithms [39]. Since we are primarily concerned with the tumor motion, we applied the deformation inversion procedure on only one voxel, that is, the tumor center of volume in the reference image.

7.3.4 Fast Computation through Algorithm Implementation on GPU

We have implemented our algorithm using compute unified device architecture (CUDA) as the programming environment. For this work, we used an NVIDIA Tesla C1060 GPU card, which has a total number of 240 processor cores (grouped into 30 multiprocessors with 8 cores each), with a clock speed of 1.3 GHz. This massively parallel computing power can be fully leveraged using a standardized library (compute unified basic linear algebra subprograms or CUBLAS) for matrix and vector operations, which are the main computational tasks for our algorithm. The average computation time for localizing the tumor from each projection ranges between 0.2 and 0.3 s on the NVIDIA Tesla C1060 GPU card, for the digital and physical respiratory phantoms as well as for real patient data.

7.4 EXPERIMENTAL RESULTS

7.4.1 Evaluations Using Digital and Physical Respiratory Phantoms

For the algorithm validation, we first used a dynamic nonuniform rational B-spline–based cardiac-torso (NCAT) phantom [40,41]. We simulated 4DCT data with 10 breathing phases. The tumor was located in the middle lobe of the right lung and had a 3D motion magnitude of 1.6 cm. We evaluated the performance of the algorithm with the dynamic NCAT phantom under regular breathing, however, with different breathing parameters from those in 4DCT. The different breathing parameters are: amplitude (diaphragm SI motion: 1 and 3 cm), period (3 and 5 s), and baseline shift (−1 and +1 cm).

We also evaluated the algorithm on the NCAT phantom under irregular breathing, where the breathing trace is from an actual lung patient.

For testing purposes, we simulated 360 cone-beam x-ray projections using the Siddon's algorithm [42] from angles uniformly distributed over one full gantry rotation. We used the end of exhale (EOE) phase as the reference image and performed DIR between the EOE phase and all other phases. The DIR algorithm used here is a fast demons algorithm implemented on GPU, which has been shown to yield an average 3D error of <2 mm on five lung cancer patients [43]. Then PCA was performed on the nine DVFs from DIR and three PCA coefficients, and eigenvectors were kept in the PCA lung motion model. Then we derived the 3D tumor location corresponding to each simulated cone-beam projection.

We found that the average 3D tumor localization error is <1 mm, which is not affected by amplitude change, period change, or baseline shift. The relative image intensity difference between the EOE and end of inhale (EOI) phases is as large as 35%. This was significantly reduced to an average of 7% in the reconstructed volumetric images by our algorithm.

The algorithm was also tested on a physical respiratory phantom, which consisted of a cork block on a platform that can be programmed to undergo translational motion. Inside the cork block were embedded several tissue-like objects including a 2.5 cm water balloon which was used as the target for localization [44]. We simulated an arc treatment during which continuous cone-beam x-ray images were acquired over a full gantry rotation and the phantom moved according to an actual lung patient breathing trace. Figure 7.2 shows the estimated and ground truth tumor position in the SI direction. The average tumor localization error is 0.8 mm.

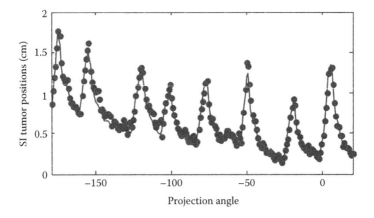

FIGURE 7.2 Tumor localization results (dots) for the physical phantom with an irregular patient breathing trace. The solid line is ground truth.

7.4.2 Evaluations Using Real Patient Data Sets

We acquired 4DCT for five lung cancer patients. The raw cone-beam x-ray images during the CBCT scan for treatment setup purposes were used retrospectively to test our algorithm. For all patients in this study, there were no implanted fiducial makers and real-time 3D location of the tumor was not available to evaluate our algorithm. Instead, we projected the estimated 3D tumor location onto the 2D imager and compared with that manually defined by a clinician. For each patient, the clinician marked the tumor in the largest continuous set of projections in which the tumor was visible. All five patients had somewhat irregular breathing during the CBCT scans. Figure 7.3 shows the localization results for two of the

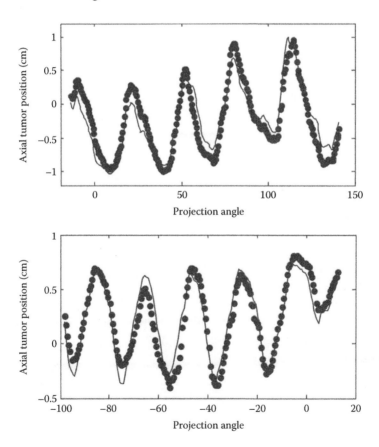

FIGURE 7.3 Tumor localization results (dots) for two patients with irregular breathing. The solid lines are ground truths. Only the direction with the largest motion (axial) is shown. The average error is 1.9 and 0.9 mm for the two patients, respectively.

patients with irregular breathing. The average tumor localization error is <2 mm for all five patients. Figure 7.4 shows the raw x-ray projections and the coronal and sagittal views of the reconstructed images at two breathing phases. Overall, the algorithm gives realistic and consistent anatomy (including tumor, diaphragm, bronchial, and vascular structures) during respiration.

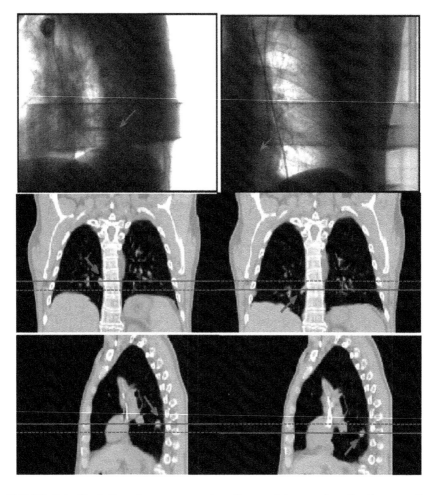

FIGURE 7.4   Real-time image reconstruction and tumor localization results. Left column: cone-beam x-ray projection image (top) and the coronal and sagittal views of the reconstructed image at an EOE phase (middle and bottom). Right column: same as left column, except at an EOI phase. Arrow indicates the tumor.

7.5 CONCLUSION AND DISCUSSIONS

We have performed in-depth analysis and comprehensive evaluations of an efficient and flexible patient-specific PCA lung motion model. Based on this, we have developed a novel algorithm to reconstruct a volume image and localize the tumor using a single x-ray projection image. The algorithm is able to handle x-ray images acquired with rotational geometry as well as fixed-angle geometry, for example, fluoroscopy. We have implemented this new algorithm on an NVIDIA Tesla C1060 GPU card and achieved real-time efficiency. We have performed comprehensive evaluations on digital and physical respiratory phantoms and achieved sub millimeter accuracy, which was not affected by moderate changes in breathing amplitude, period, and baseline drift. We have also done some preliminary testing on patient data and a localization accuracy of ~2 mm was achieved. Therefore, we have developed an accurate noninvasive real-time tumor localization and CBCT algorithm in fluoroscopy, without implanted markers.

It is possible that during the course of the treatment patients may undergo anatomical changes on an inter-fractional basis, which could make the results worse if the lung motion model is built based on 4DCT acquired during patient simulation. In this case, a PCA lung motion model built from 4DCBCT during patient setup might overcome this difficulty. Another potential issue is when the tumor gets close to the heart. Then the tumor motion will be affected not only by breathing motion, but by heart motion too, which is not represented by the PCA motion model. This is a fundamental limitation of 4DCT, which is usually synchronized to respiration, not to cardiac motion.

Although the results of five lung cancer patients confirmed the validity of our algorithm on a preliminary basis, a more comprehensive study on a larger patient population is warranted. It is worth mentioning that the accuracy of the tumor localization and image reconstruction is directly affected by the PCA lung motion model, which is in turn influenced by the quality of the training data (4DCT or 4DCBCT). Therefore, it is important to minimize the motion artifacts in the training data, for example, by optimizing the scanning protocol or coaching the patients to breathe regularly whenever possible [45]. A reliable deformable registration algorithm, especially one that can accurately model large motion (e.g., around the diaphragm), is also likely to help improve the accuracy of the PCA lung motion model. In this work, since there were no fiducial markers implanted in the lung cancer patients, clinician marked tumor positions had to be used to evaluate the tumor localization results. The accuracy of the ground truth is thus, limited due to noisy projection images as well as relatively poor soft-tissue contrast. For validation purposes, it would be beneficial to have patient data with implanted fiducial markers, from which better ground truth can be derived.

REFERENCES

1. Keall, P.J. et al., The management of respiratory motion in radiation oncology report of AAPM Task Group 76. *Med Phys*, 2006. 33(10): 3874–3900.
2. Underberg, R.W. et al., Use of maximum intensity projections (MIP) for target volume generation in 4DCT scans for lung cancer. *Int J Radiat Oncol Biol Phys*, 2005. 63(1): 253–260.
3. Timmerman, R. et al., Excessive toxicity when treating central tumors in a phase II study of stereotactic body radiation therapy for medically inoperable early-stage lung cancer. *J Clin Oncol*, 2006. 24(30): 4833–4839.
4. Timmerman, R.D. and B.D. Kavanagh, Stereotactic body radiation therapy. *Curr Probl Cancer*, 2005. 29(3): 120–157.
5. Keall, P.J. et al., Four-dimensional radiotherapy planning for DMLC-based respiratory motion tracking. *Med Phys*, 2005. 32(4): 942–951.
6. D'Souza, W.D., S.A. Naqvi, and C.X. Yu, Real-time intra-fraction-motion tracking using the treatment couch: A feasibility study. *Phys Med Biol*, 2005. 50(17): 4021–4033.
7. Cho, B. et al., First demonstration of combined kV/MV image-guided real-time dynamic multileaf-collimator target tracking. *Int J Radiat Oncol Biol Phys*, 2009. 74(3): 859–867.
8. Jiang, S.B., Technical aspects of image-guided respiration-gated radiation therapy. *Med Dosim*, 2006. 31(2): 141–151.
9. Ford, E.C. et al., Evaluation of respiratory movement during gated radiotherapy using film and electronic portal imaging. *Int J Radiat Oncol Biol Phys*, 2002. 52(2): 522–531.
10. Kubo, H.D. et al., Breathing-synchronized radiotherapy program at the University of California Davis Cancer Center. *Med Phys*, 2000. 27(2): 346–353.
11. Vedam, S.S. et al., Determining parameters for respiration-gated radiotherapy. *Med Phys*, 2001. 28(10): 2139–2146.
12. Wagman, R. et al., Respiratory gating for liver tumors: Use in dose escalation. *Int J Radiat Oncol Biol Phys*, 2003. 55(3): 659–668.
13. Mageras, G.S. and E. Yorke, Deep inspiration breath hold and respiratory gating strategies for reducing organ motion in radiation treatment. *Semin Radiat Oncol*, 2004. 14(1): 65–75.
14. Shirato, H. et al., Physical aspects of a real-time tumor-tracking system for gated radiotherapy. *Int J Radiat Oncol Biol Phys*, 2000. 48(4): 1187–1195.

15. Kriminski, S.A. et al., Comparison of kilovoltage cone-beam computed tomography with megavoltage projection pairs for paraspinal radiosurgery patient alignment and position verification. *Int J Radiat Oncol Biol Phys*, 2008. 71: 1572–1580.

16. Li, H. et al., Comparison of 2D radiographic images and 3D cone beam computed tomography for positioning head-and-neck radiotherapy patients. *Int J Radiat Oncol Biol Phys*, 2008. 71: 916–925.

17. Nelson, J.W. et al., Stereotactic body radiotherapy for lesions of the spine and paraspinal regions. *Int J Radiat Oncol Biol Phys*, 2009. 73: 1369–1375.

18. Oldham, M. et al., Cone-beam-CT guided radiation therapy: A model for on-line application. *Radiother Oncol*, 2005. 75(3): 271–278.

19. Yin, F.-F. et al., Physics and imaging for targeting of oligometastases. *Semin Radiat Oncol*, 2006. 16: 85–101.

20. Wang, Z. et al., Refinement of treatment setup and target localization accuracy using three-dimensional cone-beam computed tomography for stereotactic body radiotherapy. *Int J Radiat Oncol Biol Phys*, 2009. 73(2): 571–577.

21. Chang, Z. et al., Dosimetric characteristics of novalis Tx system with high definition multileaf collimator. *Med Phys*, 2008. 35: 4460–4463.

22. Jaffray, D.A. et al., Flat-panel cone-beam computed tomography for image-guided radiation therapy. *Int J Radiat Oncol Biol Phys*, 2002. 53(5): 1337–1349.

23. Sonke, J. et al., Respiratory correlated cone beam CT. *Med Phys*, 2005. 32(4): 1176–1186.

24. Li, T. et al., Four-dimensional cone-beam CT using an on-board imager. *Med Phys*, 2006. 33: 3825–3833.

25. Li, T. and L. Xing, Optimizing 4D cone-beam CT acquisition protocol for external beam radiotherapy. *Int J Radiat Oncol Biol Phys*, 2007. 67(4): 1211–1219.

26. Dietrich, L. et al., Linac-integrated 4D cone beam CT: First experimental results. *Phys Med Biol*, 2006. 51(11): 2939–2952.

27. Li, T., A. Koong, and L. Xing, Enhanced 4D cone-beam CT with inter-phase motion model. *Med Phys*, 2007. 34(9): 3688–3695.

28. Lu, J. et al., Four-dimensional cone beam CT with adaptive gantry rotation and adaptive data sampling. *Med Phys*, 2007. 34(9): 3520–3529.

29. Zeng, R., J.A. Fessler, and J.M. Balter, Estimating 3-D respiratory motion from orbiting views by tomographic image registration. *IEEE Trans Med Imaging*, 2007. 26(2): 153–163.

30. Gao, H. et al., 4D cone beam CT via spatiotemporal tensor framelet. *Med Phys*, 2012. 39(11): 6943–6946.

31. Seppenwoolde, Y. et al., Precise and real-time measurement of 3D tumor motion in lung due to breathing and heartbeat, measured during radiotherapy. *Int J Radiat Oncol Biol Phys*, 2002. 53(4): 822–834.

32. Vedam, S.S. et al., Quantifying the predictability of diaphragm motion during respiration with a noninvasive external marker. *Med Phys*, 2003. 30(4): 505–513.

33. George, R. et al., The application of the sinusoidal model to lung cancer patient respiratory motion. *Med Phys*, 2005. 32(9): 2850–2861.

34. Li, R. et al., Real-time volumetric image reconstruction and 3D tumor localization based on a single x-ray projection image for lung cancer radiotherapy. *Med Phys*, 2010. 37(6): 2822–2826.

35. Li, R. et al., 3D tumor localization through real-time volumetric x-ray imaging for lung cancer radiotherapy. *Med Phys*, 2011. 38(5): 2783–2794.

36. Li, R. et al., PCA-based lung motion model. In: *Proceedings of the 16th International Conference on the Use of Computers in Radiation Therapy*, 2010. Amsterdam, the Netherlands.

37. Zhang, Q. et al., A patient-specific respiratory model of anatomical motion for radiation treatment planning. *Med Phys*, 2007. 34(12): 4772–4781.

38. Bazaraa, M.S., H.D. Sherali, and C.M. Shetty, *Nonlinear Programming: Theory and Algorithms*, 2006. Hoboken, NJ: John Wiley & Sons.

39. Chen, M. et al., A simple fixed-point approach to invert a deformation field. *Med Phys*, 2008. 35(1): 81–88.

40. Segars, W.P., Development and application of the new dynamic NURBS-based cardiac-torso (NCAT) phantom. PhD dissertation, 2001. Chapel Hill, NC: University of North Carolina.

41. Segars, W.P., D.S. Lalush, and B.M.W. Tsui, Modeling respiratory mechanics in the MCAT and spline-based MCAT phantoms. *IEEE Trans Nucl Sci*, 2001. 48(1): 89–97.

42. Siddon, R.L., Fast calculation of the exact radiological path for a three-dimensional CT array. *Med Phys*, 1985. 12: 252–255.

43. Gu, X. et al., Implementation and evaluation of various demons deformable image registration algorithms on a GPU. *Phys Med Biol*, 2010. 55(1): 207–219.

44. Lewis, J.H. et al., Markerless lung tumor tracking and trajectory reconstruction using rotational cone-beam projections: A feasibility study. *Phys Med Biol*, 2010. 55(9): 2505–2522.

45. Kini, V.R. et al., Patient training in respiratory-gated radiotherapy. *Med Dosim*, 2003. 28(1): 7–11.

CHAPTER 8

GPU Denoising for Computed Tomography

Andreas Maier and Rebecca Fahrig

CONTENTS

8.1 INTRODUCTION

THE MAIN SOURCE OF noise in x-ray images is quantum noise. It follows a Poisson process that has a standard deviation of $\sqrt{N}$, where N is the number of photons that arrive at the detector pixel. Thus, the resulting signal-to-noise ratio is $(N/\sqrt{N}) = \sqrt{N}$, that is, the more photons arrive at the detector, the higher the signal-to-noise ratio. As the number of measured photons is proportional to the number of emitted photons at the source, image noise is inversely proportional to the dose that is applied to the patient. Thus, methods for the reduction of image noise also enable a reduction in patient dose. For this reason, noise reduction techniques have been investigated in the literature extensively.

8.2 NOISE IN X-RAY IMAGES

In order to understand the nature of noise reduction techniques, it is useful to have an understanding of the properties of noise in x-ray images.

8.2.1 Noise in Projection Data

The distribution of noise in an x-ray projection image follows Beer–Lambert's Law for the monochromatic case; the measured intensity I_m at the detector is found to be

$$I_m = I_0 e^{-\int f(x)dl}, \tag{8.1}$$

where

I_0 is the intensity that was emitted at the source
$f(x)$ is the object that was irradiated
l is the line that follows the path of the x-ray.

Note that

$$I_0 = N_0 \cdot E_0 \tag{8.2}$$

is composed of the number of photons N_0 and their respective energy E_0. Thus, the number of photons is proportional to the measured intensity and the monochromatic case follows exactly the Poisson distribution given by

$$P(n) = \frac{\kappa^n}{n!} \cdot e^{-\kappa} \tag{8.3}$$

where

$P(n)$ is the likelihood of observing n photons
$\kappa = N_m = (I_m/E_0)$ is the expected value of the Poisson process.

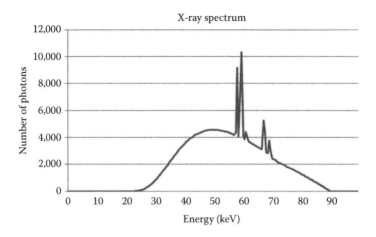

FIGURE 8.1 The intensity measured at the detector is found as integral over the source spectrum after attenuation by the object. This spectrum was simulated using CONRAD.

For the polychromatic case, the measured intensity I_p is found as an integral along energy ϵ that is present at the source I_ϵ:

$$I_p = \int I_\epsilon e^{-\int f(x)dl} d\epsilon \tag{8.4}$$

Figure 8.1 displays an example for an x-ray spectrum at 90 kVp. As a consequence, the noise is no longer a single Poisson process, but its distribution is now an integral over all Poisson processes at each individual energy ϵ. For simplification, we will assume, in the following, that our source is able to generate monochromatic x-rays.

8.2.2 Noise in Reconstructed Image Data

In order to compute noise in reconstruction domain, one has to determine line integrals p from measured intensities I_m:

$$p = \int f(x) = -\ln \frac{I_m}{I_0} \tag{8.5}$$

As I_0 is known from the x-ray tube parameters, the main effect on the noise is the application of the logarithm. For high number of photons, it is valid to approximate the Poisson process by a Gaussian distribution

Water cylinder (20 cm)   Noisy projection (0.1 mAs, 90 kVp)   Difference image

FIGURE 8.2 Noise-free and noisy projection of a water cylinder. The third image shows the difference between the two. The longer the path length in water, the higher is the noise variance. These images were created using CONRAD.

with mean value $\mu = N_m$ and standard deviation $\sigma = \sqrt{N_m}$. By application of a first-order Taylor expansion [1,2], one is able to show that the log transform results in a new mean value of $\kappa_p = -\ln((N_m \cdot E_0)/I_0)$ and $\sigma_p = 1/\sqrt{N_m}$. Figure 8.2 shows a noise-free projection, a noisy projection, and their difference. Note that the noise variance increases with object thickness.

The reconstruction process involves filtering and back projection. Back projection $b(x)$ is—depending on the imaging geometry—only a weighted sum of the projection values p_i with back projection weights c_i:

$$b(x) = \sum_i c_i p_i \qquad (8.6)$$

Even, if the effect of the filter is not considered, one immediately understands that the noise variance becomes object dependent at this point, as p_i and the noise at this detector pixel σ_{pi} are dependent on the number of photons that arrived at this detector pixel N_m. Figure 8.3 shows these effects in the reconstructed images. In the water cylinder, the most noise is found in the center of the object. In a noncircular object like the one shown on the right, streak structures appear in the noise. Thus, noise removal in the reconstructed image needs to consider the directional nature of the noise generation. The effects of filtering on the noise properties and how to reconstruct the noise variance at every pixel are beyond the scope of this book chapter and we refer to further literature [3–5].

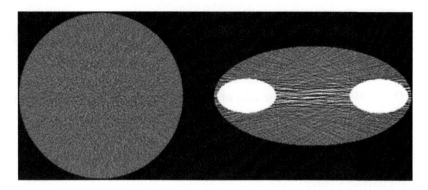

FIGURE 8.3 The left side shows the noise in the reconstruction of a water cylinder phantom. The noise is stronger in the center than in the off-center regions. The right side shows an elliptic phantom with two high intensity insets. In the center, streak noise emerges that is caused by the high attenuating structures. For both simulations N_0 = 90.000 photons @ 75 keV per pixel were used. The display window/level is [−200,200] HU. These images were created using CONRAD.

```
convolution2D (image) {
    sumWeight = 0;
    sumFilter = 0;
    step = halfWidth*2+1;
    for (j=0; j < step; j++) {
        for (i=0; I < step; i++) {
            nx = halfWidth + i;
            ny = halfWidth + j;
            g = geomClose(nx, ny, sigma_spatial)
            sumWeight += g;
            sumFilter += g * image[(x-nx)+((y-ny)*width)];
        }
    }
    return sumFilter/sumWeight;
}
```

Algorithm 8.1: Pseudo-code for a 2D convolution for application of a Gaussian kernel of odd neighborhood size, for example, $\mathcal{N}=5\times5$ using halfWidth=2. The function *geomClose* refers to the computation of $g(x, y, \sigma_r)$ with *sigma _ spatial* as σ_r, *image* denotes $p(x, y)$, and *sumWeight* and *sumFilter* are used to compute $k(x, y, \sigma_r)$ and $p'(x, y, \sigma_r)$ incrementally. Note that x and y are not arguments of our kernel function, as the kernel code is intended to be executed in parallel in a 2D grid structure with x and y as grid indexes.

8.3 DENOISING METHODS

In the following, we will give a short introduction to the most common denoising methods. We start with the simple Gaussian filter and extend this concept to edge-preserving methods. All reported methods can be employed for 2D and 3D denoising. For sake of simplicity, we will stay in 2D space in the following. Note that we only selected a subset of possible denoising algorithms that we deem as the most important selection, for example, we omit methods that require an iterative solution such as diffusion filters.

8.3.1 Gaussian Filter

The Gaussian filter is the most common way of suppressing noise in any kind of image. It is applied as convolution in projection space:

$$p'(x, y, \sigma_r) = \frac{1}{k(x, y, \sigma_r)} \sum_{x', y' \in \mathcal{N}} p(x', y') \cdot g(x', y', \sigma_r)$$

$$g(x, y, \sigma_r) = e^{-\frac{x^2 + y^2}{2\sigma_r^2}} \tag{8.7}$$

$$k(x, y, \sigma_r) = \sum_{x', y' \in \mathcal{N}} g(x', y', \sigma_r)$$

where
  $p(x, y)$ is the projection image at coordinates (x, y)
  $g(x, y, \sigma_r)$ is the Gaussian function with standard deviation σ_r
  $k(x, y, \sigma_r)$ is the mass of the kernel
  $\mathcal{N}$ is the neighborhood in which the kernel is evaluated
  $p'(x, y, \sigma_r)$ is the filtered projection image.

This kind of filtering is often included in the filter kernel of filtered back projection-type reconstruction methods [4]. Thus, explicit implementation is often not required. For small kernel sizes, the implementation in spatial domain is efficient. For larger kernels, the convolution in Fourier space is more efficient.

Nonetheless, investigation of the algorithm as graphics card implementation is interesting, as one can see how the computation is parallelized. Algorithm 8.1 lists pseudo-code for the spatial implementation on a graphics card. The code omits a loop over the coordinates (x, y) as the kernel code is designed to be executed in a parallel grid structure that matches the image dimensions. The main part of the code are the two for-loops over the double sums that are required to compute $k(x, y, \sigma_r)$ and $p'(x, y, \sigma_r)$. Note that the values of $g(x, y, \sigma_r)$ are the same in all kernel executions as the kernel is shift-invariant. In this example, we do not pre-compute the values of $g(x, y, \sigma_r)$. Depending on the graphics card hardware, this may lead to an increase or decrease in run time. If the hardware has much compute power, the implementation in Algorithm 8.1 will be more efficient, as only a single access to the *global* memory is performed. In this case, the hardware is *memory* or *bandwidth limited*; that is, the execution is limited by memory transfers. Thus, additional operations in the kernel that are computed in *local* or *shared* memory only can be done without increasing the run time as the *global* memory transfers are the dominating factor in the execution time. If the hardware is *compute limited*: that is, the computational power is already maxed out. In this case, the online computation of $g(x, y, \sigma_r)$ will add to the run time. It is therefore, advisable to pre-compute the values of $g(x, y, \sigma_r)$ as it will result in a reduced run time. Unfortunately, this behavior is problem and hardware dependent. Thus, the most efficient implementation depends on the actual use case. This effect can be exploited for various applications in the field of medical image computing [6,7].

```
bilateralFilter (image) {
   sumWeight = 0;
   sumFilter = 0;
   step = halfWidth*2+1;
   for (j=0; j < step; j++) {
      for (i=0; I < step; i++) {
            nx = halfWidth + i;
            ny = halfWidth + j;
            g1 = geomClose(nx, ny, sigma_spatial)
            g2 = intensityClose(nx, ny, x, y, sigma_int,
            image)
            sumWeight += g1*g2;
            sumFilter += g1*g2*image[(x-nx)+((y-ny)*width)];
      }
   }
   return sumFilter/sumWeight;
}
```

Algorithm 8.2: Pseudo-code for a bilateral filter. Note the similarity to Algorithm 8.1. The only difference is the introduction of a new function *intensityClose* that computes the closeness of the two intensities of the image.

8.3.2 Bilateral Filter

The bilateral filter [8] is an extension of the Gaussian filter that adds edge preservation. The concept is very easy to follow, the implementation is straightforward, and its parameterization can be determined directly from the image. These properties make the bilateral filter one of the most popular edge-preserving filtering methods. Unfortunately, the bilateral filter has a rather high computational complexity. Parallel execution hardware such as graphics cards have lessened this burden which lead to a broad application of the bilateral filter.

The definition of the bilateral filter is as follows:

$$p'_b(x,y,\sigma_r,\sigma_i) = \frac{1}{k(x,y,\sigma_r,\sigma_i)} \sum_{x',y' \in \mathcal{N}} p(x',y') \cdot g(x',y',\sigma_r) \cdot g(p(x',y')$$

$$- p(x,y),\sigma_i) \tag{8.8}$$

$$k(x,y,\sigma_r,\sigma_i) = \sum_{x',y' \in \mathcal{N}} g(x',y',\sigma_r) \cdot g(p(x',y') - p(x,y),\sigma_i)$$

where we introduce another 1D Gaussian function $g(x,\sigma_i) = e^{-(x^2/(2\sigma_r^2))}$ with a standard deviation σ_i, which describes the intensity similarity between the pixel at $p(x',y')$ and $p(x,y)$. If they are identical, their difference is zero which leads to a maximum in the Gaussian function. If the kernel is evaluated in a uniform image area, all differences will be zero which leads the normal Gaussian kernel. In the presence of an intensity difference, the pixels that have different intensities will contribute less to the filtering, that is, the edge is preserved. Doing so, the filtering is not shift-invariant and pre-computation of $g(p(x',y') - p(x,y), \sigma_i)$ is impossible.

In order to choose σ_i, it is advisable to measure the smallest edge in the image that should still be preserved. For projection images, σ_i should be chosen very small as too high values might introduce streaks in the reconstruction. In this case, 10% of the smallest edge values are advisable. In reconstruction space, σ_i can be chosen up to the value of the smallest edge. Note that this will already lead to a slight blurring of the edge [9].

Algorithm 8.2 depicts pseudo–kernel code. Again, the parallelization is performed over the image coordinates (x,y). Also in this implementation, we do not precompute the geometric closeness as in the previous example, which may lead to increased run time depending on the execution hardware and problem size.

8.3.3 Joint Bilateral Filter

The joint bilateral filter is an extension of the bilateral filter that introduces a so-called guidance image [10]. The idea is that one image shows the desired information in very noisy conditions while another one shows the correct edge information. In the original paper, the desired information is a color photograph that was taken without flash and the edge information comes from an image with flash that has suboptimal color information. Thus, the bilateral filter is adjusted in the following manner:

```
jointBilateralFilter (image, guide){
   sumWeight = 0;
   sumFilter = 0;
   step = halfWidth*2+1;
   for (j=0; j < step; j++){
       for (i=0; I < step; i++){
           nx = halfWidth + i;
           ny = halfWidth + j;
           g1 = geomClose(nx, ny, sigma_spatial)
           g2 = intensityClose(nx, ny, x, y,
               sigma_int, guide)
           sumWeight += g1*g2;
           sumFilter += g1*g2*image[(x-nx)+((y-ny)*width)];
       }
   }
   return sumFilter/sumWeight;
}
```

Algorithm 8.3: Pseudo-code for the joint bilateral filter. The only difference is that *intensityClose* now operates on a guidance image *guide*.

$$p'_j(x,y,\sigma_r,\sigma_i) = \frac{1}{k(x,y,\sigma_r,\sigma_i)} \sum_{x',y' \in \mathcal{N}} p(x',y') \cdot g(x',y',\sigma_r) \cdot g(p_g(x',y')$$

$$- p_g(x,y),\sigma_i) \tag{8.9}$$

$$k(x,y,\sigma_r,\sigma_i) = \sum_{x',y' \in \mathcal{N}} g(x',y',\sigma_r) \cdot g(p_g(x',y') - p_g(x,y),\sigma_i)$$

In this formulation, we have now introduced a guidance image $p_g(x,y)$ that is used to compute the geometric closeness. Everything else is identical to the bilateral filter. Subsequently, also Algorithm 8.3 is almost identical to Algorithm 8.2. The only change is the introduction of the guidance image. If the original image is supplied as guidance image, this implementation will result in the original bilateral filter.

While the implementation is straightforward, results are astonishing and find applications in many fields from super resolution [11], over multimodal imaging to energy-resolving detectors [9].

8.3.4 Guided Filter

The guided filter [12] is a common alternative to the joint bilateral filter. The main idea is to express the filtered image as a linear transform of the guidance image:

$$p'_{gf}(x, y) = a_k p_g(x, y) + b_k \qquad (8.10)$$

The local coefficients a_k and b_k are found as the solution to the following optimization problem:

$$a_k, b_k = \mathrm{argmin}_{a,b} \sum_{x',y' \in \mathcal{N}} \{[a p_g(x', y') + b - p(x', y')]^2 + \epsilon a^2\} \qquad (8.11)$$

These optimal coefficients are found as

$$a_k = \frac{\mathrm{Cov}_{\mathcal{N}}(p, p_g)}{\mathrm{Var}_{\mathcal{N}}(p_g) + \epsilon}$$

$$b_k = \mathrm{Mean}_{\mathcal{N}}(p) - a_k \cdot \mathrm{Mean}_{\mathcal{N}}(p_g) \qquad (8.12)$$

where
  $\mathrm{Mean}_{\mathcal{N}}(p)$ computes the mean of the image p in the local neighborhood $\mathcal{N}$
  $\mathrm{Var}_{\mathcal{N}}(p)$ computes the local variance of p
  $\mathrm{Cov}_{\mathcal{N}}(p, p_g)$ computes the covariance between p and p_g.

If we consider the case where the guidance image is identical to the filtered image and $\epsilon = 0$, we can make the following observations:

- The covariance in the numerator of a_k becomes a variance.

- If the neighborhood is completely homogeneous, this variance will become zero and therewith a_k will become 0. Thus, the filtered image will only consist of b_k which is the mean value of the neighborhood.

- If the variance is not 0, a_k will become 1 and thus, b_k will be zero. In those regions, the filtered image will consist only of the original image values.

With a value of ϵ that is higher than zero, one can adjust the filter behavior between these two extreme cases. Note that for actual application on an image, the local values of a_k and b_k have to be determined first for every pixel. Then, a low-pass filter has to be applied to all of the a_k and b_k before the final filter is computed. As mean filters can be implemented very efficiently using integral images, the guided filter can be applied fast, even if it is applied on large neighborhoods. Algorithm 8.4 describes a summary of the steps that need to be done in order to compute the guided filter. In contrast to previous algorithms, it does not describe kernel code but subsequent steps that can be executed parallel for every pixel.

```
guidedFilter (image, guide){
    compute a_k and b_k for all pixels
    apply smoothing a_k and b_k for all pixels
    compute final filter for all pixels
}
```

Algorithm 8.4: In contrast to previous algorithms, the guided filter has to be separated into several kernel executions. Thus, it cannot be described as a single kernel. The preceding code gives an abstract description on how the filter is implemented.

8.3.5 Structure Tensor

The idea of the structure tensor–based filtering is to use a structure detector—the structure tensor—to steer the denoising of the image. Before doing so, we decompose the projection stack into directional high-frequency components and into a single low-frequency component [13]. Next, a structure tensor is computed for every pixel in order to measure the amount of structure and its contributions to the different spatial directions. If a high amount of structure is detected, more weight is given to the respective directional high-pass component. If the tensor reports a low value, more weight is given to the low-pass component. In the end, the filtered image is recovered as a weighted sum of its high- and low-frequency components.

```
structureTensorFilter (image){
    compute directional high-pass components at each pixel
    compute low-pass component at each pixel
    compute structure tensor for each direction at each pixel
    low-pass filter structure tensor
    compute filter output for each pixel
}
```

Algorithm 8.5: In order to filter the image with a structure tensor, the image needs to be decomposed into directional high-frequency components and a low-pass component. Then the structure tensor is computed for each pixel and low-pass filter. In the final step, the filter output is computed as a weighted sum of low-pass and high-pass components where the weights are obtained from the structure tensor.

Again, this filter cannot be described by a single kernel. Algorithm 8.5 reports the individual steps which can be computed in parallel. For the filtering, the use of fast Fourier transforms is advisable. Note that if this code is to be executed on a graphics card on an entire projection stack with 6 directional components in floating point precision up to 24 times, the projection stack memory may be required. Thus, it is advisable to process the data in smaller blocks.

8.4 DENOISING FOR CT

The previously described methods are directly applicable to x-ray projection images or to reconstructed CT slice data in 2D or 3D. However, there have been many adaptations of these methods that aim at modeling prior information about x-ray imaging process into the denoising process. In the following, we present a small subset of the literature that is found on this broad topic. We regard these methods to be the most important ones in everyday use.

8.4.1 Projection Domain

In filtered back projection–type algorithms, streak noise as shown in the beginning of this chapter can be omitted by projection-based noise reduction. All of the previously mentioned methods can be extended to incorporate the average path length of the photons through the object. Kachelriess et al. developed a method based on triangular filters that performs filtering on the entire projection stack [14]. Zeng and Zamyatin included such ray-dependent weighting into the ramp filter of the filtered back projection algorithm and combined it with an edge-preserving filtering in reconstruction domain [15]. Furthermore, also the structure tensor can be applied in projection domain by processing the entire projection stack [13]. Schäfer Dirk et al. investigated several implementations of edge-preserving and noise-adaptive filtering and found that filtering minimizing the total variation worked best in their high-contrast examples [16].

8.4.2 Iterative Reconstruction

Although iterative reconstruction is an order of magnitude slower than analytic reconstruction methods, there is a broad variety of iterative reconstruction approaches that aim at reducing the image noise and the x-ray dose. Probably the most well known one is penalized least squares iterative reconstruction [17]. It aims at weighting each ray with its noise variance. Doing so, the noisy ray gets a reduced weight while the more reliable rays with a lower path length get more influence in each iteration. Recently, iterative methods that reduce the total variation in reconstruction space became more and more popular [18]. In these approaches, the assumption is made that the object of interest is piecewise constant which allows extreme noise suppression. If performed in an extreme way, the resulting images are often described as surrealistic or comic-like. Thus, such regularization has to be performed with caution. In general, iterative reconstruction methods are able to deliver superior image quality compared to traditional methods. However, their parameterization is difficult and requires a lot of experience and/or grid search of the parameter space which increases their run time even further. If applied in a clinical context, often only a single iteration is performed.

8.4.3 Reconstruction Domain

There is a large body of literature on noise reduction in reconstruction space. Many approaches build on the previously presented denoising methods. A quite different approach that uses wavelets for denoising was presented by

Borsdorf et al. [19]. This approach aims at estimating the noise by a dual reconstruction. The set of projection images is split in half and each set is reconstructed individually. Then both reconstructions are correlated against each other in wavelet domain and only correlated structures, that is, structures that belong to the object are preserved. Bruder et al. have shown that given a high projection number—as in the case of CT—penalized least squares regularization can be expressed entirely in image domain as a nonlinear filter [20]. Given a piecewise constant object, nonlinear filtering can also achieve reconstructions that resemble the outcomes of total variation–based iterative reconstruction methods [21]. This is in line with Manhart et al. who reported a filtered back projection–type method for perfusion C-Arm CT that deliversimage quality that is en par with comparable iterative methods [22]. The run time of the analytic method, however, was more than 23 times faster than the iterative method, although both were implemented on graphics cards.

8.5 SUMMARY

This chapter gave a comprehensive summary of denoising algorithms for GPU. We described the underlying physical processes that cause the noise from photon statistics to the noise in reconstruction domain and gave examples what kind of structure the noise can exhibit. This was followed by an introduction to denoising methods in general with a focus on GPU implementation. We started from shift-invariant Gaussian filtering, introduced the edge-preserving bilateral filter, the guided filter, and structure tensor–based image filters. In the last part of the chapter, we discussed important implementations of noise reduction methods into the reconstruction process which included projection-based, iterative, and reconstruction-based denoising methods. In order to implement fast and reliable noise reduction, a combination of projection-based and reconstruction-based filtering is probably the fastest and easiest way to incorporate noise reduction [9,15].

All examples in this book chapter have been created with CONRAD, an open-source software framework for CT Simulation and Reconstruction [23].

REFERENCES

1. M. Manhart, A. Fieselmann, Y. Deuerling-Zheng, A. Maier, and M. Kowarschik. Dynamic reconstruction with statistical ray weighting for C-arm CT perfusion imaging. *12th International Meeting on Fully Three-Dimensional Image Reconstruction in Radiology and Nuclear Medicine*, Lake Tahoe, CA, June 18, 2013, pp. 221–224.

2. J. Hsieh. Adaptive filtering approach to the streaking a artifact reduction due to X-ray photon starvation. *Medical Physics*, 25, 2139–2147, 1998.
3. A. Borsdorf. Adaptive filtering for noise reduction in x-ray computed tomography. PhD thesis, Friedrich-Alexander University Erlangen-Nuremberg, Erlangen, Germany, 2010.
4. T. Buzug. *Computed Tomography: From Photon Statistics to Modern Cone-Beam CT*. Springer, Heidelberg, Germany, 2008.
5. J. Baek and N. Pelc. Local and global 3D noise power spectrum in cone-beam CT system with FDK reconstruction. *Medical Physics*, 38, 2122–2131, 2011.
6. T. Zinsser and B. Keck. Systematic performance optimization of cone-beam back-projection on the Kepler architecture. *Proceedings of the 12th Fully Three-Dimensional Image Reconstruction in Radiology and Nuclear Medicine*, Lake Tahoe, CA, June 16–21, 2013, pp. 225–228.
7. A. Maier, H. Hofmann, C. Schwemmer, J. Hornegger, A. Keil, and R. Fahrig. Fast simulation of x-ray projections of spline-based surfaces using an append buffer. *Physics in Medicine and Biology*, 57(19), 6193–6210, 2012.
8. V. Aurich and J. Weule. Non-linear Gaussian filters performing edge-preserving diffusion. *Mustererkennung 1995. Informatik aktuell*, G. Sagerer, S. Posch, F. Kummert (eds.). Springer, Berlin, Germany, 1995, pp. 538–545.
9. M. Manhart, R. Fahrig, J. Hornegger, A. Dörfler, and A. Maier. Guided noise reduction for spectral CT with energy-selective photon counting detectors. *Proceedings of the Third CT Meeting (The Third International Conference on Image Formation in X-Ray Computed Tomography)*, Salt Lake City, UT, June 23, 2014, pp. 91–94.
10. G. Petschnigg, R. Szeliski, M. Agrawala, M. Cohen, H. Hoppe, and K. Toyama. Digital photography with flash and no-flash image pairs. *ACM Transactions on Graphics (TOG)*, 23(3), 664–672, 2004.
11. S. Farsiu, E. Michael, and M. Peyman. Multiframe demosaicing and super-resolution of color images. *IEEE Transactions on Image Processing*, 15(1), 141–159, 2006.
12. K. He, S. Jian, and T. Xiaoou. Guided image filtering. *Computer Vision–ECCV 2010*, Heraklion, Greece. Springer, Berlin, Germany, 2010, pp. 1–14.
13. A. Maier, L. Wigström, H. Hofmann, J. Hornegger, L. Zhu, N. Strobel, and R. Fahrig. Three-dimensional anisotropic adaptive filtering of projection data for noise reduction in cone beam CT. *Medical Physics*, 38(11), 5896–5909, 2011.
14. M. Kachelriess, O. Watzke, and W.A. Kalender. Generalized multi-dimensional adaptive filtering for conventional and spiral single-slice, multi-slice, and cone-beam CT. *Medical Physics*, 28(4), 475–490, 2001.
15. G.L. Zeng and A. Zamyatin. A filtered backprojection algorithm with ray-by-ray noise weighting. *Medical Physics*, 40(3), 031113, 2013.
16. S. Dirk, P. van de Haar, and M. Grass. Comparison of Gaussian and non-isotropic adaptive projection filtering for rotational 3D X-ray angiography. *Proceedings of the 12th International Meeting on Fully Three-Dimensional Image Reconstruction in Radiology and Nuclear Medicine*, University of Southern California, Los Angeles, CA, 2013, pp. 130–133.

17. J.A. Fessler. Penalized weighted least-squares image reconstruction for positron emission tomography. *IEEE Transactions on Medical Imaging*, 13(2), 290–300, 1994.
18. X. Pan, E. Sidky, and M. Vannier. Why do commercial CT scanners still employ traditional, filtered back-projection for image reconstruction? *Inverse Problems*, 25(12), 1–50.
19. A. Borsdorf, R. Raupach, T. Flohr, and J. Hornegger. Wavelet based noise reduction in CT-images using correlation analysis. *IEEE Transactions on Medical Imaging*, 27(12), 1685–1703, 2008.
20. R.R. Bruder, J. Sunnegardh, M. Sedlmair, K. Stierstorfer, and T. Flohr. Adaptive iterative reconstruction. *SPIE Medical Imaging*, 7961, 79610J–12, 2011. doi: 10.1117/12.877953.
21. R. Christian, M. Berger, H. Wu, M. Manhart, R. Fahrig, and A. Maier. TV or not TV? That is the Question. *Proceedings of the 12th International Meeting on Fully Three-Dimensional Image Reconstruction in Radiology and Nuclear Medicine*, 2013, pp. 341–344.
22. M.T. Manhart, A. Aichert, T. Struffert, Y. Deuerling-Zheng, M. Kowarschik, A.K. Maier, J. Hornegger, and A. Doerfler. Denoising and artefact reduction in dynamic flat detector CT perfusion imaging using high speed acquisition: First experimental and clinical results. *Physics in Medicine and Biology*, 59(16), 4505, 2014.
23. A. Maier, H. Hofmann, M. Berger, P. Fischer, C. Schwemmer, H. Wu, K. Müller et al. CONRAD—A software framework for cone-beam imaging in radiology. *Medical Physics*, 40(11), 111914, 2013.

CHAPTER 9

GPU-Based Unimodal Deformable Image Registration in Radiation Therapy

Sanjiv S. Samant, Soyoung Lee,
and Sonja S.A. Samant

CONTENTS

9.1 INTRODUCTION

THE EFFICACY OF RADIATION therapy, beyond tumor biology, is determined by two issues that are increasingly reliant on medical imaging: (1) accuracy in delineation of target tumor and neighboring organs at risk (OARs) and (2) accuracy of treatment delivery [1]. Accurate target contouring utilizes high contrast and spatial resolution imaging and can involve multimodal imaging. However, unimodal image registration remains prominent in clinical practice, as observed in strategies that use serial CT imaging to monitor treatment efficacy and re-treatment of patients where there is concern of excess dose to OARs. In these cases, the patient anatomy necessitates deformable image registration (DIR). While DIR is clearly necessary for extracranial imaging, it can also play a role in cranial imaging, where physiological changes can deform the anatomy [2]. Thus, fast computation strategies are needed for unimodal deformable image registration that is used in a variety of clinical settings.

Accurate treatment delivery utilizes all the modern tools of image-guided radiation therapy (IGRT): cone-beam computed tomography (CBCT) [3], onboard magnetic resonance imaging (MRI) [4], 2D–3D (radiograph-CBCT) [5], radiofrequency markers [6], and camera-based surface topography [7]. For the highest accuracy and most faithful representation of the patient anatomy with respect to treatment beam geometry, it is desirable to have 3D imaging of internal anatomy that visualizes the targeted tumor and the neighboring OARs. Only with 3D imaging can one accurately determine the delivered dose distribution. IGRT based on CBCT-CT-based patient positioning has become the standard of care for intensity-modulated radiation therapy (IMRT), volumetric-modulated arc therapy (VMAT), and stereotactic body radiation therapy (SBRT). Although CBCT and CT are both volumetric x-ray imaging modalities, CBCT-CT can constitute multimodal registration since the usual algorithms for unimodal registration lack robustness and accuracy for CBCT-CT registrations [8]; however, there is still a role for unimodal image registration. Initially, CBCT-CBCT registration [8] was proposed for patient positioning, as it allowed for a more robust and faster registration than CT-CBCT (with the assumption a reference CBCT could be established on the first day of treatment). Though the accuracy of bony anatomical landmarks is comparable for CT-CBCT (2.2 ± 0.6 mm) and CBCT-CBCT (1.8 ± 1.0 mm) registrations, this is not generally true for soft tissue landmarks [8,9]. The approach still requires that the reference

CBCT is established using a CT-CBCT registration, which could be manual, automatic intensity or automatic feature based. Unimodal intensity-based image registration is generally more robust and accurate than multimodal image registration [8,10,11]. With subsequent improved registration strategies utilizing fast intensity-based registrations, CT-CBCT has become the preferred registration strategy in the clinic. Nonetheless, with the increased feasibility of adaptive radiation therapy (ART) within a clinical setting, unimodal registration remains a useful clinical tool [12]. This strategy involves using a CT-CBCT deformation map to deform the reference planning CT to a "CT of day," thereby re-mapping planning structures and recalculating the dose distribution [9,12,13].

9.2 GPU VS CPU

Medical imaging is generally computationally intensive, which hampers maximal utilization of image processing applications in the clinic. Translational imaging research requires ever increasing computational efficiency [14–16]. The conventional approach in imaging has been to reduce the computation time by employing higher-performance central processing units (CPUs) (i.e., higher clock speed and multicore architecture and multiprocessor systems). OpenMP, an application programming interface (API) widely used for multi-CPU-based computing, is used for high-performance computing. Over the past decade, cell broadband engine [17] and field-programmable gate array (FPGA) [18–20] represented promising alternatives to graphics processing unit (GPU)-based image processing; however, today GPU is the most popular alternative to CPU for computationally intensive medical image processing [21–23]. Although the role of GPU in imaging research is well established, commercial products are still lacking.

Due to the relative ease of programming and rapid technological developments, GPUs are now preferred for accelerated image registrations in medical imaging research. Moore's Law, an observation based on 50 years of historical data and only now slowing down due to technological limits in 450 mm wafers and extreme ultraviolet lithography, holds that the density of integrated circuits doubles every 18–24 months with an accompanying increase in computational efficiency; this law drives CPU-based computations [24]. Similarly, a new Moore's Law has emerged for GPUs with performance, based on gigaflop ratings, observed to double every year, and is driven by the ease with which transistors can be added to a GPU board and the demands of a multibillion dollar video gaming industry [25].

A GPU is a highly parallel computing structure with tremendous memory bandwidth and computation horsepower, but it does not have the programming flexibility of the CPU due to its parallel data architecture; thus, it can be challenging for many types of computations, including those required for DIR [16]. In addition, significantly less dedicated processor memory is available for GPU programming than for CPU programming. Notwithstanding this, GPU computing is a viable high-performance alternative for general nongraphics computing tasks (e.g., DIR). Standard GPUs are limited to single precision (i.e., 32 bits), whereas CPUs and only the latest high-end GPUs have both single and double precision (i.e., 64 bits) [26,27].

Typically, single precision is deemed to be sufficient for DIR in medical imaging [16] and other general computations, including diffusion equations, wave equations, Poisson problems, and Navier–Stokes equations. For the Demons algorithm, which is used for benchmarking in medical imaging, there is no significant difference between a single-precision floating-point GPU computation and a double-precision floating-point CPU computation [16]. However, there are cases, such as for finite element modeling, where double precision is needed [28].

Among the GPU programming languages, CUDA is especially popular in medical imaging but is limited to NVIDIA GPUs [22,29]. OpenCL [30], similar to CUDA in its tools, is a cross-vendor language that can be used with GPUs manufactured by NVIDIA (NVIDIA Corp., Santa Clara, CA) and AMD (Advanced Micro Devices, Inc., Sunnyvale, CA). GFlops (i.e., 10^9 floating-point operations per second) is an accepted overall approximate representation of the computational capability of a processor, but it is not a complete metric for computational efficiency. For example, the concept of time per megapixels per iteration (TPMI) can be used to evaluate memory optimization in the GPU implementation [16].

Rigid registration of images uses linear transformations based on similarity, mutual information, and normalized mutual information with only a small number of global transformation parameters (i.e., three translations and three rotations) to be computed. Affine transformations, which allow for image scaling, also have six global parameters. As DIR requires determining a (x, y, z) coordinate mapping for each individual pixel,

DIR with images on order of 10^7–10^8 pixels becomes a computationally complex problem requiring intensive computing. With their inherent parallelized structure, GPUs are better suited than CPUs to this class of intensive computations. Using OpenMP for multiple CPUs, individual CPUs can be assigned to each GPU card to simultaneously carry out calculations and further increase overall computation performance. Imaging applications in radiation therapy that require fast computing for clinical tasks (e.g., image fusion, automated contouring, dose mapping, and ART) need efficient computational strategies. GPU computations, which can compress computation times by several orders of magnitude, provide a strategy for achieving near real-time computing.

9.3 IMAGE REGISTRATION ALGORITHMS

The following sections describe nonrigid registration algorithms based on an iterative strategy that has been successfully ported to GPUs.

9.3.1 B-spline-Based Free-Form Deformation

In free-form deformation, which is the most general deformation, each voxel has flexibility without explicit constrains. B-spline-based free-form deformation can be used to locally deform an image volume, and its computational efficiency can be improved with GPU implementation. A B-spline model, where the mesh of control points can be adjusted to model a deformation field, can represent a transformation model of deformable image registration. The deformation field can be described as follows:

$$T(x, y, z) = \sum_{l=0}^{3} \sum_{m=0}^{3} \sum_{n=0}^{3} B_l(u) B_m(v) B_n(w) \phi_{i+l, j+m, k+n}, \tag{9.1}$$

where

$\phi_{i+l, j+m, k+n}$ is the control point in a control points mesh with the size of $n_x \times n_y \times n_z$ and uniform spacing δ

$i = \lfloor x/n_x \rfloor - 1$

$j = \lfloor y/n_y \rfloor - 1$

$k = \lfloor z/n_z \rfloor - 1$

$u = x/n_x - \lfloor x/n_x \rfloor$

$v = y/n_y - \lfloor y/n_y \rfloor$

$w = z/n_z - \lfloor z/n_z \rfloor$.

B_l represents the lth B-spline basis function:

$$B_0(u) = \frac{(1-u)^3}{6}$$

$$B_1(u) = \frac{3u^3 - 6u^2 + 4}{6}$$

$$B_2(u) = \frac{-3u^3 + 3u^2 + 3u + 1}{6}$$

$$B_3(u) = \frac{u^3}{6}$$

Rueckert et al. modeled the cost function of B-spline-based deformable image registration as [31]

$$C(\phi) = C_{similarity}(S, M(T)) + \lambda \cdot C_{smooth}(T) \qquad (9.2)$$

where

$C_{similarity}(S, M(T))$ and $C_{smooth}(T)$ are the similarity measures and smoothing terms with respect to transformation T

ϕ represents the control points to be optimized.

In other words, the goal of solving Equation 9.2 is to find the optimal ϕ to minimize the cost function $C(\phi)$. A gradient descent strategy is used to update the control points ϕ by

$$\phi_{i+1} = \phi_i + \mu \nabla C / \|\nabla C\|. \qquad (9.3)$$

The advantage of a B-spline model is that control points can be used to affect the transformation of the neighborhood of the individual control points, thereby making the B-spline computationally efficient [31]. However, additional constraints must be considered to prevent the deformation field from folding [32]. Application of B-spline models include liver motion [33], chest PET-CT image fusion [34], cardiac modeling [35,36], and breast imaging [31]. Deformable image registration based on B-spline transformations using multi-resolution to increase robustness has been implemented on a Silicon Mechanics Hyperform HPCg R2504.v2 workstation with two Intel Xeon X5675 Six-Core 3.06 GHz processors and a NVIDIA Tesla 2070 GPU for unimodal and multimodal imaging [37].

9.3.2 Demons

The Demons algorithm proposed by Thirion was based on an analogy of thermodynamic concepts [38] in which image entities called "demons" were positioned to move the pixels based on the local characteristics of the images in a method similar to Maxwell's thermodynamics equations. In the case of free-form deformations where all the nonzero gradient pixels are selected as the demons, the displacement field $u(x)$ (the transformation between the target image $T(x)$ and the source image $S(x)$) is estimated as follows:

$$\frac{\partial u(x)}{\partial t} = \frac{(S(x+u(x))-T(x))\nabla T(x)}{\|\nabla T(x)\|^2 + (S(x+u(x))-T(x))^2}, \tag{9.4}$$

where
$S(x+u(x))$ is the deformed source image
∇ is the gradient operator.

At each iteration, the computed displacement field $u(x)$ is convoluted with a Gaussian function for regularization. The Demons algorithm was shown to outperform other methods from the evaluation study of inter-subject brain registrations, and remains a popular algorithm for unimodal imaging [38].

Equation 9.4 indicates that the Demons algorithm will not work well for multimodal image registration because the voxel intensity for the corresponding voxels of these images may not be similar. For example, if one directly applies the Demons algorithm to register cerebrospinal fluid (CSF) in brain MRI T1 images (dark) to MRI T2 images (bright), convergence in Equation 9.4 will not occur even in the presence of a good registration because the corresponding voxel intensity differences $(M^i - S)$ do not approach zero. One solution to make the Demons algorithm applicable in multimodality images is to find an intensity transformation to map the voxel intensity from one imaging modality to another (i.e., causing T1 images to resemble T2 images [39] and CT images resemble CBCT images [40]).

Deformable registration using a Demons algorithm on a GPU has been tested with 3D-CT data. Sharp et al. achieved substantial speedup using the optical flow equation and Gaussian regularization in the Brook programming environment [14]. In terms of GPU memory utilization and computational speed, Samant et al. further optimized Demons-based DIR using NVIDIA's CUDA environment [16,41]. Courty and Hellier [42] presented a GPU implementation of the Demons algorithm with 2D texture mapping for brain MRI. To achieve better registration accuracy and performance, Gu et al. also implemented and compared various versions of the Demons algorithms on GPU [21].

9.3.3 Optical Flow Method

The optical flow method was introduced by Horn and Schunck to estimate the motion between frames in an imaging sequence [43]. The fundamental assumption is that the brightness of the images is preserved (i.e., the voxel intensity of the same object does not change within two image frames for the same anatomical object). Given two images $I(x, y, z, t)$ and $I(x+\delta x, y+\delta y, z+\delta z, t+\delta t)$, the optical flow velocity u can be described as follows:

$$I(x, y, z, t) = I(x+\delta x, y+\delta y, z+\delta z, t+\delta t). \tag{9.5}$$

Equation 9.5 can be rewritten as follows:

$$\frac{\partial I(x, y, z, t)}{\partial t} = 0 \Rightarrow \frac{\partial I}{\partial x}\frac{dx}{dt} + \frac{\partial I}{\partial y}\frac{dy}{dt} + \frac{\partial I}{\partial z}\frac{dz}{dt} + \frac{\partial I}{\partial t} = 0. \tag{9.6}$$

Equation 9.6 can be simplified as

$$\nabla I \cdot u = -\frac{\partial I}{\partial t}, \tag{9.7}$$

where
optical flow velocity $u = [dx/dt, dy/dt, dz/dt]$
$\nabla I = [\nabla I_x, \nabla I_y, \nabla I_z]$ is the image intensity gradient.

Equation 9.7 is underconstrained because one equation is not sufficient for solving three unknown components of u. Smoothness constraints can be used to address this problem. Carefully choosing the smoothing parameters, such as the standard deviation and the width of the Gaussian filter, is essential for obtaining good registration results. Horn and Schunck [43] proposed a constraint that minimizes the square of the gradient magnitude of the optical flow velocity u. In terms of a similarity measure and regularization, this can be rewritten as follows:

$$E(u) = E_{sim}(u) + \lambda E_{reg}(u)$$

$$E_{sim}(u) = \left(\nabla I \cdot u + \frac{\partial I}{\partial t} \right)^2, \tag{9.8}$$

$$E_{reg}(u) = \left\| \nabla u \right\|^2$$

where

E_{sim} and E_{reg} are the similarity and regularization term, respectively

λ is the weighting factor for balancing the similarity and regularization terms.

In image registration, we usually consider u as a displacement, and no temporal information is included. Using calculus of variation, which is equivalent to a gradient decent optimization, Equation 9.8 can be solved by the following iterative scheme:

$$\begin{cases} u_x^{i+1} = u_x^i - \dfrac{\nabla I_x (\nabla I_x u_x^i + \nabla I_y u_y^i + \nabla I_z u_z^i + I_t)}{(\lambda + \nabla I_x^2 + \nabla I_y^2 + \nabla I_z^2)} \\[4mm] u_y^{i+1} = u_y^i - \dfrac{\nabla I_y (\nabla I_x u_x^i + \nabla I_y u_y^i + \nabla I_z u_z^i + I_t)}{(\lambda + \nabla I_x^2 + \nabla I_y^2 + \nabla I_z^2)}, \\[4mm] u_z^{i+1} = u_z^i - \dfrac{\nabla I_z (\nabla I_x u_x^i + \nabla I_y u_y^i + \nabla I_z u_z^i + I_t)}{(\lambda + \nabla I_x^2 + \nabla I_y^2 + \nabla I_z^2)} \end{cases} \tag{9.9}$$

where

u_x^i, u_y^i, and u_z^i are the optical flow in x, y, and z direction at iteration i.

∇I_x, ∇I_y, ∇I_z are the image gradient in x, y, and z direction.

Ostergaard et al. [8] reported that GPU implementation accelerated computation speeds by up to 48.7 times compared to CPU for thorax CT imaging using the aforementioned Horn–Schunck algorithm and also implemented a variation of this algorithm for CT-CBCT and CBCT-CBCT in head and neck imaging.

9.3.4 Evaluation of Image Registration

Deformable registration for unimodal imaging, though less complex than multimodal registration, remains an active research problem. A number of approaches have been used to validate deformable registration algorithms, including computer-generated deformations [44–46], high-contrast phantoms [46,47], and human expert-identified anatomical landmarks [48–50]. However, when a clinician must evaluate the "goodness" of a registration for an individual case, the problem is even more challenging, especially as one seeks to balance computational speed to achieve near real-time registrations with registration accuracy [46].

A qualitative evaluation of the goodness of a registration globally can be carried out between the reference image and the deformed image using difference imaging, color blending, and checkerboard pattern in conjunction with identification of common anatomical landmarks. This approach, while sufficient for rigid registration, does not necessarily reflect the

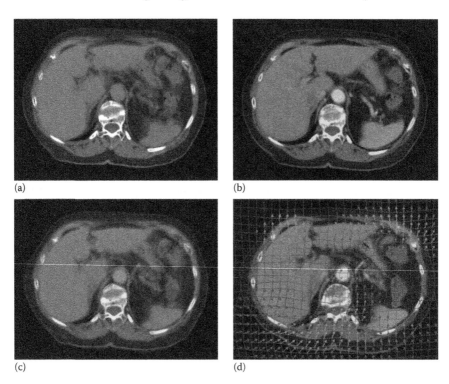

(a) (b)

(c) (d)

FIGURE 9.1 **(See color insert.)** CT images from (a) primary imaging, (b) secondary imaging, (c) fused imaging with deformed secondary imaging, and (d) superimposed vector map.

accuracy of the individual voxels, especially in regions of uniform voxel intensity. Methodologies have also been proposed for assessing different algorithms or variation of parameters for the same algorithm involving global metrics, such as normalized correlation coefficient and mutual information [51] and inverse consistency error (ICE) [52], but these provide approximate relative comparisons and do not allow one to assess the clinical validity of a specific registration.

At an intuitive level, deformation vector maps, consisting of voxel mappings between the reference and target images, can be been used to verify that homogenous regions are mapped with minimal separation between neighboring voxels of the same intensity [53]. As indicated in Figure 9.1, the vector field can be superimposed on top of the secondary or target image to assess the reasonableness of the registration (i.e., bony structures should remain rigid, whereas tissue is generally deformable while maintaining anatomical landmarks). This serves as a useful "reality check," especially when the deformation vector map is used to map organ contours between reference and target images.

9.4 CLINICAL USES OF UNIMODAL IMAGE REGISTRATION

9.4.1 Image Fusion

Image fusion represents the most popular clinical use of DIR in radiation therapy. Although this usually involves different imaging modalities, it can also involve the same modality (i.e., CT–CT, MR–MR). CT imaging acquired in different patient positions, such as CT imaging from a diagnostic study and a CT simulation prior to irradiation, is helpful for accurate target contouring. This process can be significantly accelerated with GPU computing, yet commercial products, such as Velocity (Varian Medical Systems, Palo Alto, CA) and MIM Maestro (MIM Software, Inc., Cleveland, OH), still utilize CPU computing since the deformable registration is generally done offline in current clinical settings. In Figure 9.2, reference and target images are shown along with the superimposed fused image using color blending to highlight areas of mismatch.

9.4.2 Segmentation

Accurate delineation of target volume and critical OARs is essential for precise IMRT. Manual delineation of complex anatomy is time-consuming and laborious; indeed, the length of the time for contouring depends on the disease site and physicians' clinical experience. Contouring consistency can also vary widely due to intra-observer and inter-observer variations

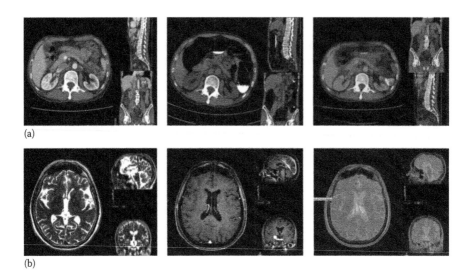

(a)

(b)

FIGURE 9.2 **(See color insert.)** Example of unimodal image registration for image fusion: (a) CT images and (b) MR images. The left most image represents the primary data set, the center image represents the secondary data set, and the right most image represents the fused data set. The CT imaging represents different patient positions, while the MR imaging represents different imaging sequences (T2 and T1 images).

[54–56]. In head and neck IMRT, manual contouring requires significant physician time, with the initial contouring typically taking 120 min for up to 30 structures [57]. To reduce contouring time and increase consistency, atlas-based autosegmentation (ABAS) using deformable registration [57,58] was introduced. The accuracy of automatic contouring based on CT images and its clinical validation have been quantified for various IMRT sites, leading to reductions in contouring times and minimization of inter-observer and inter-observer variability. This reduces the workload for treatment re-planning in cases where patient anatomy changes significantly over the course of radiation therapy, especially for the head and neck and abdomen, for which inter-fraction deformations must be assessed in the setup. As shown in Figure 9.3, the heart was segmented in the reference CT data set, and DIR was used to map the heart onto the secondary CT data set, which represents a different patient setup position.

For robust contouring, one successful autosegmentation strategy utilizes multiple atlases accompanied with the simultaneous truth and performance level estimation (STAPLE) algorithm, statistical shape model

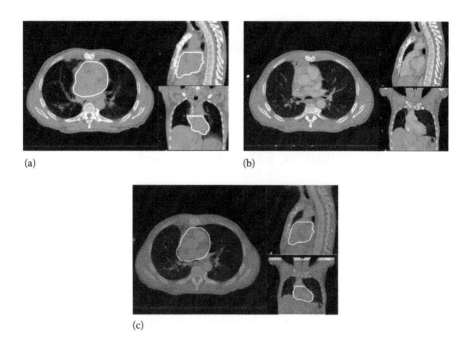

(a)

(b)

(c)

FIGURE 9.3 **(See color insert.)** Contour deformation: (a) average CT and (b) reference CT image data sets. The top two slices represent CT imaging acquired at different times with the heart contoured in the first CT, while (c) is the resulting overlay using DIR.

and statistical appearance model (SSM/SAM) method, or hybrid methods (ABAS+SSM/SAM) [58]. This multi-atlas segmentation strategy, compared to inter-observer variability, has achieved reasonable agreement between autosegmented contours and expert-delineated manual contours, with a median slicewise (i.e., 2D) Hausdorff distance of 7.6 ± 1.4 mm, a median slicewise Dice similarity coefficient of $80.2\% \pm 5.9\%$, and an overall average (i.e., 3D) Dice similarity coefficient of $77.8\% \pm 3.3\%$ over 10 head and neck IMRT cases. However, autosegmentation is known to be susceptible to image noise and low contrast differences (e.g., parotid) in CT imaging, which is the standard treatment planning modality. Autosegmentation has shown good accuracy in clinical brain imaging [58], and it offers the potential of increased clinical efficacy, such as reduced margins for hypo-fractionated prostate IMRT. Commercial software, such as Velocity and MIM Maestro, make use of atlas-based segmentation but currently do not use GPU for increased computational efficiency. Radiation therapy for oligometastatic disease, which is characterized by multiple radiation re-treatments that often involve the same organ, requires contour mapping and consideration of cumulative dose effects. These computationally intensive tasks could benefit from a GPU implementation.

GPU can be a powerful tool for tracking and motion analysis for respiratory motion and cardiac motion in 4DCT, including the estimation of the uptake rate of a contrast agent. The motion information can be consequently linked to calculate a real-time 3D conformal dose of a moving target. A GPU implementation for 3D real-time lung dosimetry has been implemented in the PlanUNC treatment planning software [59].

9.4.3 Adaptive Radiation Therapy

Advanced treatment technologies, such as IMRT and VMAT, can result in improved tumor control and reduced normal tissue toxicity with hypofractionated radiation therapy. However, delivery of an optimal conformal dose can be adversely affected by volumetric or geometric changes in a patient's anatomy during the course of radiation therapy, either

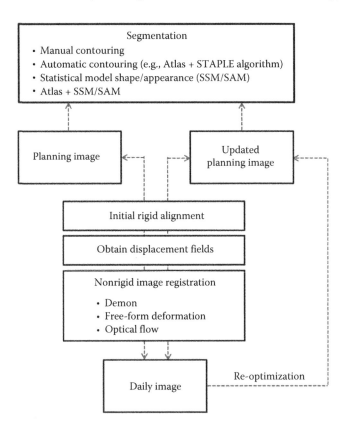

FIGURE 9.4　General image registration workflow for ART.

intra-fraction or inter-fraction motion. ART is a promising technique to adjust the treatment plan and delivery according to daily anatomical and uncorrected positioning changes. Figure 9.4 illustrates a typical workflow for ART, which incorporates the initial treatment plan into the newly updated images. Clinical implementation of ART requires many computationally intensive tasks, such as dose accumulation, treatment evaluation, re-contouring, re-optimization, and dose computations. For the modified treatment plan to be clinically implemented, these tasks must be accomplished within a few minutes. A GPU implementation of online ART makes this possible for clinical use.

A number of advanced imaging technologies have been developed for monitoring and measuring target motion in near real-time target motion or surrogate respiratory motion. Currently, the 4D-CBCT XVI (Elekta Oncology Systems Ltd., Crawley, United Kingdom) and ultrasound-based Clarity linac systems can be used to verify the tumor motion just before treatment but are limited to rigid registrations; currently they do not take advantage of GPU to reduce image reconstruction or compute deformable registration maps.

Several linear accelerators utilizing near real-time-based image guidance, such as the Vero linac (Mitsubishi Heavy Industries, Kobe, Japan; and BrainLab AG, Munich, Germany) [60] and the Cyberknife (Accuray, Sunnyvale, CA), use GPU computing. Imaging processing involves 2D–3D registrations of 2D radiographs taken intermittently during irradiation and the treatment planning CT to ensure correct target delivery in the presence of intra-fraction motion. For radiation therapy involving online MR imaging, GPU-based computing has been proposed, but it has not been implemented in commercial treatment delivery systems [61].

9.5 CONCLUSION

Use of multiple unimodal data sets, typically CT and MR, are integral to treatment planning and treatment assessment in radiation therapy. Deformable registration for general image fusion and accurate target contouring has become a valuable clinical tool. GPUs can speed up clinical tasks with the eventual goal of near real-time image processing to fully realize the potential of deformable registration. Although GPUs are primarily found in research settings, their inclusion within the purely clinical environment is expected to grow with the emergence of ART as part of the clinical paradigm.

ACKNOWLEDGMENTS

The authors would like to thank Kate Casey-Sawicki, MA, and the staff in the UF Department of Radiation Oncology Research Office for editing and preparing this manuscript for publication.

REFERENCES

1. Simpson DR, Lawson JD, Nath SK, Rose BS, Mundt AJ, Mell LK. A survey of image-guided radiation therapy use in the United States. *Cancer* 2010; 116: 3953–3960.
2. Castadot P, Lee JA, Parraga A, Geets X, Macq B, Gregoire V. Comparison of 12 deformable registration strategies in adaptive radiation therapy for the treatment of head and neck tumors. *Radiother Oncol* 2008; 89: 1–12.
3. Nithiananthan S, Brock KK, Daly MJ, Chan H, Irish JC, Siewerdsen JH. Demons deformable registration for CBCT-guided procedures in the head and neck: Convergence and accuracy. *Med Phys* 2009; 36: 4755.
4. Mutic S, Dempsey JF. The ViewRay system: Magnetic resonance-guided and controlled radiotherapy. *Semin Radiat Oncol* 2014; 24: 196–199.
5. Liu WP, Otake Y, Azizian M, Wagner OJ, Sorger JM, Armand M, Taylor RH. 2D-3D radiograph to cone-beam computed tomography (CBCT) registration for C-arm image-guided robotic surgery. *Int J Comput Assist Radiol Surg* 2014; [Epub ahead of print].
6. Kupelian P, Willoughby T, Mahadevan A, Djemil T, Weinstein G, Jani S, Enke C et al. Multi-institutional clinical experience with the Calypso System in localization and continuous, real-time monitoring of the prostate gland during external radiotherapy. *Int J Radiat Oncol Biol Phys* 2007; 67: 1088–1098.
7. Bert C, Metheany KG, Doppke KP, Taghian AG, Powell SN, Chen GT. Clinical experience with a 3D surface patient setup system for alignment of partial-breast irradiation patients. *Int J Radiat Oncol Biol Phys* 2006; 64: 1265–1274.
8. Ostergaard Noe K, De Senneville BD, Elstrom UV, Tanderup K, Sorensen TS. Acceleration and validation of optical flow based deformable registration for image-guided radiotherapy. *Acta Oncol* 2008; 47: 1286–1293.
9. Boggula R, Lorenz F, Abo-Madyan Y, Lohr F, Wolff D, Boda-Heggemann J, Hesser J, Wenz F, Wertz H. A new strategy for online adaptive prostate radiotherapy based on cone-beam CT. *Z Med Phys* 2009; 19: 264–276.
10. Li A, Kumar A. Accelerating volume image registration through correlation ratio based methods on GPUs. In: *17th Euromicro Conference on Digital System Design*, Verona, Italy. 2014; pp. 82–89.
11. Saxena V, Rohrer J, Gong L. A parallel GPU algorithm for mutual information based 3D nonrigid image registration. *Lecture Notes Comput Sci* 2010; 6272: 223–234.
12. Zhang T, Chi Y, Meldolesi E, Yan D. Automatic delineation of on-line head-and-neck computed tomography images: Toward on-line adaptive radiotherapy. *Int J Radiat Oncol Biol Phys* 2007; 68: 522–530.

13. Wu QJ, Thongphiew D, Wang Z, Mathayomchan B, Chankong V, Yoo S, Lee WR, Yin FF. On-line re-optimization of prostate IMRT plans for adaptive radiation therapy. *Phys Med Biol* 2008; 53: 673–691.

14. Sharp GC, Kandasamy N, Singh H, Folkert M. GPU-based streaming architectures for fast cone-beam CT image reconstruction and demons deformable registration. *Phys Med Biol* 2007; 52: 5771–5783.

15. Gu X, Choi D, Men C, Pan H, Majumdar A, Jiang SB. GPU-based ultra-fast dose calculation using a finite size pencil beam model. *Phys Med Biol* 2009; 54: 6287–6297.

16. Samant SS, Xia J, Muyan-Ozcelik P, Owens JD. High performance computing for deformable image registration: Towards a new paradigm in adaptive radiotherapy. *Med Phys* 2008; 35: 3546.

17. Bockenbach O, Knaup M, Kachelriess M. Implementation of a cone-beam backprojection algorithm on the cell broadband engine processor. In: Pluim J, Reinhardt J, eds., *Proceedings of SPIE, Medical Imaging: Image Processing*, vol. 6510, San Diego, CA. 2007; p. 651056.

18. Castro-Pareja CR, Jagadeesh JM, Shekhar R. FPGA-based acceleration of mutual information calculation for real-time 3D image registration. In: Kahtarnavaz N, Laplante P, eds., *Proceedings of SPIE, Real-Time Imaging VIII*, San Jose, CA. 2004; 5297: pp. 212–219.

19. Li J, Papachristou C, Shekhar R. Accelerating mutual information-based 3D medical image registration with an FPGA computing platform. In: *Proceedings of the 2005 ACM/SIGDA 13th International Symposium on Field-Programmable Gate Arrays*, Monterey, CA. 2005: p. 279.

20. Dandekar O, Walimbe V, Shekhar R. Hardware implementation of hierarchical volume subdivision-based elastic registration. In: *Conference Proceedings of the IEEE Engineering in Medicine and Biology Society*, New York, vol. 1. 2006; pp. 1425–1428.

21. Gu X, Pan H, Liang Y, Castillo R, Yang D, Choi D, Castillo E, Majumdar A, Guerrero T, Jiang SB. Implementation and evaluation of various demons deformable image registration algorithms on a GPU. *Phys Med Biol* 2010; 55: 207–219.

22. Fluck O, Vetter C, Wein W, Kamen A, Preim B, Westermann R. A survey of medical image registration on graphics hardware. *Comp Methods Progr Biomed* 2011; 104: e45–e57.

23. Shi L, Liu W, Zhang H, Xie Y, Wang D. A survey of GPU-based medical image computing techniques. *Quant Imaging Med Surg* 2012; 2: 188–206.

24. Hruska J. This is what the death of Moore's law looks like: EUV rollout slowed, 450 mm wafers halted, and an uncertain path beyond 14 nm. *ExtremeTech*; 2014. Available from: http://www.extremetech.com/computing/178529-this-is-what-the-death-of-moores-law-looks-like-euv-paused-indefinitely-450mm-wafers-halted-and-no-path-beyond-14nm (accessed on January 26, 2015).

25. Srinivasan B. Graphical processors and CUDA. 2009. Available from: http://www.umiacs.umd.edu/~ramani/cmsc662/GPU_November_10.pdf (accessed on January 26, 2015).

26. GeForce Graphics Cards. NVIDIA Corporation; 2015. Available from: http://www.nvidia.com/object/geforce_family.html (accessed on January 26, 2015).
27. AMD Graphics. Advanced Micro Devices, Inc. Available from: http://www.amd.com/en-us/products/graphics (accessed on January 26, 2015).
28. Goddeke D, Strzodka R, Turek S. Accelerating double precision FEM simulation with GPUs. In: *Proceedings of ASIM 2005, the 18th Symposium on Simulation Technique*, Erlangen, Germany, 2005.
29. Galizia A, D'Agostino D, Clematis A. An MPI–CUDA library for image processing on HPC architectures. *J Comput Appl Math* 2015; 273: 414–427.
30. Lin YT, Wang SC, Shih WL, Hsieh BKY, Lee JK. Enable OpenCL Compiler with Open64 infrastructures. In: *Proceedings of the 2011 IEEE International Conference on High Performance Computing and Communications*, Washington, DC. 2 pp. 863–868.
31. Rueckert D, Sonoda LI, Hayes C, Hill DL, Leach MO, Hawkes DJ. Nonrigid registration using free-form deformations: Application to breast MR images. *IEEE Trans Med Imaging* 1999; 18: 712–721.
32. Crum WR, Hartkens T, Hill DL. Non-rigid image registration: Theory and practice. *Br J Radiol* 2004; 77(Spec No 2): S140–S153.
33. Rohlfing T, Maurer CR, Jr., O'Dell WG, Zhong J. Modeling liver motion and deformation during the respiratory cycle using intensity-based non-rigid registration of gated MR images. *Med Phys* 2004; 31: 427–432.
34. Mattes D, Haynor DR, Vesselle H, Lewellen TK, Eubank W. PET-CT image registration in the chest using free-form deformations. *IEEE Trans Med Imaging* 2003; 22: 120–128.
35. Frangi AF, Rueckert D, Schnabel JA, Niessen WJ. Automatic construction of multiple-object three-dimensional statistical shape models: Application to cardiac modeling. *IEEE Trans Med Imaging* 2002; 21: 1151–1166.
36. McLeish K, Hill DL, Atkinson D, Blackall JM, Razavi R. A study of the motion and deformation of the heart due to respiration. *IEEE Trans Med Imaging* 2002; 21: 1142–1150.
37. Castillo E, Castillo R, Fuentes D, Guerrero T. Computing global minimizers to a constrained B-spline image registration problem from optimal l1 perturbations to block match data. *Med Phys* 2014; 41: 041904.
38. Thirion JP. Image matching as a diffusion process: An analogy with Maxwell's demons. *Med Image Anal* 1998; 2: 243–260.
39. Guimond A, Roche A, Ayache N, Meunier J. Three-dimensional multi-modal brain warping using the demons algorithm and adaptive intensity corrections. *IEEE Trans Med Imaging* 2001; 20: 58–69.
40. Nithiananthan S, Schafer S, Uneri A, Mirota DJ, Stayman JW, Zbijewski W, Brock KK, Daly MJ, Chan H, Irish JC, Siewerdsen JH. Demons deformable registration of CT and cone-beam CT using an iterative intensity matching approach. *Med Phys* 2011; 38: 1785.
41. Muyan-Ozcelik P, Owens JD, Xia J, Samant SS. Fast deformable registration on the GPU: A CUDA implementation of demons. *ICCSA* 2008; 223–233. Perugia, Italy. June 30–July 3, 2008.

42. Courty N, Hellier P. Accelerating 3D non-rigid registration using graphics hardware. *Int J Image Grap* 2008; 8: 1–18.
43. Horn BKP, Schunck BG. Determining optical flow. *Artific. Intell.* 1981; 17: 185–203.
44. Lu W, Chen ML, Olivera GH, Ruchala KJ, Mackie TR. Fast free-form deformable registration via calculus of variations. *Phys Med Biol* 2004; 49: 3067–3087.
45. Guerrero T, Zhang G, Huang TC, Lin KP. Intrathoracic tumour motion estimation from CT imaging using the 3D optical flow method. *Phys Med Biol* 2004; 49: 4147–4161.
46. Wang H, Dong L, O'Daniel J, Mohan R, Garden AS, Ang KK, Kuban DA, Bonnen M, Chang JY, Cheung R. Validation of an accelerated 'demons' algorithm for deformable image registration in radiation therapy. *Phys Med Biol* 2005; 50: 2887–2905.
47. Juang T, Das S, Adamovics J, Benning R, Oldham M. On the need for comprehensive validation of deformable image registration, investigated with a novel 3-dimensional deformable dosimeter. *Int J Radiat Oncol Biol Phys* 2013; 87: 414–421.
48. Rietzel E, Chen GT. Deformable registration of 4D computed tomography data. *Med Phys* 2006; 33: 4423–4430.
49. Boldea V, Sharp GC, Jiang SB, Sarrut D. 4D-CT lung motion estimation with deformable registration: Quantification of motion nonlinearity and hysteresis. *Med Phys* 2008; 35: 1008–1018.
50. Brock KK. Results of a multi-institution deformable registration accuracy study (MIDRAS). *Int J Radiat Oncol Biol Phys* 2010; 76: 583–596.
51. Yaegashi Y, Tateoka K, Fujimoto K, Nakazawa T, Nakata A, Saito Y, Abe T, Yano M, Sakata K. Assessment of similarity measures for accurate deformable image registration. *J Nucl Med Radiat Ther* 2012; 3: 4.
52. Varadhan R, Karangelis G, Krishnan K, Hui S. A framework for deformable image registration validation in radiotherapy clinical applications. *J Appl Clin Med Phys* 2013; 14: 4066.
53. Schreibmann E, Pantalone P, Waller A, Fox T. A measure to evaluate deformable registration fields in clinical settings. *J Appl Clin Med Phys* 2012; 13: 3829.
54. Hong TS, Tome WA, Chappell RJ, Harari PM. Variations in target delineation for head and neck IMRT: An international multi-institutional study. *Int J Radiat Oncol Biol Phys* 2004; 60: S157–S158.
55. Jeanneret-Sozzi W, Moeckli R, Valley JF, Zouhair A, Ozsahin EM, Mirimanoff RO. The reasons for discrepancies in target volume delineation: A SASRO study on head-and-neck and prostate cancers. *Strahlenther Onkol* 2006; 182: 450–457.
56. Steenbakkers R, Duppen J, Fitton I, Deurloo K, Zijp L, Eisbruch A, Nowak P, Herk Mv, Rasch C. Observer variation in delineation of nasopharyngeal carcinoma for radiotherapy, a 3-D analysis. *Int J Radiat Oncol Biol Phys* 60: S160–S161.

57. Teguh DN, Levendag PC, Voet PW, Al-Mamgani A, Han X, Wolf TK, Hibbard LS et al. Clinical validation of atlas-based auto-segmentation of multiple target volumes and normal tissue (swallowing/mastication) structures in the head and neck. *Int J Radiat Oncol Biol Phys* 2011; 81: 950–957.

58. Yang J, Beadle BM, Garden AS, Gunn B, Rosenthal D, Ang K, Frank S, Williamson R, Balter P, Court L, Dong L. Auto-segmentation of low-risk clinical target volume for head and neck radiation therapy. *Pract Radiat Oncol* 2014; 4: e31–e37.

59. Santhanam A, Willoughby TR, Meeks SL, Rolland JP, Kupelian PA. Modeling simulation and visualization of conformal 3D lung tumor dosimetry. *Phys Med Biol* 2009; 54: 6165–6180.

60. Kamino Y, Takayama K, Kokubo M, Narita Y, Hirai E, Kawawda N, Mizowaki T, Nagata Y, Nishidai T, Hiraoka M. Development of a four-dimensional image-guided radiotherapy system with a gimbaled X-ray head. *Int J Radiat Oncol Biol Phys* 2006; 66: 271–278.

61. Uecker M, Zhang S, Voit D, Karaus A, Merboldt K-D, Frahm J. Real-time MRI at a resolution of 20 ms. *NMR Biomed* 2010; 23: 986–994.

CHAPTER 10

Inter-Modality Deformable Registration

Yifei Lou and Allen Tannenbaum

CONTENTS

10.1 INTRODUCTION

DEFORMABLE IMAGE REGISTRATION (DIR) is one of the major problems in medical image processing, occurring in dose calculation [1], treatment planning [2], and scatter removal of cone-beam CT (CBCT) [3]. It is of prime importance to establish a pixel-to-pixel correspondence between two images in many clinical scenarios. For instance, the registration of a CT image to an MRI of a given patient taken at different times can provide complementary diagnostic information. For such applications, since the required deformation of the patient anatomy cannot be represented by a rigid transform, DIR is almost the sole means to establish the necessary mapping.

DIR methodologies can be generally categorized into intra-modality and inter-modality, or multi-modality. While intra-modality DIR can many times be handled by conventional intensity-based methods [4,5], solutions to inter-modality DIR problems are still far from being satisfactory. Yet, since different imaging modalities usually provide their unique viewpoints to reveal patient anatomy and delineate microscopic disease, inter-modality registration plays a key role in combining the information from multiple modalities to facilitate diagnostics and treatment of a certain disease.

A registration algorithm is usually composed of three main components: a *transformer*, a *measure*, and an *optimizer*. Typically, a similarity measure is first established to quantify how close two image volumes are to each other according to some chosen metric. Next, the transformation that maximizes this similarity metric is computed through an optimization procedure, which constrains the transformation to a predetermined class. In the present study, we focus on local deformations, and therefore, the transform is given by a displacement field. We will describe similarity measures in Section 10.2 and an optimization process in Section 10.3 and summarize the overall algorithm in Section 10.4.

The DIR algorithm is computationally expensive considering that the images to be registered are usually three dimensional, and the deformation field is of the same size as images. In addition, the processing time is often required to be minimized. To meet this clinical requirement, we have implemented our algorithm on a graphic processing unit (GPU) platform using NVIDIA's parallel computing architecture, CUDA [6]. Even though GPU's have been used for image registration, the majority of the available software focuses on rigid and Demon-type methods, the latter of which only work for intra-modal registration; see [7,8] and references therein. To the best of our knowledge, the only two existing works on GPU-based intermodal DIR are [9,10]. However, the former uses only 2D textures and is implemented with OpenGL, which is CUDA's predecessor, and less efficient. The latter is a B-spline-based method, and thus, it is unable to handle large deformation. The details of our GPU/CUDA implementation will be addressed in Section 10.5.

10.2 SIMILARITY MEASURES

As mentioned earlier, we wish to compute a displacement field $T(\vec{z}) = z - \vec{u}(\vec{z})$ that deforms one image I_1, often called the *moving* image, to match the *static* image I_2 by mapping the fixed position $\vec{z}$ in I_2 onto the corresponding point $\vec{z} - \vec{u}$ in I_1. During the process, a metric quantifying the similarity of two images is required to be predefined.

The standard strategy for inter-modal registration involves comparing intensity distributions, instead of intensity values as in the intra-modal case. Mutual information (MI) is widely used as a similarity metric, since the pioneering work of Viola and Wells [11], and independently by Maes et al. [12]; see the extensive survey in Reference 13 and other more recent work [14–17]. Next, we will give some details about MI as well as describing the Bhattacharyya distance (BD), as recently proposed in Reference 18.

10.2.1 Mutual Information

Mutual information quantifies the mutual dependence of two random variables. More precisely, let $p_1(x)$ and $p_2(y)$ be the probability density functions (PDFs) of two random variables X and Y, respectively, and let $p(x, y)$ be their joint probability distribution function. The MI of X and Y is defined as follows:

$$\mathcal{M}(X,Y) = \int_Y \int_X p(x,y) \log \frac{p(x,y)}{p_1(x)p_2(y)} dxdy. \tag{10.1}$$

It can also be expressed as a Kullback–Leibler (KL) divergence [19]. The KL divergence is a nonsymmetric measure of the difference between two probability distributions P and Q, defined as follows:

$$\mathcal{D}_{KL}(P,Q) = \int p(r) \log \frac{p(r)}{q(r)} dr, \tag{10.2}$$

where p and q are the PDFs of P and Q, respectively. Then Equation 10.1 can be interpreted as the KL divergence of $p_1(x) \times p_2(y)$ and the joint distribution $p(x, y)$:

$$\mathcal{M}(X, Y) = \mathcal{D}_{KL}(p(x, y), p_1(x) \, p_2(y)). \tag{10.3}$$

Returning to image registration, the PDF of a given image is expressed in terms of its intensity histogram. Consider, for example, the joint intensity distribution (or, for short, the joint histogram), which can be computed by

$$p(i_1, i_2; \vec{u}) = \frac{1}{V} \int_V \delta(i_1 - I_1(\vec{z} - \vec{u}), i_2 - I_2(\vec{z})) d\vec{z}$$

$$\simeq \frac{1}{V} \int_V G_\sigma(i_1 - I_1(\vec{z} - \vec{u}), i_2 - I_2(\vec{z})) d\vec{z}, \qquad (10.4)$$

where $G_\sigma = \exp(-(a^2 + b^2)/2\sigma^2)$ acts as a 2D Parzen window to approximate the δ function in the overlap region V. The integration over either i_1 or i_2 yields marginal distributions $p(i_1; \vec{u})$, $p(i_2)$, which are nothing but the PDFs of $I_1(\vec{z} - \vec{u})$ and $I_2(\vec{z})$, respectively.

To simplify notation, we assume the MI is directly defined on two images and deformation field:

$$\mathcal{M}(I_1, I_2; \vec{u}) = \iint p(i_1, i_2; \vec{u}) \log \frac{p(i_1, i_2; \vec{u})}{p(i_1; \vec{u}) p(i_2)} di_1 di_2. \qquad (10.5)$$

Note that the similarity measure compares the registered image $I_1(\vec{z} - \vec{u})$ and the static image I_2. Consequently, the MI is a functional of the deformation field $\vec{u}$. We will demonstrate below that the gradient of the MI with respect to $\vec{u}$ is not well defined, which leads to instability in the iterative process for approaching optimal solutions.

10.2.2 Bhattacharyya Distance

The Bhattacharyya distance (BD) can be derived in the same way in which MI is derived from the KL divergence. In particular, it defines a symmetric distance between probability distributions as$-\log D_{BD}$, where D_{BD} is given by

$$\mathcal{D}_{BD}(P, Q) = \int \sqrt{p(r)q(r)} \, dr. \qquad (10.6)$$

Replacing the KL divergence in Equation 10.2 with Equation 10.6, one obtains a similarity metric:

$$\mathcal{B}(X,Y) = \mathcal{D}_{BD}(p(x,y)), p_1(x)p_2(y)) \tag{10.7}$$

$$= \int \sqrt{p(x,y)p_1(x)p_2(x)p_2(y)}dxdy. \tag{10.8}$$

In an similar manner to the MI, the BD for images is defined as follows:

$$\mathcal{B}(I_1,I_2;\vec{u}) = \iint \sqrt{p(i_1,i_2;\vec{u})p(i_1;\vec{u})p(i_2)}\, di_1 di_2. \tag{10.9}$$

Both the BD and MI illustrate some type of divergence between two distributions $p(x, y)$ and $p_1(x)p_2(y)$. A major drawback of mutual information stems from the *logarithm* function involved in its computation. It is well known that the value of the logarithm function is highly sensitive to variations of its argument taking relatively small positive values. Since it is undefined at zero, it results that computing the gradient of the MI is prone to numerical errors near the origin. In contrast, the square root in the BD is continuous at zero, thus, making it less susceptible to numerical errors.

10.3 VISCOUS FLUID MODEL

Once the similarity measure is established, our goal is to find a deformation vector $\vec{u}$ to maximize this measure. We prefer a free-form deformation model, where the number of degrees of freedom of $\vec{u}$ is the same as the number of pixels in the image volumes, in contrast to the case of B-splines [20], where the deformation is defined on a number of control points and where the motion between points is propagated by interpolation. The use of B-splines may result in inaccuracy and may cause artifacts if the grid of control points is coarse. On the other hand, a dense grid fails to handle large deformations as local folding of the deformation is not allowed. In spite of their drawbacks, B-splines are still widely used in DIR with the combination of a multi-resolution approach due to its speed.

In the viscous fluid model [21], the deformation is assumed to be governed by the Navier–Stokes equation of viscous fluid motion, which is expressed as follows:

$$\nabla^2\vec{\upsilon} + \vec{\nabla}(\vec{\nabla}\cdot\vec{\upsilon}) + \vec{F}(\vec{z},\vec{u}) = 0. \tag{10.10}$$

The deformation velocity $\vec{\upsilon}(\vec{z},t)$ is related to $\vec{u}$ by

$$\vec{\upsilon} = \frac{d\vec{u}}{dt} = J(\vec{u})\vec{\upsilon} + \frac{\partial \vec{u}}{\partial t}, \tag{10.11}$$

where

$J(\vec{u})$ is the Jacobian of $\vec{u}$

$\vec{F}(\vec{z},\vec{u})$ is the force field, which drives the deformation in the appropriate direction.

Here, the force is to maximize the similarity measure, thus, parallel to the gradient of certain similarity measure with respect to the deformation $\vec{u}$.

By direct computation, the gradients of the BD and MI with respect to $\vec{u}$ share the same structure:

$$\frac{1}{V}\left[\frac{\partial G_\sigma}{\partial i_1} L(i_1,i_2,\vec{u})\right](I_1(\vec{z}-\vec{u}),I_2(\vec{z}))\nabla I_1(\vec{z}-\vec{u}), \tag{10.12}$$

where the L term for the BD and MI is given by

$$L_B(i_1,i_2;\vec{u}) = \frac{1}{2}\sqrt{\frac{p(i_1;\vec{u})p(i_2)}{p(i_1,i_2;\vec{u})}} + \frac{1}{2}\int\sqrt{\frac{p(i_2)p(i_1,i_2,\vec{u})}{p(i_2;\vec{u})}}\,di_2, \tag{10.13}$$

$$L_M(i_1,i_2;\vec{u}) = 1 + \log\frac{p(i_1,i_2;\vec{u})}{p(i_1;\vec{u})p(i_2)}. \tag{10.14}$$

The subscripts B and M correspond to the BD and MI, respectively.

To solve the Navier–Stokes Equation 10.10, the approach adopted here involves convolving the force field with the 3D Gaussian kernel Φ_s as an approximation to the solution [21,22]. In summary, the deformation field $\vec{u}$ at iteration k is given by

$$\vec{\upsilon}^k = \Phi_s\vec{F}^k \tag{10.15}$$

$$\vec{R} = \vec{\upsilon}^k - \sum_{i=1}^{3}\upsilon_i^k\left[\frac{\partial u^k}{\partial z_i}\right] \tag{10.16}$$

$$\vec{u}^{k+1} = \vec{u}^k + \delta_t \vec{R}, \qquad (10.17)$$

where
$$\vec{z} = (z_1, z_2, z_3)^T$$
$$\vec{\upsilon} = (\upsilon_1, \upsilon_2, \upsilon_3)^T.$$

The time step δ_t is chosen adaptively during iteration and set to

$$\delta_t = \frac{\delta_u}{\max \|\vec{R}\|}, \qquad (10.18)$$

with δ_u (in voxels) being the maximal voxel deformation that is allowed in one iteration. The numerical errors involved in computing the gradient of the MI can cause large gradient values, that is, $\max \|\vec{R}\|$ may be large. It follows from Equation 10.18 that one should use a smaller δ_t in order to make the iterative algorithm more stable. Consequently, the MI formulation needs more iterations to converge as compared to the BD.

To preserve the topology of the deformed template, *re-gridding* is performed when there exists $\vec{z}$ such that the Jacobian of $\vec{z} - \vec{u}$ becomes less than a predetermined positive value. As long as the determinant of the Jacobian is larger than this predetermined value, the deformation field is guaranteed to be invertible, and thus, the topology is preserved. In [21], this value is chosen to be 0.5, which is also used in our case. Re-gridding is performed in such a way that the current deformed image is set to a new template and the incremental deformation field is set to zero. In the end, the total deformation is the concatenation of the incremental deformation fields associated with each propagated template.

10.4 ALGORITHM

The flow chart of this algorithm is illustrated in Figure 10.1. In addition to what has been discussed earlier, we also employ a multi-resolution scheme to increase speed and robustness. Specifically, the deformation field obtained at the lower resolution is fed back to the initial value for the higher resolution. As the algorithm is implemented on a GPU, the data transfer between CPU and GPU is also included in the flow chart. Step 6 is a main component of our algorithm, which corresponds to registering two images at a certain resolution. Further details of Step 6 are provided on the right of Figure 10.1.

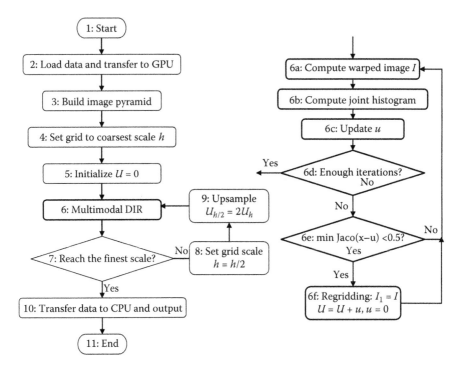

FIGURE 10.1 Flow chart of the intermodal DIR algorithm on the GPU.

10.5 GPU IMPLEMENTATION

A number of computationally intensive tasks involved in our algorithm share a common feature, namely, applying a single operation to a different part of the data elements. For this type of task, it is straightforward to accomplish it in a data-parallel fashion, namely, having all GPU threads running the same operation, one for a given subset of the data. Such a parallel scheme is particularly suitable for the single instruction multiple data (SIMD) structure of a GPU, and high computation efficiency can be therefore, achieved. In particular, Steps 6c and 6f in the flow chart (Figure 10.1) can be accomplished by simply launching one thread on GPU for each pixel to compute its gradient and to update its displacement field.

Yet, other steps are not quite so easily parallelizable or could be further improved due to the GPU's special properties. Consider, for example, Step 6f. The update of the displacement field is simply an interpolation process. The aforementioned method is a naive way of implementation. Since the GPU's development has been closely linked to handling computer graphics, which rely heavily on interpolation calculations, the GPU has been given dedicated hardware support to compute interpolation. This is done through the use of *texture memory*, which is a special type of memory space on the GPU. Not only does it support fast data access due to its associated memory cache, it also supports hardware linear interpolation, which is sufficient for image registration considered here. In short, the use of texture memory can greatly simplify coding while ensuring efficiency.

We will elaborate on how to implement histogram and convolution on the GPU in the subsequent sections.

10.5.1 Histogram Computation

Computing the joint histogram is an essential step in the inter-modal registration process. Consider two images $I_1(\vec{z})$ and $I_2(\vec{z})$, whose intensities range in a known interval $[a, b]$. Let us partition the interval $[a, b]$ into a set of N bins of equal size. A joint histogram is simply a 2D array of size $N \times N$ defined according to Equation 10.4. In a CPU implementation, it is straightforward to compute the joint histogram according to Equation 10.4, where we sequentially loop over all coordinates $\vec{z}$ and increase $p(i_1, i_2)$ by unity at $i_1 = I_1(\vec{z})$ and $i_2 = I_2(\vec{z})$ followed by a Parzen window smoothing. However, it is difficult to parallelize this sequential process on a GPU. If we were simply having each GPU thread responsible for one coordinate $\vec{z}$ and update $p(i_1, i_2)$, we would encounter a memory conflict problem due to the potential simultaneous update of a same memory address holding $p(i_1, i_2)$ from different GPU threads. One thread has to wait for another thread whenever this conflict occurs, thus, significantly reducing the computational efficiency.

Previous work on GPU-based histogram computation is reviewed in Reference 23, while our implementation relies on a GPU library, THRUST [24]. It is a powerful library of parallel algorithms and data structures. This library enables developers to write succinct codes to perform GPU-accelerated sort, scan, transform, and reduction operations with orders of magnitude faster than the latest multi-core CPUs.

To use the THRUST libary, we label the 2D histogram bins (i_1, i_2) by a 1D index $i = i_1 * N + i_2$ and create a vector $P = (I_1(\vec{z}), I_2(\vec{z})), \forall \vec{z}$ of dimension $2 \times N_p$, where N_p is the number of voxels $\vec{x}$ on two images. A thrust function is then used to transform each pair in P to the corresponding histogram indices, yielding a vector P_I of length N_p with index:

$$\text{histogram index}(\vec{z}) = \text{bin index of } I_1(\vec{z}) \times \text{\# of histogram bins}$$

$$+ \text{bin index of } I_2(\vec{z}).$$

We then sort the histogram index vector P_I using a sorting function *thrust::sort* from THRUST, so that P_I is in ascending order. Finally, we compute a histogram $p(i)$ by using a THRUST function to count the number of entries in P_I that are equal to i. Note $p(i)$ is exactly the joint histogram to be computed after transforming the linear index i to its equivalent row and column subscripts i_1 and i_2, respectively. During this process, the key operations invoking THRUST functions are executed on the GPU in parallel, ensuring computational efficiency.

Once a joint histogram $p(i_1, i_2)$ is obtained, the marginal distributions can be computed by summing $p(i_1, i_2)$ over the i_1 or i_2 coordinate. Summation is another calculation that is hard to parallel. We employ a technique called *parallel reduction* [25]. We illustrate this process for 256 bins, as usually the intensity range is $[a, b] = [0, 255]$. The task is to sum over 256 elements, and we first divide them into two 128-element regions. Each of 128 threads will add its element in the second half to its element in the first half, writing back to the first half. After synchronizing all threads, using *__syncthreads*() on the GPU, we have 128 elements remaining. Then we loop again on this resulting region of summing these 128 items down to 64 items, and so on, until we have one element remaining at the very end of the iterations, which is the final output. This approach reduces the number of sequential updates from N to $\log_2 N$ for $N = 256$ in this case.

10.5.2 Convolution

In Equation 10.15, the update of deformation velocity $\vec{\upsilon}$ involves convolving the force term $\vec{F}$ by a Gaussian kernel Φ_s. For Gaussian-type kernels that can be separated as the product of two Gaussian kernels along two directions, 2D convolution can be carried out by the combination of 1D convolution of the image vertically and then horizontally. In tests, it has been observed that *convolutionSeparable* [26] is more than twice as fast as a purely texture-based implementation of the same algorithm running on the same hardware. This can be explained by the number of elementary computation steps. Suppose an image is of size $N \times N$, and kernel is $k \times k$. The computational complexity of separable convolution is $O(2kN^2)$, while nonseparable implementation has complexity of $O(k^2N^2)$.

Another popular way to implement convolution is in the frequency domain using the fast Fourier transform (FFT) based on the fact that the convolution of two signals equals pointwise multiplication in the frequency domain. The CUFFT library offers a straightforward way to implement convolution in terms of the FFT, *convolutionFFT2D* [27]. The complexity of the FFT approach is $O(4N^2 \log_2(N))$, since each pixel is processed in $\log_2(N)$ for FFT and repeats four times (both dimensions and both forward and backward transforms).

By simply comparing the complexities, we can see that FFT-based approach is faster for nonseparable and/or large separable kernels. A detailed comparison between spatial and frequency implementations is studied in Reference 28, which gives the same conclusion. However, both implementations on CUDA are targeted for 2D convolutions. In our registration problem, the kernel size is much smaller compared to the image size, and the Gaussian kernel is separable. Consequently, we adopt the same idea in *convolusionseparable* to program a 3D separable convolution for the need to deal with 3D medical data. In particular, for a separable Gaussian kernel, we only need to store the 1D version, since it is homogeneous along the three directions. In addition, the best place to store the kernel is in the GPU's *constant memory*, since the kernel is fixed once defined, and constant memory offers the fastest access. We then perform convolution dimension by dimension three times, just as *convolutionseparable* in 2D.

It is possible to encounter large nonseparable kernels. For example, the voxel resolution might be at different levels at the three axes, and hence, it is better to apply a different amount of smoothing. As suggested in [28], a "large kernel" refers to 33 pixels for 2D images and 51 pixels for 3D. If this is the case, we would adopt the FFT approach or a decomposition method as in Reference 29.

10.6 EXPERIMENTS

The DIR algorithm using both the BD and MI was implemented in C++ and CUDA and executed on a personal laptop with 1.80 GHz CPU and an NVIDIA Tesla C1060 GPU. Both similarity measures are found to require approximately the same amount of time regardless of which platform is employed. Furthermore, the parallel GPU implementation achieves a speedup by a factor of 50 compared to the one serial CPU. The deformable registration of two volumes of sizes $256 \times 256 \times 60$ is found to require 6 s. Registration is performed with three multi-resolution levels, with a maximum of 50 iterations on each level, and with 256 histogram bins.

For comparison, BRAINSFit [30], NiftyReg [31], and ANTs [32] were tested on the same data under identical hardware configurations. All of these methods have the option to use MI as the similarity measure. In addition, BRAINSFit and NiftyReg are both B-spline based implementations, while ANTs implements a diffeomorphic deformation. BRAINSFit is used here as a plugin in 3D Slicer [33] and is found to require about 8 min. NiftyReg contains both CPU and GPU implementations of the method [30], which takes about 15 min and 35 s (CPU) and 2 min 18 s (GPU). ANTs, a diffeomorphic registration method, requires about 1 h and 4 min.

We demonstrate our algorithm on a real patient case of registering a CT image to cone-beam CT (CBCT). CT-CBCT registration is frequently encountered in current clinical practice. By establishing voxel correspondence between these two modalities, the treatment plan can conveniently transfer the organ contours from the planning CT image to the daily CBCT image.

The CT image used here was scanned in a 4-slice GE LightSpeed RT scanner (GE Healthcare, Milwaukee, WI) for treatment plan purposes, while the CBCT image was from the onboard imager system on a Varian Trilogy linear accelerator (Varian Medical System, Palo Alto, CA) before one treatment. Both CT and CBCT are of size $256 \times 256 \times 68$, while we only visualize one 2D slice in Figure 10.2.

Registration results are visualized using a checkerboard display tool. In this technique, the two images to be compared are merged together in a checkerboard pattern where the black region is filled by one image and the white region is filled by another. At the transition zone, this manner of display makes the mismatch between anatomical structures in the two images visible. Ideally, when two images are fully aligned, there is no difference in the boundary between checkerboard squares. Figure 10.2 shows that MI and BD have the same performance in this simple example where the intensity difference between CT and CBCT is not so dramatic.

10.7 CONCLUSION

In this chapter, we have presented an efficient GPU-based implementation of 3D inter-modal deformable image registration. In our experiments, we found that it takes about 6 s to register two image volumes of

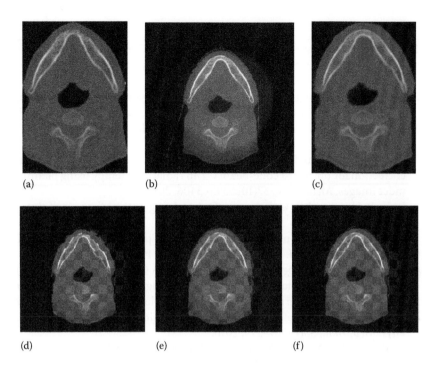

FIGURE 10.2   Real patient CT-CBCT registration. The top row shows (a) moving image (CT), (b) static image (CBCT), and (c) the registered CT by MI. The bottom row is about checkerboard comparison, which are from left to right (d) CT + CBCT, (e) registered CT by BD + CBCT, and (f) registered CT by MI + CBCT.

size $256 \times 256 \times 60$ for 50 iterations using three multi-resolution levels. The resulting registrations meet clinical requirements in terms of both time and accuracy. In Reference 34, we have provided the users with source codes and a sample data set in order to test our algorithm. Some of the future work will be focused on the visualization of the registration process during the iterations, in order to allow experts to quickly check the plausibility of the registration, and make any possible adjustments.

ACKNOWLEDGMENTS

This work was supported in part by the National Center for Research Resources through NIH under Grant P41-RR-013218, by the National Institute of Biomedical Imaging and Bioengineering through NIH under Grant P41-EB-015902 (Neuroanalysis Center at Brigham and Women's Hospital), AFOSR, and ONR. We would like to thank NVIDIA for providing GPU cards for this project.

REFERENCES

1. D. Maarten, M. van Zijtvetd, and B. Heijmen. Correction of cone beam CT values using a planning CT for derivation of the "dose of the day". *Radiother. Oncol.*, 85(2):195–200, 2007.

2. Q. J. Wu, D. Thongphiew, Z. Wang, B. Mathayomchan, V. Chankong, S. Yoo, W. R. Lee, and F. F. Yin. On-line re-optimization of prostate IMRT plans for adaptive radiation therapy. *Phys. Mel. Biol.*, 53(3):673–691, 2008.

3. T. Y. Niu, M. S. Sun, J. Star-Lack, H. W. Gao, Q. Y. Fan, and L. Zhu. Shading correction for on-board cone-beam CT in radiation therapy using planning mdct images. *Med. Phys.*, 37(10):5395–5406, 2010.

4. B. K. P. Horn and B. G. Schunck. Determining optical flow. *Artif. Intell.*, 17:185–203, 1981.

5. J. P. Thirion. Image matching as a diffusion process: An analogy with Maxwell demons. *Mel. Image Anal.*, 2:243–260, 1998.

6. NVIDIA. *CUDA C Programming Guide*. NVIDIA Corporation, p. 120 (2011).

7. X. Gu, H. Pan, Y. Liang, R. Castillo, D. Yang, D. J. Choi, E. Castillo, A. Majum-dar, T. Guerrero, and S. B. Jiang. Implementation and evaluation of various demons deformable image registration algorithms on a GPU. *Phys. Med. Biol.*, 55(1):207–219, 2010.

8. R. Shams, P. Sadeghi, R. Kennedy, and R. Hartley. Parallel computation of mutual information on the GPU with application to real-time registration of 3D medical images. *Comp. Methods Prog. Biomed.*, 99(2):133–146, 2010.

9. V. Saxena, J. Rohrer, and L. Gong. A parallel GPU algorithm for mutual information based 3d nonrigid image registration. In *Proceedings of the 16th international Euro-Par Conference on Parallel Processing: Part II, Euro-Par'10*, pp. 223–234, 2010.

10. G. Vetter, C. Guetter, C. Xu, and R. Westermann. Non-rigid multi-modal registration on the GPU. In *Proceedings of SPIE*, vol. 6512, 2007.

11. P. Viola and W. M. Wells. Alignment by maximization of mutual information. *Int. J. Comp. Vision*, 24:137–154, September 1997.

12. F. Maes, A. Collignon, D. Vandermeulen, G. Marchal, and P. Suetens. Multimodality image registration by maximization of mutual information. *IEEE Trans. Med. Imaging*, 16:187–198, 1997.

13. J. P. W. Pluim, J. B. A. Maintz, and M. A. Viergever. Mutual-information-based registration of medical images: A survey. *IEEE Trans. Med. Imaging*, 22(8):986–1004, 2003.

14. J. Chappelow, B. N. Bloch, N. Rofsky, E. Genega, R. Lenkinski, W. DeWolf, and A. Madabhushi. Elastic registration of multimodal prostate MRI and histology via multiat-tribute combined mutual information. *Med. Phys.*, 38(4):2005–2018, 2011.

15. D. Loeckx, P. Slagmolen, F. Maes, D. Vandermeulen, and P. Suetens. Nonrigid image registration using conditional mutual information. *IEEE Trans. Med. Imaging*, 29(1):19–29, 2010.

16. Q. R. Razlighi, N. Kehtarnavaz, and A. Nosratinia. Computation of image spatial entropy using quadrilateral markov random field. *IEEE Trans. Image Proc.*, 18:2629–2639, 2009.
17. Z. Yi and S. Soatto. Nonrigid registration combining global and local statistics. In *Proceedings of the IEEE Conference on Computer Vision and Pattern Recognition*, pp. 2200–2207, June 2009.
18. Y. Lou, A. Irimia, P. Vela, M. C. Chambers, J. Van Horn, P. M. Vespa, and A. Tannen-baum. Multimodal deformable registration of traumatic brain injury MR volumes via the Bhattacharyya distance. *IEEE Trans. Bioeng.*, 60(9):2511–2520, April 2013.
19. S. Kullback and R. A. Leibler. On information and sufficiency. *Ann. Math. Stat.*, 22(1):79–86, 1951.
20. D. Rueckert, L. I. Sonoda, C. Hayes, D. L. G. Hill, M. O. Leach, and D. J. Hawkes. Nonrigid registration using free-form deformations: Application to breast MR images. *IEEE Trans. Med. Imaging*, 18:712–721, 1999.
21. E. D'Agostino, F. Maes, D. Vandermeulen, and P. Suetens. A viscous fluid model for multimodal non-rigid image registration using mutual information. In *Proceedings of the International Conference on Medical Image Computing and Computer-Assisted Intervention-Part II, MICCAI*, pp. 541–548, 2002.
22. M. Bro-Nielsen and C. Gramkow. Fast fluid registration of medical images. In *Proceedings of the Fourth International Conference on Visualization in Biomedical Computing*, pp. 267–276, 1996.
23. A. Eklund, P. Dufort, D. Forsberg, and S. M. LaConte. Medical image processing on the GPU—Past, present and future. *Med. Image Anal.*, 17:1073–1094, 2013.
24. J. Hoberock and N. Bell. Thrust: A parallel template library, 2010. http://thrust.github.io/ (accessed on May 29, 2015).
25. M. Harris. Optimizing parallel reduction in CUDA. NVIDIA Developer Technology 2.4, 2007.
26. V. Podlozhnyuk. Image convolution with CUDA. NVIDIA Corporation white paper, June 2097.3, 2007.
27. V. Podlozhnyuk. FFT-based 2D convolution. NVIDIA white paper, 2007.
28. O. Fialka and M. Cadik. FFT and convolution performance in image filtering on GPU. In *Tenth International Conference on Information Visualization, 2006, IV 2006.*, pp. 609–614, July 2006.
29. P. Karas and D. Svoboda. Convolution of large 3D images on GPU and its decomposition. *EURASIP J. Adv. Signal Process.*, 2011:120, 2011.
30. H. Johnson, G. Harris, and K. Williams. BRAINSFit: Mutual information registrations of whole-brain 3D images, using the insight toolkit. http://hdl.handle.net/1926/1291, 2007. (Accessed on May 29, 2015).
31. M. Modat, G. R. Ridgway, Z. A. Taylor, M. Lehmann, J. Barnes, D. J. Hawkes, N. C. Fox, and S. Ourselin. Fast free-form deformation using graphics processing units. *Comp. Methods Prog. Biomed.*, 98(3):278–284, 2010.

32. B. B. Avants, C. L. Epstein, M. Grossman, and J. C. Gee. Symmetric diffeomorphic image registration with cross-correlation: Evaluating automated labeling of elderly and neurodegenerative brain. *Med. Image Anal.*, 12:26–41, 2008.

33. A. Fedorov, R. Beichel, J. Kalpathy-Cramer, J. Finet, J.-C. Fillion-Robin, S. Pujol, C. Bauer et al. 3D slicer as an image computing platform for the quantitative imaging network. *Magn. Reson. Imaging*, 30(9):1323–1341, 2012.

34. Y. Lou and A. Tannenbaum. Multimodal deformable image registration via the Bhattacharyya distance. submitted to *IEEE Trans. Image Proc.*, 2011.

CHAPTER 11

CT-to-Cone-Beam CT Deformable Registration

Xin Zhen and Xuejun Gu

CONTENTS

11.1 INTRODUCTION

Oᴺʟɪɴᴇ ᴀᴅᴀᴘᴛɪᴠᴇ ʀᴀᴅɪᴀᴛɪᴏɴ ᴛʜᴇʀᴀᴘʏ (ART) allows real-time treatment adaptations based on the current patient anatomy and geometry. In a typical online ART process, a computed tomography (CT) image is usually acquired prior to the treatment course for treatment planning purposes. Before each treatment fraction, a cone-beam CT (CBCT) image is then obtained, on which the treatment plan is redesigned to account for setup errors, deformations of tumor and other organs, and the change of their relative locations. Deformable image registration (DIR) technique plays an important role in this process in establishing a correspondence between voxels in the CT and the CBCT for various purposes, for instance, transferring the organ contours from the planning CT images to the daily CBCT images. It is, hence, desirable to have an accurate and robust DIR algorithm to facilitate this step.

Among the existing DIR methods, demons [1] have been proven to be a fast and robust algorithm and a number of its variants have been developed [2–5]. However, the demons algorithm assumes that there exists intensity consistency between two images to be registered. Therefore, although demons can successfully deal with images of the same modality (e.g., CT-CT registration), considerable registration error may present, when it comes to intermodality registration problems (e.g., CT-MRI registration), where corresponding points in the two images to be registered do not necessarily pertain to the same intensity level [6–9].

CT-CBCT DIR is considered to be an intermodality DIR problem. Although CT and CBCT are reconstructed under the same physical principles, the intensity (Hounsfield Units [HU]) consistency between CT and CBCT images is violated due to many reasons. First, almost all current commercial systems reconstruct CBCT images using the well-known FDK [10] algorithm. As a fundamental limitation of this algorithm, CBCT quality degrades with the increase of cone angle [11]. Second, scatter contamination also leads to severe cupping, streak artifacts, and degrade of image contrast. In a typical clinical CBCT system for radiotherapy, scatter-to-primary ratio (SPR) may even exceed 100% [12], although many methods have been proposed to correct scatter artifacts [13–20], it is still an open problem. Third, the gantry mounted bowtie filter in CBCT system may wobble, as the gantry rotates, which can result in crescent artifacts [21,22]. Moreover, there are also other factors that contribute to the intensity inconsistency between CT and CBCT, for example, different level of noise, beam-hardening effects, and motion [23–27].

Despite the difficulties caused by the different image intensities, and the availability of many existing algorithms for multimodal image DIR (cited papers), it is still desirable to use demons-type algorithm in DIR due to its simplicity and hence high efficiency and robustness [28,29]. Castadot et al. [30] compared 12 deformable registration strategies in adaptive radiation therapy for the treatment of head and neck tumors, and found that the multiresolution demons or using multiresolution demons as an initialization of a subsequent single-resolution level-set registration are the best DIR strategies and are effective in deforming CT images of a same patient at different phases of his/her treatment course. As such, a variety of intermodality demons methods have been proposed. Some researchers incorporate more reliable statistical similarity metrics into demons, such as normalized cross-correlation (NCC) or normalized mutual information (NMI), to measure the similarity between corresponding anatomical points. For instance, Modat et al. [31] implemented a diffeomorphic demons using the analytical gradient of NMI in a conjugate gradient optimizer. Lu et al. [7] has also proposed a variational approach for multimodal image registration based on the diffeomorphic demons algorithm by replacing the standard demons similarity metric with pointwise mutual information in the energy function. However, it is still unclear what alternative metrics is robust for this CT-CBCT DIR problem and how to incorporate it into demons style algorithms. On the other hand, some researchers focused on estimating the intensity relation between CT and CBCT images and combining intensity correction with geometrical transformation. Within this category, Guimond et al. [6] investigated the functional transformation that maps the intensities of one image to those of another, and implemented the intensity correction prior to each iteration. However, the intensity mapping of a polynomial form estimated globally based on the entire image may not be accurate enough and hence may result in errors in subsequent registration process. Recently, Hou et al. [9] attempted to correct image intensities by aligning the cumulative histograms of the two images. Nithiananthan et al. [8] also tried to correct intensities at each demons iteration by estimating linear transformations for intensity mapping for a few segmented tissue types. Nevertheless, all these works essentially assume that there exists a global mapping between the CT and the CBCT intensities. This, however, may not hold, when CBCT is contaminated by artifacts that usually have local patterns.

In this work, we propose and evaluate a modified demons algorithm embedded with a simultaneous intensity correction step, called *Deformation with Intensity Simultaneously Corrected* (DISC). Rather than estimating a global mathematical transformation model between CT and CBCT images, our method corrects CBCT intensity of each voxel at every iteration step of demons by matching the first and the second moments of the voxel intensities inside a patch around this voxel with those in the CT image. Quantitative evaluations of our method are performed by using both a simulation data set and six clinical head-and-neck cancer patient data sets. It is found that DISC can handle CBCT artifacts and the intensity inconsistency issue and therefore improves the registration accuracy when compared with the original demons.

11.2 THE ORIGINAL DEMONS ALGORITHM

Suppose we would like to register the moving images $I_m(x)$ and static image $I_s(x)$, where a vector field $v(x)$ relates these two images by $I_m(x + v(x)) = I_s(x)$. The vector field $v(x)$ is solved in an iterative fashion and at each iteration the increment of the vector field $dr(x)$ is determined based on the image intensity at the voxel x. There are six different variants of the demons algorithm that have been studied by Gu et al. [29]. Take the double force demons [3,4] as an example, the demons algorithm iteratively performs the following steps. First, calculate the increment of the moving vector (or the displacement vector) at all voxel points. Specifically, $dr = (dx, dy, dz)$ at a voxel in the double force demons is as follows:

$$dr^{(k+1)} = \frac{(I_m^{(k)} - I_s)\nabla I_s}{(I_m^{(k)} - I_s)^2 + |\nabla I_s|^2} + \frac{(I_m^{(k)} - I_s)\nabla I_m^{(k)}}{(I_m^{(k)} - I_s)^2 + |\nabla I_m^{(k)}|^2}, \qquad (11.1)$$

where
  the superscript indexes the iteration step
  $I_m^{(k)}$ is the intensity of the moving image at the kth iteration
  I_s is the original static image

Second, smooth the resulting incremental vector field dr by convolving it with a Gaussian kernel. Third, add the incremental deformation field to the global deformation field $v(x)$ and update the moving image. This process is iteratively performed until convergence.

11.3 THE DISC ALGORITHM

The DISC algorithm solves the CT-CBCT intensity inconsistency problem by integrating a novel local intensity correction step into the demons framework. This step estimates an intensity transformation at a voxel x based on voxels in small cubic volumes centering at x in both CT and CBCT (termed *patches*). Then the intensity of the CBCT image is adjusted at each voxel using the estimated voxel-dependent transformations.

Suppose at an intermediate step k of the registration process, the moving image is deformed into $I_m^{(k)}$. We would like to estimate two parameters a and b at each voxel x, such that the CBCT intensity $I_s(x)$ is corrected into $a(x)I_s(x) + b(x)$. This is achieved by comparing the patches centering at x on $I_m^{(k)}$ and I_s. Let us consider two patches of size n: $\mathbf{M}\{m_1, m_2, ..., m_n\}$ and $\mathbf{S}\{s_1, s_2, ..., s_n\}$, both centered at voxel x in $I_m^{(k)}$ and I_s, respectively. It is our objective to find a linear intensity mapping:

$$\mathbf{S}' = a(x)\mathbf{S} + b(x), \tag{11.2}$$

such that the intensity distributions of $\mathbf{S}'$ are in agreement with $\mathbf{M}$. This can be achieved by matching the moments of intensity distributions of the two groups of voxels, and $a(x)$ and $b(x)$ for the voxel x are given by R [32]

$$a(x) = \frac{\text{STD}(\mathbf{M})}{\text{STD}(\mathbf{S})}, \tag{11.3}$$

$$b(x) = \text{E}(\mathbf{M}) - a(x)\text{E}(\mathbf{S}), \tag{11.4}$$

where $\text{STD}(\cdot)$ is the standard deviation operator. Let $\mathbf{A}$ and $\mathbf{B}$ denote two vectors of length N (number of voxels in I_m or I_s) with entries a and b, respectively. Strictly speaking, the estimation of $a(x)$ and $b(x)$ in Equations 11.3 and 11.4 are valid only when the two voxels $I_m^{(k)}(x)$ and $I_s(x)$ are at the same anatomical location. Yet, due to the apparent violation caused by the image deformation, this estimation is not always reliable. In practice, we only estimate those a and b when $I_m^{(k)}(x)$ and $I_s(x)$ belong to the same tissue class and limit a and b in a certain range to avoid false correction. Specifically, we first define a mask:

$$\xi(x) = \begin{cases} 1, & \text{where } \left| I_m^{(k)}(x) - I_s(x) \right| > \text{HU}_0 \\ 1, & \text{where } a(x) < a_{min} \text{ or } a(x) > a_{max}, \\ 0, & \text{elsewhere} \end{cases} \tag{11.5}$$

where
   HU_0 is the threshold in Hounsfield Unit to identify if two voxels in CT and CBCT images belong to the same tissue class
   a_{min} and a_{max} are the lower and upper bounds of a

The way of choosing the parameters HU_0, a_{min}, and a_{max} is detailed in Reference 32. We then estimate $a(x)$ and $b(x)$ values for voxels with $\xi = 1$ by computing a weighted average of all a and b values of voxels inside a small cubic region T centering at x:

$$a(x) = \frac{\sum_{i \in T} \omega_i a_i}{\sum_{i \in T} \omega_i},$$ (11.6)

$$b(x) = \frac{\sum_{i \in T} \omega_i b_i}{\sum_{i \in T} \omega_i},$$ (11.7)

$$\omega_i = \exp\left[\frac{-(I_s(i) - I_s(x))^2}{h^2} \right]$$ (11.8)

where
   ω_i is the weighting factor determined by the intensity values of $I_s(i)$
   h is a parameter that adjusts to what extent we would like to enforce the similarity

This interpolation/extrapolation step may need to be performed for multiple times since for some voxels, the values of ξ may be 1 for all voxels in their neighbor T.

As soon as a and b are available at all voxels, an intensity correction is performed to update the image intensity of the original static image $I_s(x)$ to yield an intensity-corrected static image $I_s'(x) = a(x)I_s(x) + b(x)$, which is thus ready for the displacement calculation using the original demons algorithm.

11.4 IMPLEMENTATION OF DISC

Before starting the DIR procedure, a global intensity transformation is first performed to shift the CBCT intensity by a constant (denoted as ΔHU) to match the mean intensities of CT and CBCT. This is to correct the average HU difference between the two images to a certain extent.

Then, a multiscale strategy is adopted so as to reduce the magnitude of the displacement with respect to voxel size. The iteration starts with the lowest resolution images, and the moving vectors obtained at a coarser level are upsampled to serve as initial solution at a finer level. In this study, we considered two different resolution levels. Further downsampling was found not to improve registration accuracy nor efficiency.

One of the most commonly used stopping criterion to judge whether the moving image has been correctly deformed to the static image is the cross-correlation coefficient [3,5,28,33]. However, such a similarity metric does not work well in this CT-CBCT DIR context. We therefore use a convergence criterion based on the difference between successive deformation fields. We define a relative norm $l^{(k)} = \sum |d\mathbf{r}^{(k+1)}| / \sum |\mathbf{r}^{(k)}|$, and use $l^{(k-10)} - l^{(k)} \leq \varepsilon$, where $\varepsilon = 1.0 \times 10^{-4}$ as our stopping criterion. This measure is found to have a closer correspondence with spatial accuracy than correlation coefficient as DIR is stopped when there is no "force" to push voxels any more [29].

In this work, we use the compute unified device architecture (CUDA) with an NVIDIA GPU card as the implementation platform. In order to efficiently parallelize DISC in the CUDA environment, the data-parallel portions of the algorithm are identified and grouped into the following kernels: (1) an intensity correction kernel to compute and interpolate **A** and **B**; (2) a Gaussian filter kernel to smooth images and moving vectors; (3) a gradient kernel to calculate the gradient of images; (4) a moving vector kernel to calculate and update moving vectors; (5) an interpolation kernel to deform images with moving vectors; and (6) a comparison kernel to stop the program based on the stopping criteria.

11.5 EVALUATION

11.5.1 Synthetic Data: MC Simulation

To validate our algorithm, we have generated a test data set based on two CT images of a head-and-neck cancer patient. The first CT image is called planning CT acquired before the treatment while the second CT image was acquired halfway in the treatment course for replanning purpose and is called treatment CT here. Then, a CBCT image with realistic image artifacts is synthesized using the treatment CT image. DISC is applied to perform DIR between the planning CT image and the synthesized CBCT image and to correct the CBCT intensity. This approach offers us the ground truth for the evaluation of DISC: the treatment CT can be regarded as the scatter-free CBCT image and the deformation vector field between the planning and treatment CT images obtained using the original demons algorithm is the ground truth deformation vector field.

To synthesize a realistic CBCT image using the treatment CT image, we first convert the CT image into a digital phantom by assigning each voxel with a density value and a material type. CBCT projection images at 360 equally spaced directions covering an entire 2π angular range are then calculated using an in-house developed software tool (called gDRR) [34] under a realistic projection geometry for a Varian On-board-Imaging system (OBI) (Varian Medical Systems, Inc., Palo Alto, CA). In this package, the primary component in a projection image is calculated by a ray-tracing algorithm, while the scatter component is obtained by Monte Carlo simulations followed by an image smoothing process to suppress noise. Both the primary and the scatter calculations consider a variety of effects occurred in a realistic CBCT scan including the energy spectrum, the source fluence map, and the detector response. Once the projections are generated, an FDK reconstruction algorithm is invoked, yielding the CBCT image with exactly the same anatomy structures as in the treatment CT image but with all major CBCT artifacts such as scatter.

11.5.2 Clinical Data

The performance of DISC is further assessed using clinical CT and CBCT data of six head-and-neck cancer patients. Each patient has a planning CT image and a CBCT image. The CBCT images were acquired 1–7 weeks after the first fraction of treatment on a Varian OBI system integrated in a Trilogy™ linear accelerator (Varian Medical Systems, Inc., Palo Alto, CA) using full-fan mode with a full-fan bow-tie filter on site. The planning CT images are cropped and resampled to match the dimension and resolution of the CBCT image after rigid registration. Both CT and CBCT images are downsampled to half of their original size in the transverse plane. Hence, the image resolution for both CT and CBCT images after rigid registration is $256 \times 256 \times 68$ (Cases 1, 2, 5, 6) or $256 \times 256 \times 52$ (Cases 3, 4), and the voxel size is $0.94 \times 0.94 \times 2.5$ mm^3.

11.5.3 Quantification of Registration Performance

Three similarity metrics are used in this work to quantify the DIR results. They are chosen based on two considerations. First, the metric should be observer independent. Second, the metric should be insensitive to intensity inconsistency.

The first metric is normalized mutual information (NMI), ranging from 0 to 1 with 1 representing the highest image similarity. The second metric is called feature similarity index (FSIM) [35,36], which tries to model the mechanism of the human visual system by capturing the main image features such as the phase congruency of the local structure and the image gradient magnitude. Detailed definition and description of FSIM are given in Reference 35. In this work, FSIM is calculated at each pair of corresponding transverse slices between two three-dimensional data sets, and the average value and standard deviation are calculated. The FSIM score varies between 0 and 1 with 1 representing the most image similarity. The third metric is $RMSE_{edge}$: the root mean square error (RMSE) between two edge images [32], when two images are perfectly aligned, $RMSE_{edge}$ should be zero.

11.6 RESULTS

For clarity, following symbols are used to represent different images used in the algorithm evaluation: $CT_{original}$ and $CBCT_{original}$ are the CT and CBCT images before registration, respectively. $CT_{deformed}^{demons}$ and $CT_{deformed}^{DISC}$ are the deformed CT images using the original demons algorithm and the DISC algorithm, respectively. $CBCT_{corrected}$ is the intensity-corrected CBCT image using DISC. In the simulation study, $CT_{original}$ refers to the planning CT before registration; $CBCT_{original}$ is the synthesized CBCT using the treatment CT before registration. $CBCT_{no\ artifacts}$ is the treatment CT and regarded as the primary part of $CBCT_{original}$ before registration; $CT_{deformed}^{demons}(\rightarrow CBCT_{no\ artifacts})$ and $CT_{deformed}^{demons}(\rightarrow CBCT_{original})$ are the deformed CT images to match $CBCT_{no\ artifacts}$ and $CBCT_{original}$ using the original demons algorithm, respectively. $CT_{deformed}^{DISC}(\rightarrow CBCT_{original})$ is the deformed CT image to match $CBCT_{original}$ using DISC.

11.6.1 Synthetic Data

Figure 11.1 shows the results for the simulation case. Figure 11.1a is the planning CT image before registration; Figure 11.1b is the synthesized CBCT image before registration, in which intensity variations due to scatter artifacts can be clearly observed inside the yellow dashed circle. Figure 11.1c is the scatter-free primary component of $CBCT_{original}$ before registration. The comparison of vertical intensity profiles through the image center between $CBCT_{original}$ and $CBCT_{no\ artifacts}$ is shown in Figure 11.1g.

Registration between $CT_{original}$ and $CBCT_{no\ artifacts}$ (Figure 11.1a and c) using the original demons algorithm can be regarded as the ground truth which yields $CT_{deformed}^{demons}(\rightarrow CBCT_{no\ artifacts})$ (Figure 11.1d). When the original demons algorithm is applied to deform the original CT to the original CBCT, we can see that in $CT_{deformed}^{demons}(\rightarrow CBCT_{original})$ (Figure 11.1e) soft tissue and bone inside the yellow dashed circle are significantly distorted after registration due to the intensity inconsistency mainly caused by scatter artifacts. We can also see the significant intensity difference between $CT_{deformed}^{demons}(\rightarrow CBCT_{no\ artifacts})$ and $CT_{deformed}^{demons}(\rightarrow CBCT_{original})$ in the intensity profile comparison in Figure 11.1h. In contrast, DISC can yield correct result, as shown in Figure 11.1f and h.

11.6.2 Clinical Cases

The performance of the DISC algorithm is further assessed on six head-and-neck cancer patient cases. An example result (Case 2) is shown in

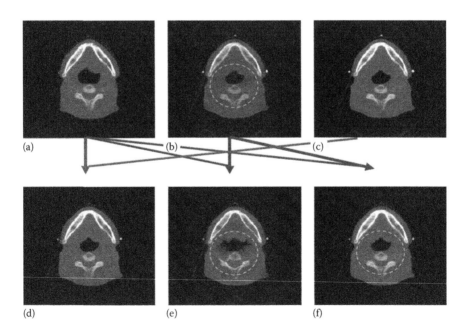

FIGURE 11.1 Simulation and registration results. (a) $CT_{original}$; (b) $CBCT_{original}$; (c) $CBCT_{no\ artifacts}$; (d) $CT_{deformed}^{demons}(\rightarrow CBCT_{no\ artifacts})$; (e) $CT_{deformed}^{demons}(\rightarrow CBCT_{original})$; (f) $CT_{deformed}^{DISC}(\rightarrow CBCT_{original})$. The dashed circles indicate the regions which severely suffer from the scatter artifact. The arrows indicate DIR between different image sets. *(Continued)*

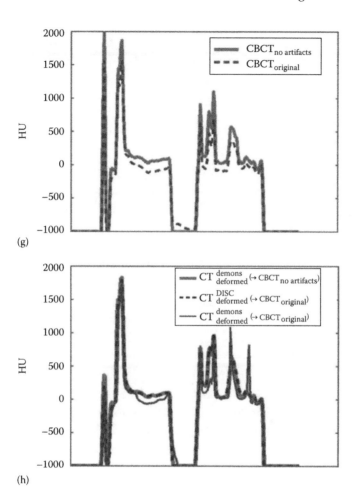

(g)

(h)

FIGURE 11.1 (*Continued*) Simulation and registration results. (g) and (h) are the corresponding vertical intensity profiles through the image center. (Reprinted from Zhen, X. et al., *Phys. Med. Biol.*, 57(21), 6807, 2012.)

Figures 11.2 and 11.3. Misalignment is evident before registration (Figure 11.3a-1 through a-3). The results from the original demons algorithm are shown in Figures 11.2c and 11.3b-1 through b-3. As we can see, the anatomical structures in the deformed CT image are significantly distorted due to intensity inconsistency. This effect is more severe in regions that are dominated by artifacts (as indicated by arrows in Figures 11.2c and 11.3b-1 through b-3). The DISC algorithm (Figures 11.2d and 11.3c-1 through c-3),

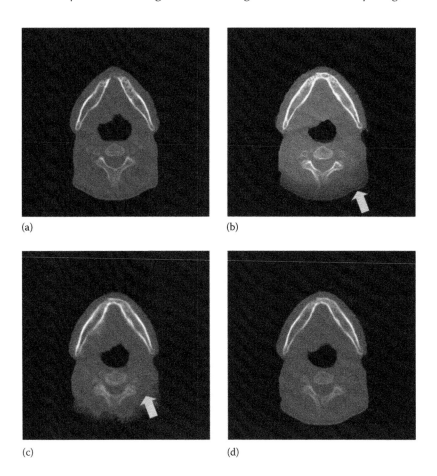

FIGURE 11.2 (a) $CT_{original}$; (b) $CBCT_{original}$; (c) $CT_{deformed}^{demons}$; (d) $CT_{deformed}^{DISC}$. The arrows indicate the region contaminated by artifacts (b) and the region with geometrical distortion (c). (Reprinted from Zhen, X. et al., *Phys. Med. Biol.*, 57(21), 6807, 2012.)

on the other hand, yields undistorted CT image that matches well with the original CBCT, which can also be seen from the comparison between $CT_{deformed}^{DISC}$ and $CBCT_{corrected}$ as shown in Figure 11.3d-1 through d-3.

In terms of DIR accuracy, the DISC algorithm is quantitatively evaluated using NMI, FSIM, and Canny edge RMSE between $CT_{original}$ and $CBCT_{original}$, $CT_{deformed}^{demons}$ and $CBCT_{original}$, $CT_{deformed}^{DISC}$ and $CBCT_{original}$, and $CT_{deformed}^{DISC}$ and $CBCT_{corrected}$. For all six cases, the average NMI increases from 0.62 ± 0.02 to 0.63 ± 0.02, the average FSIM increases from 0.91 ± 0.04 to 0.94 ± 0.02, and the average edge RMSE decreases from 0.24 ± 0.03 to 0.21 ± 0.03, when comparing DISC with the original demons algorithm.

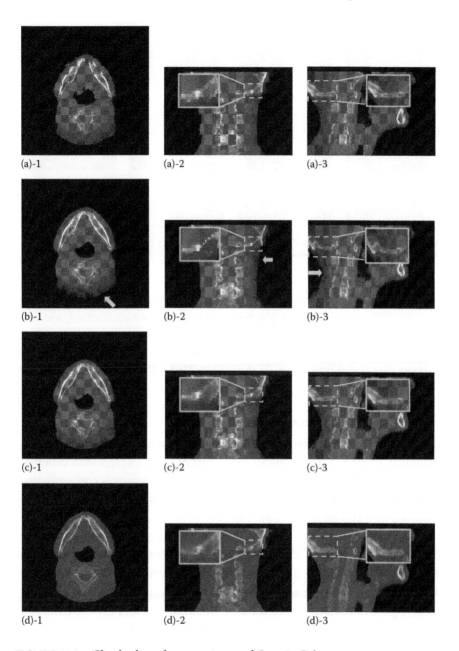

FIGURE 11.3 Checkerboard comparisons of Case 2. *Columns*: transverse, coronal and sagittal images, respectively. (a) $CT_{original}$ and $CBCT_{original}$; (b) $CT_{deformed}^{demons}$ and $CBCT_{original}$; (c) $CT_{deformed}^{DISC}$ and $CBCT_{original}$; (d) $CT_{deformed}^{DISC}$ and $CBCT_{corrected}$. The insets show the zoomed-in views, and the arrows indicate the regions contaminated by artifact or with geometrical distortion. (Reprinted from Zhen, X. et al., *Phys. Med. Biol.*, 57(21), 6807, 2012.)

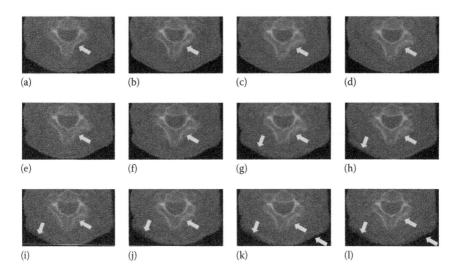

FIGURE 11.4   Effect of patch size on the deformed CT images. (a) through (l) are part of $CT^{DISC}_{deformed}$ obtained with patch size of $3 \times 3 \times 3$, $5 \times 5 \times 3$, $7 \times 7 \times 3$, $9 \times 9 \times 3$, $11 \times 11 \times 3$, $13 \times 13 \times 5$, $15 \times 15 \times 5$, $17 \times 17 \times 5$, $19 \times 19 \times 5$, $21 \times 21 \times 5$, $23 \times 23 \times 7$, $25 \times 25 \times 7$. The arrows indicate the regions of deterioration as the patch size increase. (Reprinted from Zhen, X. et al., *Phys. Med. Biol.*, 57(21), 6807, 2012.)

11.6.3 Effect of Patch Size

The patch size in DISC has a considerable impact on the registration performance. Figure 11.4a through l show part of the $CT^{DISC}_{deformed}$ with different patch sizes in an example clinical case. This part of image is contaminated by artifacts most severely. As the patch size increases, the bones become distorted and the edges become blurred. When the patch size gets even larger, soft tissue region is distorted as well (Figure 11.4g through l). This effect is also observed in other clinical cases. The reason is that, as the patch size increases, it is more likely to have different structures included in the same patch, resulting in errors in intensity correction. Therefore, we use $3 \times 3 \times 3$ patch size in this work for all simulation data and clinical data, and the patch size is kept constant at each image resolution level.

11.6.4 Computational Efficiency

All the experiments in this study were conducted on an NVIDIA Telsa C1060 card with a total number of 240 processors of 1.3 GHz. It is also equipped with 4 GB DDR3 memory, shared by all processors. In the CUDA implementation of intensity correction kernel in DISC, the computation corresponding to each voxel is handled by a thread, which loops over all voxels in the patch centered at the voxel. Consequently, the computational time is highly dependent on the patch size. The most time-consuming step in the calculation is the iterative interpolation/extrapolation procedure to get a and b values for voxels with $\xi = 1$. Currently, there is no efficient way for sparse data interpolation. Compared to ~20 s needed with the original demons algorithm, the DISC algorithm takes about 69 s for a patch size of $3 \times 3 \times 3$ and an image size of $256 \times 256 \times 68$.

11.7 DISCUSSION AND CONCLUSIONS

A fast, accurate, and robust CT-CBCT DIR algorithm is a key step in ART. Though many methods had been studied specifically for intermodality image registration, which might be applied directly to CT-CBCT DIR, it is still tempting to use a more accurate and efficient intensity-based algorithm, such as demons, to deal with the DIR problem for CT and CBCT images. Other than modifying the DIR similarity metric, which is not straightforward in demons, we presented in this chapter a patch-based intensity correction method that is performed in conjunction with demons, namely DISC, and evaluated it comprehensively with simulated data and clinical patient data. By incorporating the intensity correction at every iteration step, the DISC algorithm can robustly estimate the spatial transformation while guaranteeing the intensity consistency between CT and CBCT image.

It is technically possible to use different patch sizes at different parts of the image. One practical approach is to adjust the patch size according to the local image content. For instance, a relatively large patch size can be used in those homogeneous regions, while a small patch size can be adopted in those regions with more image details. The patch size can be thus determined adaptively by calculating the standard deviation of a specific region, which might help the convergence of the algorithm. This idea will be studied in the future work.

Note that in Case 2, the $CT_{original}$ has been truncated. This is because the CBCT has a limited field of view (FOV) due to the limited size of the image detector. In this work, we first rigidly register the CT with this truncated CBCT and then crop the CT outside the FOV, which results in a cropped CT image after registration. Undoubtedly, this cropped CT image cannot be used for dose calculation afterward in ART, as it may introduce significant errors. This truncation problem is not unique to DISC but a common issue for all DIR algorithms. Yang et al. [37] proposed to assign those missing voxels outside of the FOV with not-a-number (NaN) value. This simple method makes it possible to avoid the cropping of the CT image. We are developing a more mathematically strict method to modify DISC for the DIR problem between CT image and truncated CBCT image.

Most previous studies focused on CBCT images without severe artifacts, in which case simple histogram matching [9] or polynomial transformation model [8] might be sufficient to describe the intensity relationship between the CT and CBCT image. In our DISC algorithm, intensity correction is performed on a voxel-by-voxel basis and the correction parameters are estimated by considering the voxels nearby. It is advantageous to do so, because usually, obtaining an ideal mathematical model to describe the intensity relation is not easy, if not impossible, when considering the existence of severe artifacts in CBCT images. While estimating the intensity transformation locally, one is able to make use of the prior information from CT image, and correct the intensity in CBCT image voxelwise. By this means, we try to equalize the intensities distribution (the sample moments) inside each small patch with its corresponding patch in the other image, which is physically sensible. Embedding this additional correction step into the original demons, the intensity consistency requirement can be guaranteed.

Another merit of the DISC method is that it can yield the intensity-corrected CBCT image at the end of the DIR. This intensity-corrected CBCT image can be regarded as a CBCT image produced by the primary radiation. Hence, the DISC method, potentially, can be applied to the area of CBCT scatter removal and hence other clinically relevant tasks, for example, radiation dose calculation based on CBCT.

The intensity correction method in DISC can be easily incorporated into other intensity-based DIR algorithms, such as optical flow-based algorithms [38], viscous fluid algorithms [39], fast free form with calculus of variations [40], etc., where intensity consistency between two images is assumed. This method is effective in correcting the intensity of CBCT images because of the modality similarity between CT and CBCT images. For more challenging multimodality DIR tasks, such as CT to MR DIR, our intensity correction method needs to be modified since same anatomical structures in the MR image may have totally different or even opposite image intensity with those in the CT image.

Another potential application of this intensity correction method might be in the field of image segmentation. Medical images often suffer from various imaging artifacts and the image intensity may not be homogeneous in regions where image homogeneity should exist. This issue may result in the failure of some intensity-driven segmentation methods [41,42]. To solve this problem, our intensity correction method may be incorporated into a more sophisticated framework based on the combination of simultaneous registration, segmentation, and intensity correction using artifact-free prior images.

REFERENCES

1. Thirion, J.P., Image matching as a diffusion process: An analogy with Maxwell's demons. *Medical Image Analysis*, 1998. **2**(3): 243–260.
2. Pennec, X., P. Cachier, and N. Ayache, Understanding the "Demon's Algorithm" 3D non-rigid registration by gradient descent. In *Proceedings of Second International Conference on Medical Image Computing and Computer-Assisted Intervention (MICCAI'99) (Lecture Notes Computer Science)*, Sophia Antipolis, France, 1999, pp. 597–605. http://webdocs.cs.ualberta.ca/~dana/readingMedIm/papers/pennec99understanding-1.pdf.
3. Wang, H. et al., Validation of an accelerated 'demons' algorithm for deformable image registration in radiation therapy. *Physics in Medicine and Biology*, 2005. **50**(12): 2887–2905.
4. Rogelj, P. and S. Kovacic, Symmetric image registration. *Medical Image Analysis*, 2006. **10**(3): 484–493.
5. Yang, D. et al., A fast inverse consistent deformable image registration method based on symmetric optical flow computation. *Physics in Medicine and Biology*, 2008. **53**(21): 6143–6165.
6. Guimond, A. et al., Three-dimensional multimodal brain warping using the demons algorithm and adaptive intensity corrections. *IEEE Transactions on Medical Imaging*, 2001. **20**(1): 58–69.
7. Lu, H. et al., Multi-modal diffeomorphic demons registration based on pointwise mutual information. In *Proceedings of the 2010 IEEE International Conference on Biomedical Imaging: From Nano to Macro*. Rotterdam, the Netherlands: IEEE Press, 2010, pp. 372–375. http://ieeexplore.ieee.org/xpl/articleDetails.jsp?arnumber=5490333.
8. Nithiananthan, S. et al., Demons deformable registration of CT and cone-beam CT using an iterative intensity matching approach. *Medical Physics*, 2011. **38**(4): 1785.
9. Hou, J. et al., Deformable planning CT to cone-beam CT image registration in head-and-neck cancer. *Medical Physics*, 2011. **38**(4): 2088.
10. Feldkamp, L.A., L.C. Davis, and J.W. Kress, Practical cone-beam algorithm. *Journal of the Optical Society of America A*, 1984. **1**(6): 612–619.
11. Schulze, R. et al., Artefacts in CBCT: A review. *Dentomaxillofacial Radiology*, 2011. **40**(5): 265–273.
12. Siewerdsen, J.H. and D.A. Jaffray, Cone-beam computed tomography with a flat-panel imager: Magnitude and effects of x-ray scatter. *Medical Physics*, 2001. **28**(2): 220.

13. Siewerdsen, J.H. et al., A simple, direct method for x-ray scatter estimation and correction in digital radiography and cone-beam CT. *Medical Physics*, 2006. **33**(1): 187–197.

14. Rinkel, J. et al., A new method for x-ray scatter correction: First assessment on a cone-beam CT experimental setup. *Physics in Medicine and Biology*, 2007. **52**(15): 4633.

15. Maltz, J.S. et al., Algorithm for X-ray scatter, beam-hardening, and beam profile correction in diagnostic (kilovoltage) and treatment (megavoltage) cone beam CT. *IEEE Transactions on Medical Imaging*, 2008. **27**(12): 1791–1810.

16. Zhu, L. et al., Scatter correction for cone-beam CT in radiation therapy. *Medical Physics*, 2009. **36**(6): 2258.

17. Poludniowski, G. et al., An efficient Monte Carlo-based algorithm for scatter correction in keV cone-beam CT. *Physics in Medicine and Biology*, 2009. **54**(12): 3847.

18. Yan, H. et al., Projection correlation based view interpolation for cone beam CT: Primary fluence restoration in scatter measurement with a moving beam stop array. *Physics in Medicine and Biology*, 2010. **55**(21): 6353–6375.

19. Meyer, M., W.A. Kalender, and Y. Kyriakou, A fast and pragmatic approach for scatter correction in flat-detector CT using elliptic modeling and iterative optimization. *Physics in Medicine and Biology*, 2010. **55**(1): 99–120.

20. Sun, M. et al., Correction for patient table-induced scattered radiation in cone-beam computed tomography (CBCT). *Medical Physics*, 2011. **38**(4): 2058–2073.

21. Giles, W. et al., Crescent artifacts in cone-beam CT. *Medical Physics*, 2011. **38**(4): 2116.

22. Zheng, D. et al., Bow-tie wobble artifact: Effect of source assembly motion on cone-beam CT. *Medical Physics*, 2011. **38**(5): 2508–2514.

23. Zhu, L., J. Wang, and L. Xing, Noise suppression in scatter correction for cone-beam CT. *Medical Physics*, 2009. **36**(3): 741–752.

24. Hsieh, J. et al., An iterative approach to the beam hardening correction in cone beam CT. *Medical Physics*, 2000. **27**(1): 23–29.

25. Grimmer, R. and M. Kachelriess, Empirical binary tomography calibration (EBTC) for the precorrection of beam hardening and scatter for flat panel CT. *Medical Physics*, 2011. **38**(4): 2233–2240.

26. Li, T. et al., Motion correction for improved target localization with on-board cone-beam computed tomography. *Physics in Medicine and Biology*, 2006. **51**(2): 253–267.

27. Lewis, J.H. et al., Mitigation of motion artifacts in CBCT of lung tumors based on tracked tumor motion during CBCT acquisition. *Physics in Medicine and Biology*, 2011. **56**(17): 5485–5502.

28. Sharp, G.C. et al., GPU-based streaming architectures for fast cone-beam CT image reconstruction and demons deformable registration. *Physics in Medicine and Biology*, 2007. **52**(19): 5771–5783.

29. Gu, X. et al., Implementation and evaluation of various demons deformable image registration algorithms on a GPU. *Physics in Medicine and Biology*, 2010. **55**(1): 207–219.

30. Castadot, P. et al., Comparison of 12 deformable registration strategies in adaptive radiation therapy for the treatment of head and neck tumors. *Radiotherapy and Oncology*, 2008. **89**(1): 1–12.

31. Modat, M. et al., *Diffeomorphic Demons Using Normalized Mutual Information, Evaluation on Multimodal Brain MR Images*. San Diego, CA: SPIE, 2010. http://discovery.ucl.ac.uk/19174/1/19174.pdf.

32. Zhen, X. et al., CT to cone-beam CT deformable registration with simultaneous intensity correction. *Physics in Medicine and Biology*, 2012. **57**(21): 6807–6826.

33. Samant, S.S. et al., High performance computing for deformable image registration: Towards a new paradigm in adaptive radiotherapy. *Medical Physics*, 2008. **35**(8): 3546.

34. Jia, X. et al., A GPU tool for efficient, accurate, and realistic simulation of cone beam CT projections. *Medical Physics*, 2012. **39**(12): 7368–7378.

35. Zhang, L. et al., FSIM: A feature similarity index for image quality assessment. *IEEE Transactions on Image Processing*, 2011. **20**(8): 2378–2386.

36. Yan, H. et al., A comprehensive study on the relationship between the image quality and imaging dose in low-dose cone beam CT. *Physics in Medicine and Biology*, 2012. **57**(7): 2063–2080.

37. Yang, D. et al., Technical note: Deformable image registration on partially matched images for radiotherapy applications. *Medical Physics*, 2010. **37**(1): 141.

38. Lucas, B.D. and T. Kanade, An iterative image registration technique with an application to stereo vision. In *Proceedings of the Seventh International Joint Conference on Artificial Intelligence*. Vancouver, British Columbia, Canada: Morgan Kaufmann Publishers, Inc., 1981. **2**: 674–679. http://www.ri.cmu.edu/pub_files/pub3/lucas_bruce_d_1981_2/lucas_bruce_d_1981_2.pdf.

39. D'Agostino, E. et al., A viscous fluid model for multimodal non-rigid image registration using mutual information. *Medical Image Analysis*, 2003. 7(4): 565–575.

40. Lu, W. et al., Fast free-form deformable registration via calculus of variations. *Physics in Medicine and Biology*, 2004. **49**(14): 3067–3087.

41. Li, Y. et al., Simultaneous segmentation and inhomogeneity correction in magnetic resonance images. *Conference Proceedings: IEEE Engineering in Medicine and Biology Society*, 2011. **2011**: 8045–8048.

42. Shahvaran, Z. et al., Variational level set combined with Markov random field modeling for simultaneous intensity non-uniformity correction and segmentation of MR images. *Journal of Neuroscience Methods*, 2012. **209**(2): 280–289.

CHAPTER 12

Reconstruction in Positron Emission Tomography

Franck P. Vidal and Jean-Marie Rocchisani

CONTENTS

12.1 INTRODUCTION

THE CORE PRINCIPLE OF nuclear medicine is to administer a radioactive substance (also called radiotracer or simply tracer) to patients. The tracer can be injected, inhaled, or under certain circumstances, ingested. Tissues will absorb the tracer in proportion to a physiological process. The normal distribution of the tracer is subject to be affected by a physio-pathological process. Examples of such processes include bone fractures, the growth of malignant tumors, and low blood flow in the heart. As it reflects organ metabolism, this modality is frequently called functional imaging. Tomography in this context is called emission tomography (ET). The reconstruction corresponds to the recovery of the three-dimensional (3D) distribution of the tracer in the body. Reconstructed volumes in ET will therefore highlight if/where a given physiological process occurs. There are two kinds of emission tomography technologies in nuclear medicine. Single-photon emission computed tomography (SPECT) makes use of a gamma emitter, that is, photons, as a radiotracer. Positron emission tomography (PET) makes use of positron emitters. It is becoming the main modality in nuclear medicine departments. This is why, in this chapter, we focus on this modality only. The most common clinical applications of ET are in oncology, and to a lesser extent neurology and cardiology.

Anatomical 3D imaging, such as computerized tomography (CT) or magnetic resonance imaging (MRI), is located in radiology departments; functional 3D imaging, that is, ET, requires dedicated equipment and approvals for manipulation and administering radioactive tracers and is located in nuclear medicine departments. Those imageries are complementary. Physiological and metabolic changes can and do occur without any change in anatomical structures. Anatomical imaging may not be sufficient to diagnose staging diseases or to monitor response to therapy. In practice, modern PET scanners embed CT or MRI capabilities. The aim is to obtain registered anatomical and physiological data.

Figure 12.1 shows an example of a PET-CT scan examination of the lungs. One can notice that the anatomical data offer a relatively high signal-to-noise ratio (SNR) and a rather good spatial resolution (a typical CT slice is 512×512 pixels with a millimetric pixel size). In contrast, the physiological data are very noisy and have a much lower resolution (a typical ET slice is about 128×128 pixels with at least 4 mm pixel size). Direct volume rendering [1] is used to fuse the bony structures and soft tissues from the CT data with areas of high level of radioactivity (in this figure

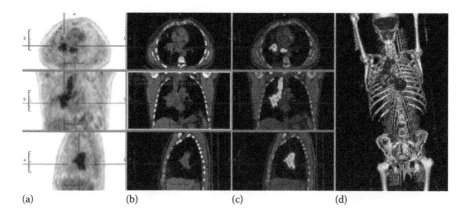

(a) (b) (c) (d)

FIGURE 12.1 PET-CT scan of the lungs: the radiotracer is absorbed by tissue in proportion to a physiological process. In this example, PET highlights areas of cancerous cells in the lung. In oncology it is used to detect tumors. (a) PET; (b) CT; (c) Fusion PET-CT; (d) Volume rendering of PET-CT. (c, d) are usually displayed in color to distinguish PET from CT data.

a lung, the heart, and the bladder) from the ET volume. It highlights the area of low blood flow in the heart that was not visible in the CT images.

The aim of this chapter is to present a brief overview of the volume reconstruction methods in PET. The necessary background knowledge specific to PET imaging is given in Section 12.2. A review of the most common rebinning techniques is provided in Section 12.3. Section 12.4 gives an introductory overview of the reconstruction methods. Analytical reconstruction methods are studied in Section 12.5, and iterative methods in Section 12.6. The chapter ends with a conclusion.

12.2 PHYSICS REVIEW AND DATA ACQUISITION

12.2.1 Physics of PET Imaging

Figure 12.2 illustrates the PET acquisition process. It makes use of physiological-like molecules labeled with positron emitters as radiotracers. Positron is a moving material particle (also denoted as β^+) that will interact with matter. After interactions, a positron may combine with an electron (β^- particle). It produces two photons of 511 keV emitted in opposite directions at almost 180°. They are detected in *coincidence*, in other words almost at the same time. The photons of a single pair are often called *coincidence photons*. When two photons are detected within a predefined time window (2τ), the PET imaging system records which detectors have been activated,

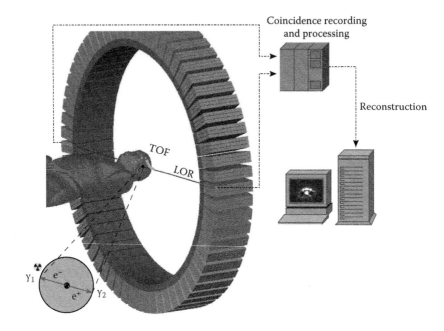

FIGURE 12.2 Schema of the PET data acquisition process.

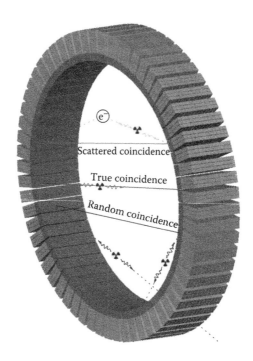

FIGURE 12.3 Events in PET.

that is, the position of the photons within the imaging system. The line between the detectors that have been activated for a given pair is called line of response (LOR).

The performance of a PET system depends on the energy and time resolutions of detectors [2]. During their travel through the body, photons can hit an electron. Some of the photon energy can be transferred to the electron. In this case, the photon energy will decrease and its trajectory is deviated. This is the Compton scattering effect (also called inelastic scattering). To improve the sensitivity of the system, photons with energy less than 511 keV are accepted. *Scattered coincidences* will then be recorded (Figure 12.3). To increase the number of recorded coincidences, the time window τ can also be increased. Two events issued from different annihilation processes could be considered as a single coincidence. They correspond to *random coincidences*, also shown in Figure 12.3. The data recorded by the PET scanner are then processed using tomography reconstruction to produce a map of radioactivity concentration. Incorrect LORs due to scattering and random events degrade the image quality.

12.2.2 Attenuation, Absorption, and Scattering

During the path of a photon through matter, it can be (1) transmitted, when no interaction between the photon and matter occur; (2) absorbed, when it transfers its energy to the tissue; or (3) scattered, when its direction changes (and it loses energy in case of Compton scattering). The result of absorption and scattering is the disappearance of photons from their point of emission, which is called attenuation.

Due to attenuation, only one of the coincidence photons for many annihilation events will reach the detector. These are called *single events* and should be discarded.

12.2.3 Electronic Collimation

Registering detected photons as pairs of coincidence photons (or LORs) corresponds to *electronic collimation*. LORs are intrinsically fully 3D. No mechanical collimation is therefore required. For these reasons, PET has a much higher sensitivity and a better spatial resolution than SPECT. Note that early PET scanners made use of tungsten septum collimators to ensure that the photon pair for each annihilation is located on the same plane. In this case, the reconstruction was performed slice by slice using exactly the same reconstruction code as SPECT.

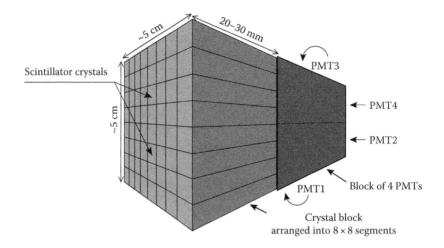

FIGURE 12.4   PET scintillation block detector.

12.2.4 Detectors

To detect a pair of annihilation photons simultaneously, the detectors are placed in a ring. Its diameter is about 80 cm and its thickness is about 20 cm. The ring is divided into blocks. A block usually consists of a segmented block of scintillation crystals paired with a set of extremely sensitive light detectors (Figure 12.4). The scintillator is 20–30 mm deep and is made of bismuth germanate (BGO), lutetium orthosilicate (LSO), or lutetium-yttrium oxyorthosilicate (LYSO). It absorbs the photon and restitutes its energy with as many visible photons. The light is then converted into an electric signal that is amplified in a photomultiplier tube (PMT), or in solid state detector in recent devices.

12.2.5 Time-of-Flight

Today's PET scanners have a relatively high time resolution, which allows us to record the arrival time of each photon. This extra information corresponds to time-of-flight (TOF). It makes possible to compute an estimate of where an annihilation occurred on an LOR with a given probability of accuracy. Figure 12.5 illustrates that TOF does not totally remove the need of tomographic reconstruction. The photon going to the right will travel a distance $d_r = D_{lor}/2 - \Delta_d$, it will reach the detector in d_r/c seconds, with c the speed of light, D_{lor} the LOR's length, and Δ_d the distance from the location of the annihilation to the center of the LOR. The photon going to the left will travel a distance $d_l = D_{lor}/2 + \Delta_d$, it will reach the detector

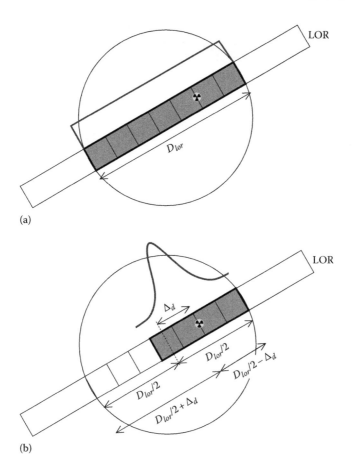

FIGURE 12.5 Illustration of the elements contributing to a LOR: (a) conventional PET without TOF; (b) PET with TOF.

in d_l/c seconds. The difference in arrival time is $\Delta_t = 2\Delta_d/c$. Δ_t is known as the scanner measures it. D_{lor} is also known as it corresponds to the LOR. Δ_d can therefore be estimated. Its accuracy depends on the timing resolution of the system. Modern reconstruction tools use this information to improve the accuracy of the final image (Figure 12.6). In oncology, TOF helps clinicians to better detect lesions in a noisy background [3–5].

12.3 REBINING IN 2D AND 3D SINOGRAMS

The final step is the reconstruction. PET scanners produce a very high number of LORs, which is significantly larger than the number of voxels in the reconstructed volume. There are redundancies in the data.

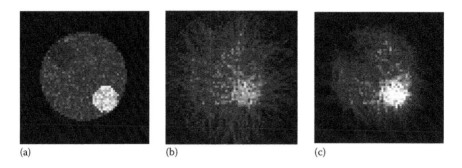

(a)   (b)   (c)

FIGURE 12.6   Impact of TOF on the back projection of simulated LORs into the image space: (a) phantom; (b) conventional PET without TOF; (c) PET with TOF.

It leads to the need of high storage capacities and it increases the processing time (data correction and tomography reconstruction). For these reasons, the first generation of algorithms required sinograms as input data. The conversion of raw list-mode data into a sinogram is called *rebinning* [6]. It is also convenient as it enables the use of standard reconstruction methods that have been originally developed for CT or SPECT.

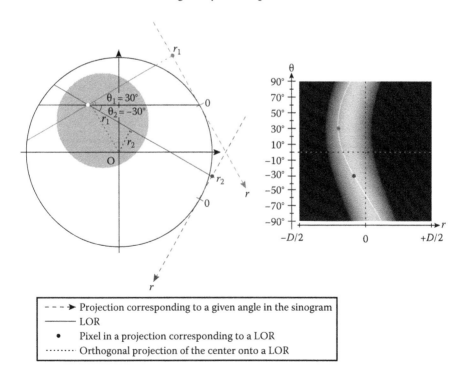

FIGURE 12.7   Conversion from coincidence events to 2D sinograms.

Figure 12.7 shows how this conversion can be achieved in 2D (for illustration purposes, we consider that the two coincidence photons of a pair are located on the same transaxial plane). Care must be given to sampling to avoid the loss of resolution. Sampling is needed along:

- The horizontal axis of the sinogram that matches the minimum distance between an LOR and the center point of the system (see oriented distances r in Figure 12.7).

- The vertical axis of the sinogram that matches the angle between a LOR and the horizontal plane of the system (see angles θ between the LORs and the horizontal axis in Figure 12.7).

A bin of the sinogram is then identified by its (r, θ) coordinates. To reconstruct a 3D volume, a 3D stack of 2D transaxial sinograms will be built. Each 2D slice of the sinogram stack will be used to reconstruct successive transaxial slices. A bin of the sinogram stack is now identified by its (r, θ, z) coordinates. The two coincidence photons of a pair are located in the system at (x_1, y_1, z_1) and (x_2, y_2, z_2), respectively. In a true 3D system, it is very unlikely that they are located on the same transaxial plane, in other words z_1 and z_2 are probably different. To build the sinogram stack, such oblique LORs are also taken into account. It allows us to take into account all the LORs that have been recorded, hence to retain sensitivity.

The simplest method to achieve this is called single-slice rebinning (SSRB) [7]. The count corresponding to an oblique LOR is assigned to the slice of the sinogram stack that coincides to the transaxial slice midway between the two detected photons of the LOR($\overline{z}_{1,2} = (z_1 + z_2)/2$) (Figure 12.8a). When the annihilation location is relatively close to the scanner axis, the plane is assigned quite accurately. When it is far from the axis, the accuracy decreases significantly. It leads to blur and distortion in the reconstructed volume.

To reduce these drawbacks, other methods have been proposed such as the popular multislice rebinning (MSRB) method [8]. In SSRB, a single bin is incremented for each LOR. In MSRB, bins located on slices between z_{low} and z_{high} are incremented for each LOR (Figure 12.8b):

$$z_{low} = \overline{z}_{1,2} - \frac{R|z_2 - z_1|}{t_{1,2}} \tag{12.1}$$

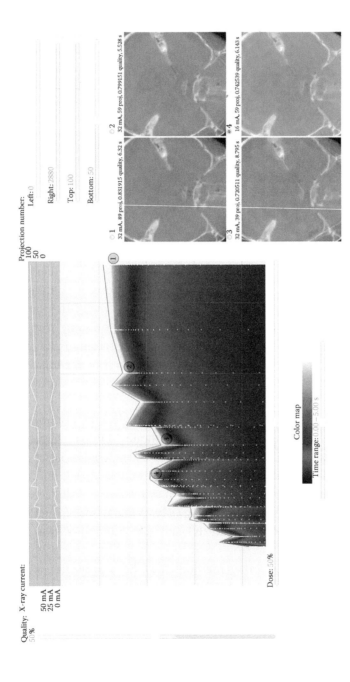

FIGURE 4.5 The quality *vs.* dose plot. The numbers inserted into the plot refer to the image matrix (the light box) on the right. Image 1 is the image of the highest quality but requires a high dose. Images 3 and 4 require much less dose but have streak artifacts or blurred features, respectively. Image 2 seems to be a good compromise—lower dose than image 1 but still offering good quality. It could serve as a starting point for further exploration of the plot, where the user would mouse-click at desirable plot locations and insert the corresponding images into the light box. (From Zheng, Z. et al., *Phys. Med. Biol.*, 58(21), 7857, 2013. With permission.)

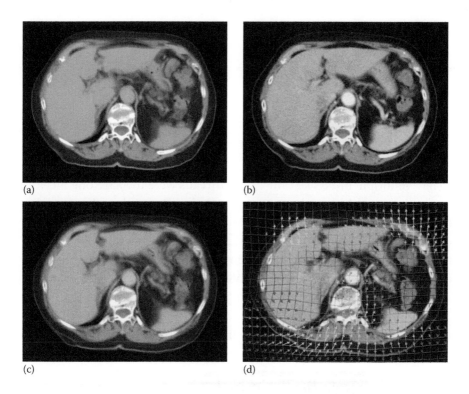

FIGURE 9.1 CT images from (a) primary imaging, (b) secondary imaging, (c) fused imaging with deformed secondary imaging, and (d) superimposed vector map.

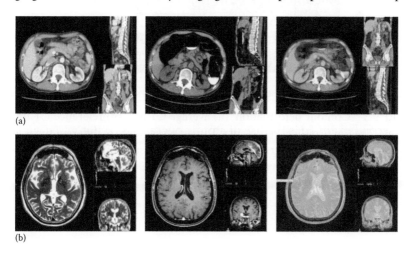

FIGURE 9.2 Example of unimodal image registration for image fusion: (a) CT images and (b) MR images. The left most image represents the primary data set, the center image represents the secondary data set, and the right most image represents the fused data set. The CT imaging represents different patient positions, while the MR imaging represents different imaging sequences (T2 and T1 images).

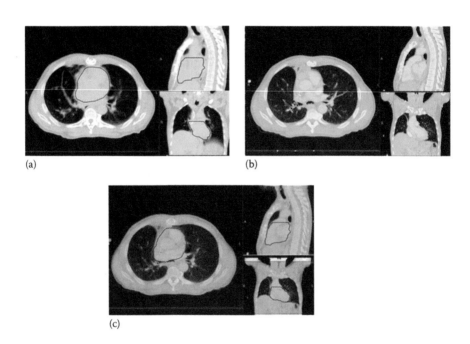

FIGURE 9.3 Contour deformation: (a) average CT and (b) reference CT image data sets. The top two slices represent CT imaging acquired at different times with the heart contoured in the first CT, while (c) is the resulting overlay using DIR.

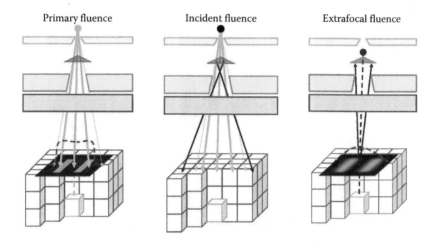

FIGURE 13.3 Diagram of the separation of incident fluence into a primary fluence and extrafocal fluence. The incident photon fluence (center) includes photons (blue) from the primary source (black) and photons induced (purple) by
interactions with parts of the linear accelerated head, such as the flattening filter
(green). Fluence intensities are modulated by the primary (yellow) and upper
(orange) and lower (red) jaws. The incident photon fluence is modeled by using
a point-like "primary" fluence model (left) and an area-like "extrafocal" fluence
model (right). Both models project the fluence computed at a reference plane
through the patient volume (dashed lines) in order to avoid computing the fluence to each voxel (blue and yellow cubes). An example of the fluence for a highly
modulated MLC field and a diagram of the open field intensity profile (dashed
red line) have been included for each model. The purple arrows in the extrafocal
fluence diagram (right) represent the limiting of the visible source area by the
upper collimating jaws (orange).

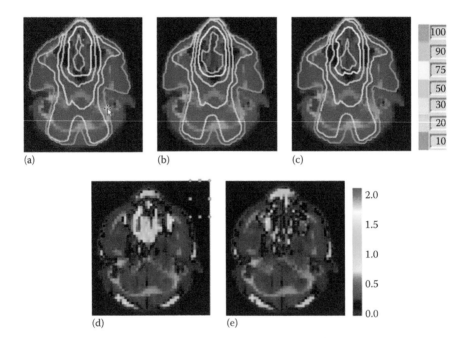

FIGURE 14.2 Dose distributions for Case H3 calculated with the MCSIM (a), g-FSPB (b), and g-DC-FSPB (c) algorithms in the X-Y plane through the isocenter. The γ-index distributions are shown in (d) for g-FSPB and (e) for g-DC-FSPB dose distributions in the same plane. (Reproduced from Gu, X. et al., *Phys. Med. Biol.*, 56(11), 3337, 2011.)

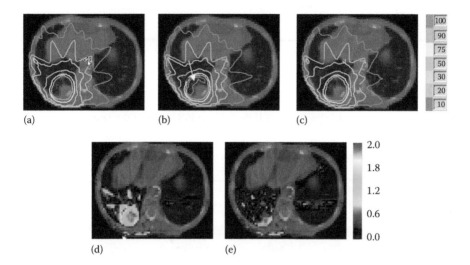

(a) (b) (c)

100
90
75
50
30
20
10

2.0
1.8
1.2
0.6
0.0

(d) (e)

FIGURE 14.3 The dose distribution of Case L2 calculated with MCSIM (a), g-FSPB (b), and g-DC-FSPB (c) in the X-Y plane through the isocenter. The γ-index distributions in the same plane are illustrated in (d) for g-FSPB and (e) for g-DC-FSPB. (Reproduced from Gu, X. et al., *Phys. Med. Biol.*, 56(11), 3337, 2011.)

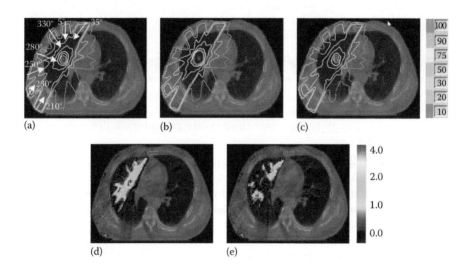

(a) (b) (c)

100
90
75
50
30
20
10

4.0
2.0
1.0
0.0

(d) (e)

FIGURE 14.4 Dose distributions for Case L4 in the X-Y plane through the isocenter calculated with the MCSIM (a), g-FSPB (b), and g-DC-FSPB (c) algorithms. γ-index distributions for the g-FSPB (d) and g-DC-FSPB (e) algorithms are given in the same plane. (Reproduced from Gu, X. et al., *Phys. Med. Biol.*, 56(11), 3337, 2011.)

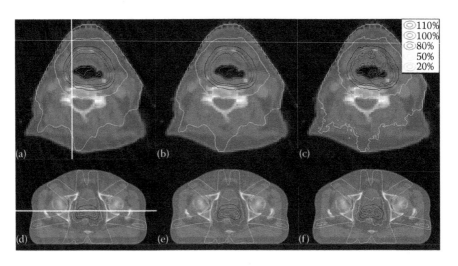

FIGURE 22.4 Isodose lines for a HN case (a through c) and a prostate case (d through f). The three columns are ground truth doses, denoised doses, and noisy doses. *(Continued)*

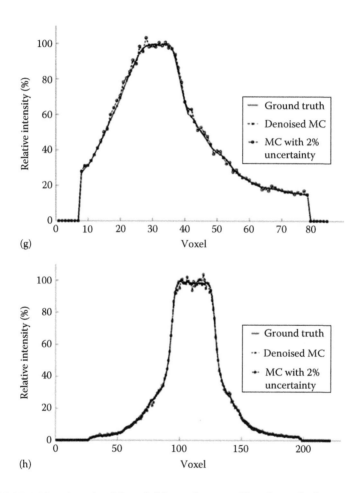

FIGURE 22.4 (*Continued*)   (g) and (h) are dose profiles along the lines indicated in (a) and (d).

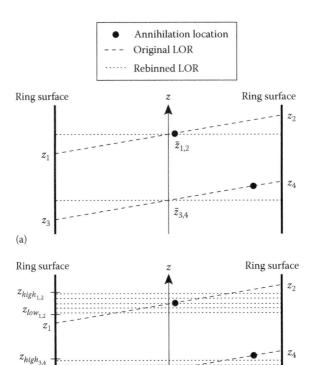

FIGURE 12.8   Illustration of rebinning methods: (a) using SSRB; (b) using MSRB.

$$z_{high} = \bar{z}_{1,2} - \frac{R|z_2 - z_1|}{t_{1,2}} \tag{12.2}$$

where
$$t_{1,2} = \sqrt{(x_2 - x_1)^2 + (y_2 - y_1)^2}$$
R controls the spread of the rebinning across several slices

If R is null, MSRB and SSRB produce exactly the same results. In Reference 8, R is equal to the radius of the transverse field of view, that is, the maximum value of r. The increment ($incr$) for each bin is

$$incr = \frac{maxincr}{(I(z_{high}) - I(z_{low})) + 1} \tag{12.3}$$

where

  $I(z)$ is the axial index in the sinogram stack of bins that contain any
  point at axial coordinate z

  maxincr is a positive constant value

MSRB is more accurate than SSRB, but it is known to be less stable for noisy data.

A further alternative to limit the trade-off in accuracy and in stability is the Fourier rebinning (FORE) algorithm [9]. Its goal is to provide an inversion formula in the Fourier transform (FT) domain. Counts from oblique sinograms are reassigned to slices of the transaxial sinogram stack using the frequency–distance relation [10]. It separates the contributions of the events located close to the detectors from the ones located close to the axis of the scanner. In some ways it is considered as a *virtual TOF principle.*

12.4 OVERVIEW OF RECONSTRUCTION METHODS

The following sections describe image reconstruction methods from PET-acquired data, which is represented by events recorded in coincidence along an LOR. An elementary LOR is given when the two photons produced by the annihilation of a positron are detected in coincidence. The aim of those methods is to recover the distribution of the radioactive activity of the tracer labeled with positrons. Assuming that the displacement of a positron is small in the matter before it annihilates, this distribution may give an approximation of the distribution of the labeled molecule. Hence this will lead to an approximation of the concentration of the radiotracer that represents a physiological parameter of interest. As a β^+ annihilates in an unknown location along the LOR, a direct reconstruction is not possible even with the state-of-the-art PET-TOF scanners. In fact, many detections in coincidence are needed for the reconstruction process. Due to the random nature of annihilations, tomographic reconstruction will provide an average of the spatial distribution of a radiotracer. The precision increases as the number of recorded LOR increases. Typically in a modern head scanner, this number is in the range of 10^9 pairs of coincidence. This will necessitate long acquisition times that are not always practical.

During the acquisition, the number of coincident detections is stored for each line of detection. This line is the path between the centers of the two involved detectors. The number of counts assigned to a given LOR is proportional to a line integral of β^+ decays along that LOR.

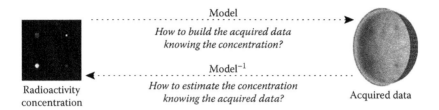

FIGURE 12.9   Model inversion.

Figure 12.9 illustrates the principle of tomography reconstruction applied to PET imaging. The spatial distribution of radioactivity is unknown, and this is what we are looking for as only the acquired data (LORs) are known. The reconstruction of the radioactivity concentration consists of the inversion of the model from the acquired data.

Different reconstruction methods have been known about for more than 40 years. They can be separated into direct or analytical methods and iterative methods. Only recently, however, a few iterative methods based on a statistical point of view have been introduced routinely. This is because this class of methods requires a huge computation power, even if they provide more accurate reconstructions. In the following sections, we present these two classes of methods.

12.5 ANALYTICAL RECONSTRUCTION METHODS

12.5.1 2D Reconstruction

Analytical reconstruction methods initially developed for CT reconstruction were also used in PET imaging. The reconstruction of a slice is the inversion of measurement equations directly (Figure 12.10). Data are usually organized as 2D sinograms (see Section 12.3). For 2D reconstruction,

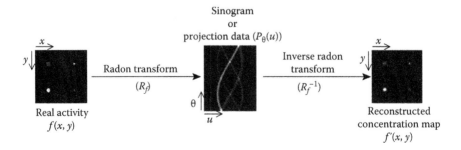

FIGURE 12.10   Analytic reconstruction.

only data organized as direct sinograms (i.e., projections perpendicular to the scanner axis) are considered. SSRB, MSRB, or FORE can be used to generate the sinograms to take into account oblique LORs. There is one 2D sinogram per 2D slice. Each slice is processed independently. 2D slices are then stacked to build the final 3D volume.

12.5.1.1 Line Integral Model and 1D Projection

The projection data consist of the observed data organized as a 2D sinogram:

$$P_\theta(u) = \iint f(x,y)\delta(u - x\cos\theta - y\sin\theta)\,dx\,dy \qquad (12.4)$$

where

$P_\theta(u)$ is the projection data at angle θ at Pixel abscissa u

$f(x, y)$ is the radioactive concentration at Image coordinate (x, y) (Figure 12.11)

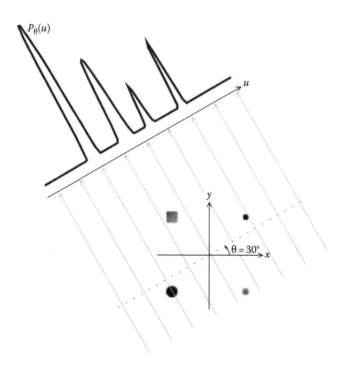

FIGURE 12.11 $P_\theta(u)$ the projection data in Figure 12.10 at angle $\theta = 30°$ and at Pixel abscissa u, and $f(x, y)$ the radioactive concentration at Image coordinate (x, y).

$P_\theta(u)$ for different values of θ and u is what is known. It corresponds to the Radon transform (or forward projection) (R_f) of the real activity $(f(x, y))$:

$$R_f(t,\theta) = P_\theta(t) \tag{12.5}$$

The real activity is unknown and the tomography reconstruction aims to recover the activity. It is performed by using the Inverse Radon transform to create an estimated activity map $(f'(x, y))$ from the projections. There are several methods to invert the Radon transform. Only the most popular methods are presented in the following sections.

12.5.1.2 2D Back Projection

The inverse transformation can be performed using a 2D back projection (also called unfiltered back projection). It is the formal inverse of the projection process. For each projection P_θ and for each value u, the quantity $P_\theta(u)$ is spread back in the image $f'(x, y)$ space along the ray corresponding to the relevant line integral. The resulting image is called laminogram. Figure 12.12 shows the result of the back projection when different numbers of angles are used. Star artifacts are observed when only a few angles are used. When a high number is used, a halo is located around bright objects. Edges are blurred due to the spreading of values across the whole image space.

12.5.1.3 Fourier Slice Theorem

The back projection is usually performed using the Fourier slice theorem (also called central slice theorem) [11]. The 1D FT of a projection for a given angle is equal to a slice of the 2D FT of the original image at the

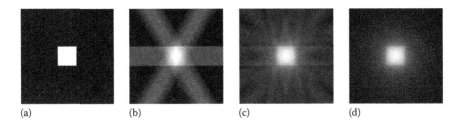

(a)　　　　　　(b)　　　　　　(c)　　　　　　(d)

FIGURE 12.12 Back projections of a square: (a) original image; (b) FBP with 3 projections; (c) FBP with 7 projections; (d) FBP with 180 projections.

same angle. The complete 2D FT of the image is reconstructed from these 1D FTs. Then, the image is obtained by inverting its FT.

12.5.1.4 2D Filtered Back Projection

Filtered back projection (FBP) is probably the most popular analytical method in tomography reconstruction. The unknown activity $f(x, y)$ may be recovered by summing filtered projections:

$$f(x,y) = \int_0^\pi \left(\int_{-\infty}^\infty P_\theta(\alpha) \cdot h(t-\alpha)d\alpha \right) d\theta$$

where

$P_\theta(u)$ is the projection

$h(t)$ is a filter called the *ramp filter*:

$$h(t) = \int_{-\omega}^{+\omega} |\omega| e^{j2\pi\omega t} d\omega$$

To recover edges in the reconstructed image, projections are filtered to amplify high-frequency components. This filter has to be damped for high frequencies, for example, by multiplying it with a window. As an example, the Hanning filter is obtained by multiplying the ramp filter and a Hanning window. It is commonly used. Its effect is to restore high frequencies in the final image without amplifying the noise too much. The filtering is usually performed in the FT domain before the inversion of the FT. Figure 12.13 shows that the reconstructed image is not so blurred. Another effect of the filtering is that high frequencies in the star artifacts are amplified. FBP can only be used when a sufficient number of angles are considered.

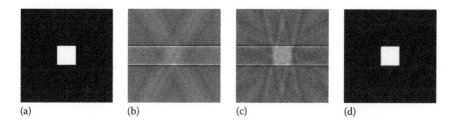

(a)   (b)   (c)   (d)

FIGURE 12.13 Filtered back projections of a square: (a) original image; (b) FBP with 3 projections; (c) FBP with 7 projections; (d) FBP with 180 projections.

12.5.2 3D Reconstruction

The FBP principle can be directly extended to 3D if projection data are available for all the possible angles of oblique LORs. However, due to the limitation in scanner geometries this is not feasible and another filter, the Colsher's filter, is used in the FBP deconvolution step [12]. For 3D analytical reconstruction, the data are organized in 2D projections as Figure 12.14

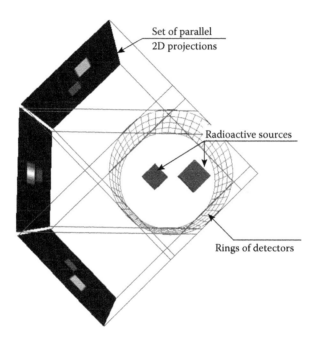

FIGURE 12.14   Parallel 2D projections for 3D analytical reconstruction.

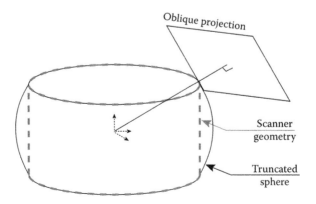

FIGURE 12.15   Oblique projection around a truncated sphere.

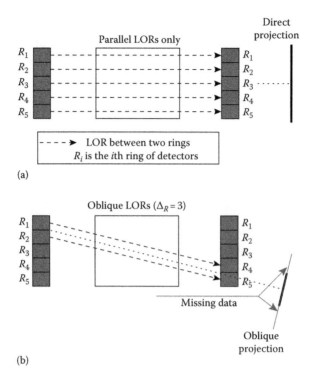

FIGURE 12.16   2D projections for 3D analytical reconstruction: (a) direct 2D projections, without missing data; (b) oblique 2D projections with missing data.

illustrates. It is also possible to build projections of oblique LORs at different angles. Figure 12.15 shows how oblique projections are located in relation to a truncated sphere that represents the scanner geometry. The effect of this incomplete sphere is illustrated in Figure 12.16. Oblique projections are truncated. To address this issue, the 3D reprojection (3D-RP) algorithm is commonly used [13]. It is composed of four successive steps:

1. Initial 2D FBP reconstruction to build an estimate of the volume.

2. Forward projections from the estimated volume to build missing data.

3. Combine the missing data with the measured data.

4. 3D FBP from the complete set of 2D projections with the Colsher's filter.

3D-RP is the most widely used 3D analytical reconstruction method.

One of the main drawbacks of analytical methods is the difficulty of taking into account the specificities of PET imaging during the reconstruction, such as scanner properties, Compton scattering, attenuation, etc. It can be performed offline before the reconstruction by cleaning the projection data. However, analytical methods have now been replaced in the clinical routine by the iterative methods presented in the next section.

12.6 ITERATIVE RECONSTRUCTION METHODS

Iterative algorithms, as opposed to analytical ones, are based on a discrete representation of the measured data as well as the unknown image to reconstruct. In this representation, it is possible to take into account some properties of the measured data (such as their statistical properties) and a priori knowledge about the way the data has been obtained (the scanner design, some aspects of the particle physics). They follow a general scheme, which consists in estimating a new image at a certain step by combining (1) the image estimated at the previous step and (2) the data generated from this estimate (Figure 12.17). The process is repeated until a given criterion is satisfied, such as a small enough difference between the projection data of the estimated image and the measured projections. The main components of an iterative algorithm are as follows:

1. A model of the physics of the acquisition process describing the deterministic aspects of the system (such as the impulse response of detectors, the statistical properties of data as the probability distribution)

2. A model of the a priori distribution of the radiotracer

3. An iterative algorithm to compute an estimate of the solution

4. A stopping criterion for that algorithm.

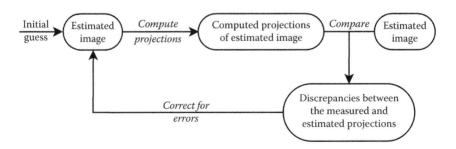

FIGURE 12.17 Iterative algorithm principles.

Image and projection data are discretized as a vector λ and a vector y, respectively. λ_j represents the activity of the radiotracers in Voxel j. y_i represents the number of events recorded in the LOR of Index i.

12.6.1 Algebraic Reconstruction Technique

Algebraic Reconstruction Technique (ART) is a popular iterative reconstruction method in tomography. It was first proposed in the 1970s [14]. The LOR-i is the path across a certain number of voxels j. The number of events recorded in LOR-i, y_i, is given by the decays occurring in each voxel of LOR-i. Due to the fact that the path in each voxel is unequal, a weight A_{ij} is given to the radioactivity for each voxel. Hence,

$$y_i = \sum_{voxel j} A_{ij} \cdot \lambda_j \tag{12.6}$$

The unknown solution and data are in relation with the linear system:

$$y = A \cdot \lambda \tag{12.7}$$

A is an operator that represents the tomographic process, also known as system matrix. It usually embeds various effects, such as the geometry of projection, the photon attenuation, and Compton scattering. λ is unknown and this is what we are trying to estimate. y is known as it is the measured data provided by the imaging system.

The solution of this linear system is given by the scheme presented below. λ^n is the estimated image at Step n. y^n is its projection computed using the system matrix (A). A new estimate λ^{n+1} is computed as follows:

$$\lambda^{n+1} = \lambda^n + (y_k - y_k^n) \cdot \frac{A_{kj}}{\sum_{voxel j} A_{kj}^2} \tag{12.8}$$

Each equation of the linear system is independently processed in a sequential. It introduces inconsistencies in the set of equations that results in salt and pepper noise. To address this, simultaneous iterative reconstruction technique (SIRT) and simultaneous algebraic reconstruction technique (SART) have been proposed [15,16]. They aim to correct errors simultaneously in all the equations of the linear system. Today, algebraic methods are depreciated. Statistical methods are now commonly used in modern PET scanners [17].

12.6.2 Statistical Method

Statistical methods, such as maximum-likelihood expectation-maximization (ML-EM), take into account the Poisson nature of the radioactive decay process [18]. In other words, the projection model deals with the probability for a decay occurring in Voxel j to be detected inside LOR-i. A_{ij}, the system matrix, will represent this probability.

The number of events y_i recorded in LOR-i is known by its expectation (or mean) and related to the unknown tracer distribution y_j by

$$\text{Expectation}\{y_i\} = \langle y \rangle = \sum_{voxel j} A_{ij} \cdot \lambda_j \tag{12.9}$$

It can be written as

$$\langle y \rangle = \langle A_i \cdot \lambda \rangle \tag{12.10}$$

Detected events are commonly described with a Poisson statistics:

$$p(y_i) = \frac{e^{-\langle A_j, \lambda \rangle} \cdot (A_j, \lambda) y_i}{y_i!} \tag{12.11}$$

with mean $\langle y \rangle$, and variance

$$\sigma^2 = \langle y \rangle = \sum_{voxel j} A_{ij} \cdot \lambda_j \tag{12.12}$$

The aim is to compute the tracer distribution λ that has the highest probability to have generated the projection data y. The probability function that is used is the likelihood function, which is derived from the Poisson statistics. This likelihood function $L(\lambda)$ is the conditional probability to measure the projection y knowing the image λ, and is expressed as

$$L(\lambda) = p(y_i \mid \lambda) = \pm \prod_i p(y_i \mid \lambda) \tag{12.13}$$

As $p(y_i|\lambda)$ is a Poisson process,

$$L(\lambda) = \prod_i \frac{e^{-A_i, \lambda} \cdot (A_i, \lambda)^{y_i}}{y_i!} \tag{12.14}$$

The most likely image λ will maximize $L(\lambda)$ so that all the $p(y_i|\lambda)$ are maximal as well. This leads to the reconstruction formula of Shepp and Vardi [18]:

$$\lambda_j^{n+1} = \frac{\lambda_j^n}{\sum_i A_{ij}} \cdot \sum_i \left(A_{ij} \cdot \frac{y_i}{\sum_i A_{ij} \cdot \lambda_j^n} \right) \qquad (12.15)$$

The reconstruction is performed iteratively using the EM algorithm. Each iteration is made of two successive steps:

1. *Expectation step*: it uses the current parameter estimates to reconstruct the Poisson process.

2. *Maximum likelihood step*: it updates the parameters of the estimates using the results of Step 1.

Figure 12.18 illustrates the ML-EM reconstruction for several iteration numbers.

The EM algorithm is notoriously slow to converge. Ordered subset-expectation maximization (OS-EM) speeds up the reconstruction process of the original ML-EM algorithm [19]. It has become the reference reconstruction method in PET. Its principle is to reduce the amount of projections used at each iteration of the EM algorithm by subdividing the projections in K subgroups. The projections of a subgroup are uniformly distributed around the volume to reconstruct.

One of the main problems of ML-EM and its derivatives, including OS-EM, is the difficulty to choose a good stopping criterion [20]. ML-EM

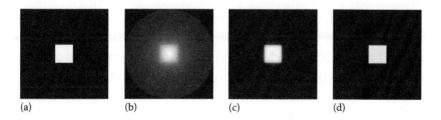

(a) (b) (c) (d)

FIGURE 12.18 ML-EM reconstructions of a square: (a) original image; (b) reconstruction with 1 iteration; (c) with 10 iterations; (d) with 40 iterations.

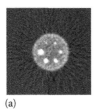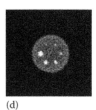

(a)   (b)   (c)   (d)

FIGURE 12.19   Reconstructions of a real acquisition of the Jasczak phantom: (a) FBP reconstruction; (b) MLEN reconstruction with 1 iteration; (c) with 16 iterations; (d) with 64 iterations.

initially converges toward an acceptable estimation of tracer distribution. Then, when the number of iterations increases the reconstruction becomes noisy. Figure 12.19 illustrates this fact. This is also the case of the OS-EM algorithm. In addition to this, OS-EM is not guaranteed to converge. These drawbacks may counterbalance the fact that it is a fast algorithm.

Other algorithms have been proposed to improve the OS-EM for convergence while keeping fastness such as Row action maximum likelihood algorithm (RAMLA) [21], or improved ML-EM for noisy data by introducing regularization during the reconstruction as in the maximum a posteriori (MAP) reconstruction.

12.7 CONCLUSION AND DISCUSSION

The use of PET in oncology is increasing exponentially, particularly for radiotherapy treatment planning and treatment response monitoring [22,23]. This chapter presents an overview of the fundamentals of PET imaging. The related physics has been reviewed, and the most popular reconstruction algorithms have been discussed. They can be classified thus: direct/analytical methods and iterative methods, of which the latter is known to be the most computationally intensive. The use of iterative methods based on a statistical point of view is now the preferred choice of volume reconstruction in PET imaging. As PET hardware improves (e.g., better spacial resolution, time-of-flight, etc.) and more computational power is available (e.g., using massively parallel processors such as graphics processing units [GPUs]), it is becoming possible to overcome computation time limitations and integrate more complex physically based corrections in the clinical routine.

REFERENCES

1. Vidal, F. P., F. Bello, K. W. Brodlie, D. A. Gould, N. W. John, R. Phillips, and N. J. Avis, Principles and applications of computer graphics in medicine. *Computer Graphics Forum*, March 2006. 25(1): 113–137.
2. Badawi, R. D., P. K. Marsden, B. F. Cronin, J. L. Sutcliffe, and M. N. Maisey, Optimization of noise-equivalent count rates in 3D PET. *Physics in Medicine and Biology*, 1996. 41(9): 1755.
3. Kadrmas, D. J., M. E. Casey, M. Conti, B. W. Jakoby, C. Lois, and D. W. Townsend, Impact of time-of-flight on PET tumor detection. *Journal of Nuclear Medicine*, 2009. 50(8): 1315–1323.
4. Lois, C., B. W. Jakoby, M. J. Long, K. F. Hubner, D. W. Barker, M. E. Casey, M. Conti, V. Y. Panin, D. J. Kadrmas, and D. W. Townsend, An assessment of the impact of incorporating time-of-flight information into clinical PET/CT imaging. *Journal of Nuclear Medicine*, 2010. 51(2): 237–245.
5. Surti, S., J. Scheuermann, G. El Fakhri, M. E. Daube-Witherspoon, R. Lim, N. Abi-Hatem, E. Moussallem, F. Benard, D. Mankoff, and J. S. Karp, Impact of time-of-flight PET on whole-body oncologic studies: A human observer lesion detection and localization study. *Journal of Nuclear Medicine*, 2011. 52(5): 712–719.
6. Fahey, F. H., Data acquisition in PET imaging. *Journal of Nuclear Medicine Technology*, 2002. 30(2): 39–49.
7. Daube-Witherspoon, M. E. and G. Muehllehner, Treatment of axial data in three-dimensional PET. *Journal of Nuclear Medicine*, 1987. 28(11): 1717–1724.
8. Lewitt, R. M., G. Muehllehner, and J. S. Karp, Three-dimensional image reconstruction for PET by multi-slice rebinning and axial image filtering. *Physics in Medicine and Biology*, 1994. 39(3): 321.
9. Defrise, M., P. E. Kinahan, D. W. Townsend, C. Michel, M. Sibomana, and D. F. Newport, Exact and approximate rebinning algorithms for 3-D PET data. *IEEE Transactions on Medical Imaging*, April 1997. 16(2): 145–158.
10. Edholm, P. R., R. M. Lewitt, and B. Lindholm, Novel properties of the Fourier decomposition of the sinogram. *Proceedings of SPIE*, 1986. 0671: 8–18.
11. Kak, A. C. and M. Slaney, *Principles of Computerized Tomographic Imaging*. Society of Industrial and Applied Mathematics, Philadelphia, PA. 2001. http://bookstore.siam.org/cl33/.
12. Colsher, J. G., Fully-three-dimensional positron emission tomography. *Physics in Medicine and Biology*, 1980. 25(1): 103.
13. Kinahan, P. E. and J. G. Rogers, Analytic 3D image reconstruction using all detected events. *IEEE Transactions on Nuclear Science*, February 1989. 36(1): 964–968.
14. Gordon, R., R. Bender, and G. T. Herman, Algebraic Reconstruction Techniques (ART) for three-dimensional electron microscopy and X-ray photography. *Journal of Theoretical Biology*, 1970. 29(3): 471–481.
15. Gilbert, P., Iterative methods for the three-dimensional reconstruction of an object from projections. *Journal of Theoretical Biology*, 1972. 36(1): 105–117.

16. Andersen, A. H. and A. C. Kak, Simultaneous Algebraic Reconstruction Technique (SART): A superior implementation of the ART algorithm. *Ultrasonic Imaging*, 1984. 6(1): 81–94.

17. Qi, J. and R. M. Leahy, Iterative reconstruction techniques in emission computed tomography. *Physics in Medicine and Biology*, 2006. 51(15): R541.

18. Shepp, L. A. and Y. Vardi, Maximum likelihood reconstruction for emission tomography. *IEEE Transactions on Medical Imaging*, 1982. 1(2): 113–122.

19. Hudson, H. M. and R. S. Larkin, Accelerated image reconstruction using ordered subsets of projection data. *IEEE Transactions on Medical Imaging*, 1994. 13(4): 601–609.

20. Bissantz, N., B. A. Mair, and A. Munk, A statistical stopping rule for MLEM reconstructions in PET, in *Nuclear Science Symposium Conference Record, 2008 (NSS'08)*, Dresden, Germany. IEEE, 2008, pp. 4198–4200. http://dx.doi.org/10.1109/NSSMIC.2008.4774207.

21. Browne, J. and A. R. De Pierro, A row-action alternative to the EM algorithm for maximizing likelihood in emission tomography. *IEEE Transactions on Medical Imaging*, October 1996. 15(5): 687–699.

22. Gregoire, V. and A. Chiti, PET in radiotherapy planning: Particularly exquisite test or pending and experimental tool? *Radiotherapy and Oncology*, 2010. 96(3): 275–276.

23. Freeman, L. M. and M. D. Blaufox, Letter from the editors: PET/CT in radiation oncology. *Seminars in Nuclear Medicine*, 2012. 42(5): 281–282.

CHAPTER 13

Implementation of Convolution Superposition Methods on a GPU

Todd R. McNutt and Robert A. Jacques

CONTENTS

13.1 CONCEPTUAL CONVOLUTION SUPERPOSITION MODEL

13.1.1 Overview

The convolution superposition algorithm was originally designed for the computation of radiation dose delivered to patients undergoing external beam radiation therapy. The algorithm was originated by the works of Mackie, Battista, and Ahnesjo in the late 1980s [1–6]. Further refinements and extensions were researched by Papanikolaou, McNutt, Liu, and others in the mid-1990s [7–16]. The algorithm became the primary dose computation method in several three-dimensional (3D) and intensity-modulated radiotherapy (IMRT) treatment planning systems, including Pinnacle[3], Helax/Oncentra, XiO, and Tomotherapy [18]. Some of these systems are also migrating their implementations to the graphics processing unit (GPU). The work described in this chapter represents more recent implementations of the algorithm specifically designed for the GPU, which further enhances the performance but, more importantly, significantly improves the accuracy by eliminating approximations made in the earlier CPU implementations.

13.1.2 Basic Description of the Method

The basic convolution superposition algorithm is depicted in Figure 13.1. It starts with a representation of the radiation exiting the linear accelerator that will be incident on the patient (incident fluence). This incident fluence is transported through a representation of the patient (typically defined by CT) to determine the Total Energy Released per unit MAss (TERMA) at every location in the patient. This TERMA volume represents where the primary radiation has interacted in the patient, but not where the energy imparted has been absorbed in the patient. To find the absorbed dose in the tissue, the TERMA volume is convolved or superposed with a dose deposition kernel. Dose deposition kernels represent the spread of absorbed dose generated by radiation interacting at a single point in the patient.

For both the TERMA computation and the superposition, ray tracing techniques are used to step through the voxel volumes to determine density and attenuation information from the patient's CT data.

13.1.3 Approximations Made in CPU-Based Implementations

In the central processing unit (CPU)-based algorithms, ray tracing was the dominant computation affecting the performance. Approximations were made to minimize the required computations for performance.

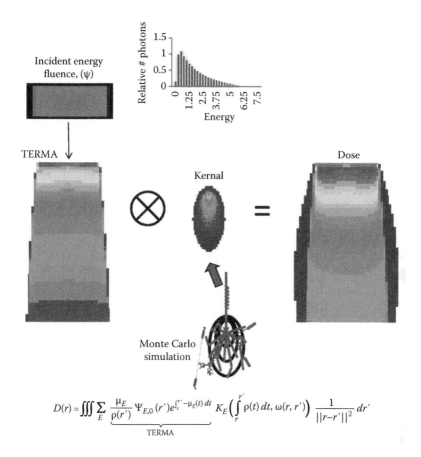

$$D(r) = \iiint \sum_E \underbrace{\frac{\mu_E}{\rho(r')} \Psi_{E,0}(r') e^{\int_{s}^{r'} -\mu_E(t)\,dt}}_{\text{TERMA}} K_E\left(\int_r^{r'} \rho(t)\,dt, \omega(r, r')\right) \frac{1}{||r-r'||^2}\,dr'$$

FIGURE 13.1 The basic convolution/superposition algorithm process: An incident fluence with a known energy spectrum is projected in the beam's direction through a density representation to generate a volume representing the total energy interacting (TERMA) in the patient. A superposition of this TERMA with the Monte Carlo-generated dose deposition kernel determines where the energy is ultimately absorbed in the patient representing the delivered dose from the beam.

During the TERMA computation, ideally, the attenuation and TERMA of each energy bin of the beam's spectrum is computed across a step along the ray from where it enters each density voxel to where it exits as depicted in Figure 13.2 for the exact radiological path. This calculation requires multiple exponentials to be evaluated at each step due to both the number of individual energies and the variable step size across the voxels for random ray directions through the voxel geometry. To minimize the computations, one can precompute the polyenergetic attenuation for a fixed step

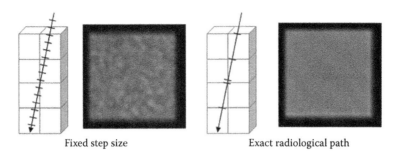

Fixed step size   Exact radiological path

FIGURE 13.2 Diagrams of fixed step size and exact radiological path ray tracing. To the right of each diagram is the first slice of the TERMA grid computed with each approach showing the mottle introduced to unequal numbers of fixed steps for each voxel. High performance exponential calculations on the GPU enable efficient exact radiological path ray tracing.

size for the range of densities. Then for each fixed step length along a ray, the computation is reduced to a single multiplication to get the TERMA and to attenuate the energy fluence. The downside is that it introduces inaccuracies in the hardening of the beam through heterogeneous densities, and introduces sampling mottle using fixed step lengths along rays that are not perfectly aligned with voxel boundaries.

During the superposition calculation rays emanate from each voxel of the dose grid. This offers the opportunity to share ray tracing computations by simply incrementing the voxel indices of a ray trace to create the parallel ray trace for all voxels. This approximation significantly reduced the required computation time; however, it assumes the dose deposition kernel's orientation also remains parallel across the volume, which is not true in a divergent radiation beam. Modifications to the algorithm to account for kernel tilting remain as approximations and add back computational complexity.

In addition to kernel tilting approximations, the kernel also hardens with depth. Most CPU-based algorithms do not account for kernel hardening and assume a constant kernel across all depths.

GPU architecture allows for the ray tracing [19] to be done in parallel, thus significantly reducing the computational burden on performance. These implementations provide the opportunity to remove the approximations present in traditional CPU implementations and thus improve the accuracy of the convolution/superposition algorithm. This chapter primarily focuses on the GPU-based method developed by Jacques et al. [20–23].

13.2 INCIDENT FLUENCE DEFINITION

All dose algorithms start with a representation of the radiation exiting the linear accelerator that will be incident on the patient as shown in Figure 13.3. The incident fluence is represented by a 2D pixelized spatial distribution where the energy spectrum and relative energy fluence are known at each pixel. For a single incident fluence distribution, it is assumed that all particles in the distribution come from a single source point, which defines their direction. To accommodate secondary or scatter sources, a second or "dual" source model is used with a second incident fluence using a different source position to define the particle directions of the secondary source.

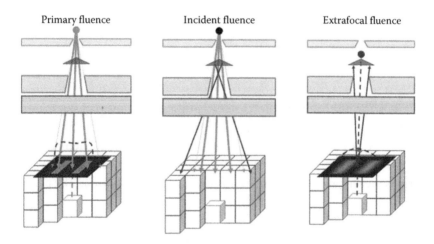

Primary fluence   Incident fluence   Extrafocal fluence

FIGURE 13.3 **(See color insert.)** Diagram of the separation of incident fluence into a primary fluence and extrafocal fluence. The incident photon fluence (center) includes photons (blue) from the primary source (black) and photons induced (purple) by interactions with parts of the linear accelerated head, such as the flattening filter (green). Fluence intensities are modulated by the primary (yellow) and upper (orange) and lower (red) jaws. The incident photon fluence is modeled by using a point-like "primary" fluence model (left) and an area-like "extrafocal" fluence model (right). Both models project the fluence computed at a reference plane through the patient volume (dashed lines) in order to avoid computing the fluence to each voxel (blue and yellow cubes). An example of the fluence for a highly modulated MLC field and a diagram of the open field intensity profile (dashed red line) have been included for each model. The purple arrows in the extrafocal fluence diagram (right) represent the limiting of the visible source area by the upper collimating jaws (orange).

The second component of the incident fluence is the energy spectrum. Most models use a primary central axis energy spectrum with a mathematical model of how the energy spectrum changes with off-axis radius for modeling the off-axis softening of the beam. For dual source models, the extrafocal scattered radiation has a separate energy spectrum representing the energy distribution of the extrafocal radiation.

Ultimately, the algorithm needs to know what the energy fluence and the associated energy spectrum at each pixel of the incident fluence plane for a given treatment beam for each source (primary or extrafocal) used in the computation. Modeling of the incident fluence for a particular linear accelerator and collimation system can be done in several ways, from deriving it from Monte Carlo simulations to modeling it with empirical methods based on the geometry of the machine head. These models consider the geometries of the multileaf collimation system, beam modifiers, focal spot blurring, and spectral variations across the field.

13.3 TOTAL ENERGY RELEASED PER UNIT MASS

TERMA represents the amount of energy removed from the primary radiation beam through photon interactions per unit mass of the material irradiated.

TERMA is the product of the mass attenuation coefficient and the energy fluence at a given location in the patient summed over all photon energies present at that location in the patient. For any point in the patient, the TERMA can be calculated knowing the source position, the material between the source and the TERMA point, and the incident energy fluence and spectrum of photons along that direction. During the process, it is important to accurately handle beam hardening through heterogeneous medium and avoid digital sampling artifacts.

In traditional methods, given an incident fluence of photons and a beam direction, TERMA is calculated by tracing rays from the source position through the patient volume for each incident fluence pixel, or "forward projected" as shown in Figure 13.4. Once the ray enters the patient, the TERMA is calculated, and the energy fluence contained in the ray is attenuated for the next step. This process is repeated for each step along the ray until it exits the patient volume.

This forward projection model is not ideal for implementation on a GPU. The main problem is that multiple rays write to the same voxel in the TERMA volume which causes race conditions in parallel processing. For the GPU implementation, it is best to break up the algorithm to have

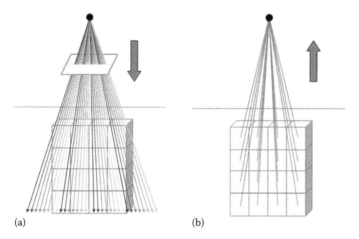

(a)   (b)

FIGURE 13.4   Diagrams illustrating the difference between the (a) forward projected and (b) back projected TERMA algorithms. The back projected model is well suited for GPU implementation as each ray can be a thread that is solely responsible for writing to one voxel of the TERMA grid.

memory writes that are independent between multiple processes to avoid the race condition. A more appropriate model is to use back projection where each process will gather the influence of the ray trace on a single voxel where TERMA is to be computed. This back projected GPU model improves on the TERMA calculation by treating each voxel in the patient volume independently. Each TERMA voxel becomes a thread on the GPU and calculates the ray tracing back to the source. Then for each step along the ray, each energy bin of the spectrum is attenuated independently resulting in an accurate determination of the hardened fluence and hence TERMA at the particular voxel in the patient. In this process, it is also possible to store the incident fluence to TERMA conversion factor which is handy for repeat calculations of the same beam with varying incident fluence such as during IMRT optimization.

A peripheral benefit to the GPU is the hardware accelerated exponential calculation that can be used during the ray tracing with little added computational cost. This operation allows for the variable exact radiological path steps to be taken during the ray trace. Figure 13.5 shows how this model improves the modeling of beam hardening through the patient. With some CPU implementations, precomputed fluence attenuation tables are used that make the assumption that all of the tissue above the point of TERMA has the same density as the voxel that is being computed,

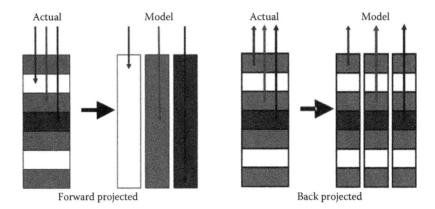

FIGURE 13.5 CPU implementations approximated spectral hardening because they continually attenuate a running energy fluence along the ray trace. The vertical arrows reflect the actual position of the voxel, and where the model assumes it is in the calculation where darker material is lower density. When each TERMA voxel has its own thread on the GPU, exact beam hardening can be included more efficiently by attenuating each energy bin independently along the exact radiological path ray trace for each voxel in the TERMA grid.

so hardening is approximated using a density scaled depth of that material. And the approximation is carried along the forward projected ray, as it carries the running energy fluence as the ray is traversed. In the back projected GPU method, each voxel has its own ray trace and each energy bin of the spectrum can be attenuated along the ray producing accurate hardening of the beam through the material traversed.

The end result of the TERMA computation is a volume encompassing the patient that contains the TERMA at each voxel inside the patient. This is the energy released from the beam and not where the energy is absorbed in tissue (dose).

13.4 DOSE DEPOSITION KERNELS

Dose deposition kernels represent where energy is absorbed in water from energy being released at a point. They are calculated with Monte Carlo calculations by forcing interactions at the center of a sphere for a monoenergetic photons entering from a single direction. The Monte Carlo simulation tallies the energy absorption in a spherical phantom from the particle interactions. For a spectrum of energies, the kernels can be summed over energy. Kernels may also be represented by analytical formulas.

Just as the TERMA calculation should handle beam hardening, so should the kernel. The effective energy of the primary beam changes with depth, and thus, the actual kernel should also change. Most CPU methods inherently assume an invariant kernel with both lateral position and depth. In reality, the kernel should tilt with divergence of the beam and should harden with depth. Furthermore, it gets much more complex in the presence of tissue heterogeneities.

To accommodate heterogeneities, the kernels are stretched in proportion to density in the direction from the interaction point to the dose deposition point as an approximation to the how the radiation may be effected by the heterogeneity.

13.5 SUPERPOSITION

The superposition process is the process where we determine the energy absorbed or dose in the patient from the TERMA. We could place a kernel centered at each TERMA voxel and spread the energy to the surrounding voxels. Then the method would have to go to every voxel in the TERMA grid and spread the absorbed energy before it determined the dose to a single point. This model is inefficient and poorly suited for GPU implementation. If we considered the spreading of each TERMA voxel as a thread on the GPU, then each thread would be writing to memory that another thread may also write to, creating a race condition where one thread may overwrite memory that another thread had written.

To avoid the race conditions, a better model is to gather the absorbed energy to each voxel of the dose grid. This is achieved by inverting the kernel and having each thread write only to the dose voxel that it is responsible for computing. The algorithm then disperses rays from the dose deposition point and determines the contribution of dose to the voxel from the TERMA along the casted rays as shown in Figure 13.6.

Due to the primary beam diverging from a single source, the orientation of the kernel should change at positions off of the central axis of the beam. This alignment of the kernel with the divergent beam position is referred to as kernel tilting. For performance, CPU methods share ray tracing by shifting the ray trace by voxel indices for each dose voxel. This shared ray model is not able to tilt the kernels without significant performance loss. Kernel tilting, however, is easily handled on the GPU because each thread is responsible for gathering the energy and contains its own set of rays, which can be tilted to accommodate divergence in the beam without any cost to the performance of the algorithm.

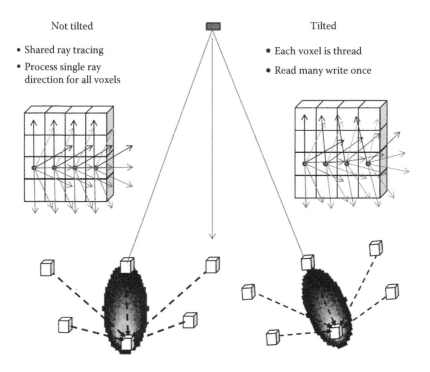

FIGURE 13.6 Nontilted kernels can share ray tracing by simply shifting the x, y, z indexes into the volume greatly reducing the computations for CPU implementations. Kernel tilting on the GPU is of little consequence as each voxel has its own thread and it does all the ray tracing for the dose accumulation independently from the other threads.

The primary beam spectrum also hardens with depth, and so should the polyenergetic kernel. The GPU-based implementation model by Jacques et al. approximated this by breaking up the spectrum into two or four subspectra and summing the full dose computation for each subspectra to obtain the total dose from the entire spectrum. Effectively this approximates kernel hardening as the kernel for each individual subspectra has relatively less hardening with depth. The sum of the dose computations thus has a larger contribution from the higher energy kernel at depth, as the higher energies attenuate less during the TERMA computation.

Lastly, Jacques et al. showed that modifying the density scaling model of kernel rays to include an empirical asymmetric filter of the scaling density along the superposition ray improved the accuracy of the algorithm in the presence of heterogeneities [24]. Figure 13.7 shows the results of this method in comparison to Monte Carlo benchmarks. The figure shows

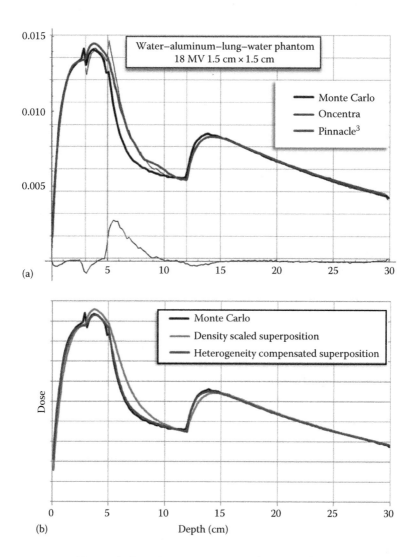

FIGURE 13.7 (a) Depth dose for two commercial C/S methods vs Monte Carlo for the ICCR benchmark showing the error in dose near heterogeneities for small fields. (b) GPU accelerated superposition with normal density scaling and heterogeneity compensated scaling showing accuracy improvements very close to the Monte Carlo results for a 1.5 cm × 1.5 cm 18 MV beam on a phantom of water–aluminum–lung–water slabs from left to right in the plots.

both the effects of the kernel hardening and density scaling improvements. It demonstrates that the convolution/superposition can rival the accuracy of Monte Carlo for normal human density variations when the CPU-based approximations for performance are removed in the GPU implementations of the method.

13.6 SOURCE MODELING

The final modeling and commissioning a convolution/superposition algorithm requires adjusting the set of source modeling parameters to accurately match a set of measured data acquired in a scanning water tank characterizing the beam. Figure 13.8 displays the agreement between measured and computed depth dose and cross beam profiles for a 6 MV beam from an Elekta accelerator. In this case, the model is a dual source incident fluence with an energy spectrum subdivided in two parts for kernel hardening. The incident fluence model includes the collimation geometry including a primary collimator and a set of jaws and multileaf collimator. For the primary source, it has an energy spectrum, an arbitrary radial function representing the change in fluence induced by the flattening filter and model for off-axis softening, and Gaussian-shaped focal spot blurring. The secondary source has a separate spectrum, a Gaussian-shaped source distribution located near the flattening filter and virtual source position closer to the patient than the primary source to represent broader scatter divergence. These parameters are adjusted to best fit the measured data as shown in Figure 13.8.

13.7 ALGORITHM PERFORMANCE

Determining the exact performance gains with GPU-based algorithms over CPU algorithms is not practical but publications have reported gains as much as 100 times faster. More important is perhaps the gains in accuracy. For a 64^3 voxel dose grid, computation times for the GPU ranged from 150 to 400 ms when using between 1 and 4 spectral groupings for kernel hardening with 82 total kernel rays. For 128^3, the times increased to between 1.8 and 5.6 s. In comparison to the Pinnacle system for similar beams, this represented between 60 and 17 times faster when Pinnacle was run on an AMD Opteron 254 (2 cores, 2.8 GHz). Due to the tilted kernel, it is also possible to use fewer superposition rays in the GPU implementation without loss of accuracy in comparison with the nontilted kernel algorithm used by Pinnacle which further improves the performance gains to more than 100 times.

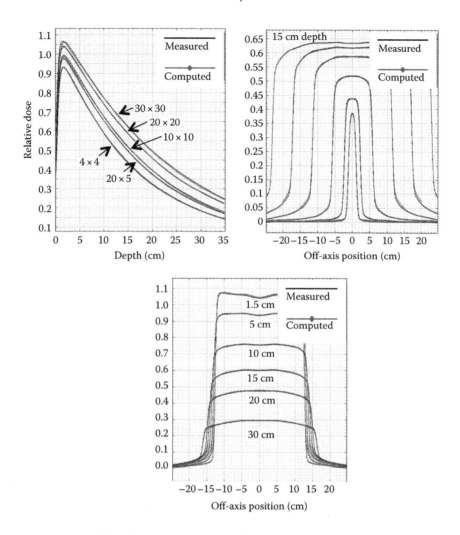

FIGURE 13.8 Plots showing agreement of the GPU superposition algorithm for measured data for an Elekta Infinity 6 MV beam. Top: depth doses and cross beam profiles (15 cm depth) for a range of field sizes; Bottom: cross beam profiles at a range of depths for a 25 cm field.

13.8 FURTHER DISCUSSION

Traditionally the convolution/superposition algorithm has not been used when several iterations of dose computation are required such as during intensity-modulated treatment optimization due to performance reasons. The GPU model offers some advantages. First of all, the raw performance may enable it to be used more frequently during hybrid dose computation-based IMRT optimization strategies. Additionally, the back projected TERMA model allows for the prestorage of TERMA to incident fluence factors that can enable instantaneous recomputation of TERMA from a change in the intensity of an incident fluence pixel. Further exploration into the GPU implementation may allow for improved dose computation accuracy during IMRT and modulated arc treatment planning.

13.9 OTHER GPU IMPLEMENTATIONS

This chapter focuses on a single implementation of the convolution/superposition algorithm to highlight the advantages that the GPU brings. The described algorithm was developed by going back to the principles of the algorithm and redesigning it for the GPU architecture. There are a few other implementations in the literature by Hissoiny et al. [25], Lu et al. [26], Zhou et al. [27], and Chen et al. [28], and it is important to understand the detailed implementations of them to appreciate both their limitations and advantages in terms of accuracy and performance.

REFERENCES

1. A. Ahnesjo, Andreo P., and Brahme A., Calculation and application of point spread functions, *Acta. Oncol.* 26, 49–56 (1987).
2. T. R. Mackie, Scrimger J. W., and Battista J. J., A convolution method of calculating dose for15-MV x-rays, *Med. Phys.* 12, 188–196 (1985).
3. T. R. Mackie, Ahnesjo A., Dickof P., and Snider A., Development of a convolution/superposition method for photon beams, *The Use of Computers in Radiation Therapy*, pp. 107–110 (1987).
4. T. R. Mackie, Bielajew A. F., Rogers D. W. O., and Battista J. J., Generation of photon energy deposition kernels using the EGS Monte Carlo code, *Phys. Med. Biol.* 33, 1–20 (1988).
5. A. Ahnesjo, Collapsed cone convolution of radiant energy for photon dose calculation in heterogeneous media, *Med. Phys.* 16, 577–592 (1989).
6. R. Mohan, Chui C., and Lidofsky L., Energy and angular distributions of photons from medical linear accelerators, *Med. Phys.* 12, 592–597 (1985).
7. N. Papanikolaou, Mackie T. R., Meger-Wells C., Gehring M., and Reckwerdt P., Investigation of the convolution method for polyenergetic spectra, *Med. Phys.* 20, 1327–1336 (1993).
8. T. R. McNutt, Mackie T. R., Reckwerdt P. J., Papanikolaou N., and Paliwal B. R., Calculation of portal dose using the convolution/superposition method, *Med. Phys.* 23–24 (1996).
9. T. R. McNutt, Mackie T. R., Reckwerdt P., and Paliwal B. R., Modeling dose distributions from portal dose images using the convolution/superposition method, *Med. Phys.* 23–28 (1996).

10. H. H. Liu, Mackie T. R., and McCullough E. C., Correcting kernel tilting and hardening in convolution/superposition dose calculations for clinical divergent and polychromatic photon beams, *Med. Phys.* 24, 1729–1741 (1997).

11. H. H. Liu, Mackie T. R., and McCullough E. C., A dual source photon beam model used in convolution/superposition dose calculations for clinical megavoltage x-ray beams, *Med. Phys.* 24, 1960–1974 (1997).

12. H. H. Liu, Mackie T. R., and McCullough E. C., Calculating dose and output factors for wedged photon radiotherapy fields using a convolution/superposition method, *Med. Phys.* 24, 1714–1728 (1997).

13. T. R. McNutt, Mackie T. R., and Paliwal B. R., Analysis and convergence of the iterative convolution/superposition dose reconstruction technique for multiple treatment beams and tomotherapy, *Med. Phys.* 24–29, 1465–1476 (1997).

14. M. Miften, Wiesmeyer M., Monthofer S., and Krippner K., Implementation of FFT convolution and multigrid superposition models in the FOCUS RTP system, *Phys. Med. Biol.* 45, 817–833 (2000).

15. T. R. Mackie, Reckwerdt P. J., McNutt T. R., Gehring M., and Sanders C., Photon dose computations, *Teletherapy: Proceedings of the 1996 AAPM Summer School* (1996).

16. A. Ahnesjo and Aspradakis M., Dose calculations for external photon beams in radiotherapy, *Phys. Med. Biol.* 44, R99–R155 (1999).

17. D. W. O. Rogers and Mohan R., Questions for comparison of clinical Monte Carlo codes, *XIII International Conference on the Use of Computers in Radiation Therapy* (ICCR) (2000).

18. T. R. McNutt, Dose calculations: Collapsed cone convolution superposition and delta pixel beam, Philips white paper.

19. M. de Greef, Crezee J., van Eijk J. C., Pool R., and Bel A., Accelerated ray tracing for radiotherapy dose calculations on a GPU, *Med. Phys.* 36, 4095–4012 (2009).

20. T. McNutt and Jacques R., Real-time dose computation for radiation therapy using graphics processing unit acceleration of the convolution/superposition dose computation method—Johns Hopkins University, US Patent, Pub. US 2011/0051893 A1.

21. R. Jacques, Taylor R., Wong J., and McNutt T., Towards real-time radiation therapy: GPU accelerated superposition/convolution, *Comput. Methods Programs Biomed.* (2009). doi:10.1016/j.cmpb.2009.07.004.

22. R. Jacques, Smith D., Tryggestad E., Wong J., Taylor R., and McNutt T., GPU accelerated real time KV/MV dose computation, *Proceedings of the XVIth ICCR* (2010).

23. R. A. Jacques, Wong J., and McNutt T., Real-time dose computation: GPU-accelerated source modeling and superposition/convolution, *Med. Phys.* 38, 294–305 (2011).

24. R. Jacques and McNutt T., An improved method of heterogeneity compensation for the convolution/superposition algorithm, *International Conference on Computer in Radiotherapy*, Melbourne, Victoria, Australia (May 6–9, 2013).

25. S. Hissoiny, Ozell B., and Després P., Fast convolution-superposition dose calculation on graphics hardware, *Med. Phys.* 36, 1998–2005 (2009).

26. W. Lu, Olivera G. H., Chen M., Reckwerdt P. J., and Mackie T. R., Accurate convolution/superposition for multi-resolution dose calculation using cumulative tabulated kernels, *Phys. Med. Biol.* 50, 655–680 (2005).

27. B. Zhou, Yu C. X., Chen D. Z., and Hu X. S., GPU-accelerated Monte Carlo convolution/superposition implementation for dose calculation, *Med. Phys.* 37, 5593–5603 (2010).

28. Q. Chen, Chen M., and Lu W., Ultrafast convolution/superposition using tabulated and exponential kernels on GPU, *Med. Phys.* 38, 1150–1161 (2011).

Photon and Proton Pencil Beam Dose Calculation

Xuejun Gu

CONTENTS

14.1 INTRODUCTION

Pᴇɴᴄɪʟ ʙᴇᴀᴍ-ʙᴀsᴇᴅ ᴅᴏsᴇ ᴄᴀʟᴄᴜʟᴀᴛɪᴏɴ has been widely adopted in modern treatment planning system, because of its simple calculation scheme and acceptable accuracy in most clinical application. This simple and accurate dose calculation model will have extensive applications in online adaptive radiotherapy (ART), where the attractive concept of real-time adaptation of the treatment to daily anatomical variation [1–16] requires extremely fast and accurate dose calculation engines.

Graphics processing unit (GPU), a massive parallel computing architecture, has been widely used in radiotherapy community to accelerate computationally intensive tasks [11,13–15,17–22]. Especially, much effort has been devoted to utilize GPU to speed up dose calculation algorithms, including Monte Carlo (MC) simulation and superposition/convolution (S/C) [23–26]. GPU follows a single instruction multiple data (SIMD) design in its execution scheme. Such design allows achieving high computation efficiency when all data are processed together with the same execution path. Pencil beam-based dose calculation is essentially a convolution method, which can be mathematically formulated as GPU favorable matrix-vector operations [27].

In this chapter, we cover both photon and proton dose calculation with the pencil beam model. We also describe the physics model of pencil beam and detail the GPU implementation of dose calculation algorithm. The performance of the GPU implementation is also evaluated with clinical cases.

14.2 PHYSICS MODEL OF FINITE-SIZE PENCIL BEAM

Pencil beam methods originally stem from the convolution approach. Gustafsson et al. [28] expressed the radiation dose calculation with a very general pencil beam formulation as

$$D(\mathbf{r}) = \int_s \int_E \int_\Omega \int \int \sum_m \Psi^m_{E,\Omega}(\mathbf{s}) \frac{P^m}{\rho}(E,\Omega,\mathbf{s},\mathbf{r}) \mathrm{d}^2\Omega \mathrm{d}E \mathrm{d}^2\mathbf{s} \qquad (14.1)$$

where

$\Psi^m_{E,\Omega}(\mathbf{s})$ is the energy fluence differential in energy E and direction Ω for beam modality m

$(P^m/\rho)(E, \Omega, \mathbf{s}, \mathbf{r})$ is the corresponding pencil kernel for energy deposition per unit mass at $\mathbf{r}$ due to primary particles entering the patient at $\mathbf{s}$

The pencil kernel involved in Equation 14.1 is not practical, since it varies with primary particle interactive with patient surface $\mathbf{s}$ and also the deposition position $\mathbf{r}$. A simplified pencil beam model can be expressed in a Cartesian coordinate system as

$$D(x,y,z) = \int\int \Psi(x',y')K(x-x',y-y',z)\,dx'dy' \qquad (14.2)$$

where
  $\Psi(x, y)$ is the fluence map
  K is the pencil beam kernel

The pencil kernel is usually derived from results of Monte Carlo calculation either directly as monoenergetic pencil kernels or as a superposition of point kernels.

With the development of multileaf collimator, more and more intensity-modulated treatment (IMRT) are adopted in radiation therapy. Generating an IMRT plan invokes twice dose calculation, one before plan optimization and one final dose calculation after plan optimization. The one before plan optimization requires calculating dose deposition coefficients which determine the dose contribution to a dose point from a pixel of fluence map, namely beamlet. Assuming that one radiation beam can be geometrically divided into identical finitely sized unit fluence elements, Bourland and Chaney [29] initially developed an finite-size pencil beam (FSPB) concept to exploit the particular geometry of the multileaf collimator used in IMRT field shaping. Basic assumptions in an FSPB model include the following: (1) the broad beam from a point source can be divided into identical beamlets and (2) the dose to a point is the integration of the contribution dose from all beamlets to that point. The accuracy of the FSPB model has been improved in past decades and is sufficient to be used in most clinical situation [30–32].

14.2.1 Basic Formulation

Mathematically, the dose deposited at a point $P(x, y, z)$ can be written as the summation of the dose contributed by all beamlets:

$$D(x,y,z) = \sum_i D_i^p(x,y,z)f_i, \qquad (14.3)$$

where
  f_i denotes the photon fluence (or beamlet intensity) for the beamlet i
  $D_i^p(x,y,z)$ is the dose distribution of the ith beamlet to position $P(x,y,z)$, which is equivalent to beamlet kernel K in Equation 14.2

Jelen et al. [32] developed a beamlet dose distribution model with the following formula:

$$D_i^p(x,y,z) = A(t,\theta) \cdot \left(\frac{SAD}{d}\right)^2 \cdot F(x',y',\omega(d),u_x(d),u_y(d),x_0,y_0). \quad (14.4)$$

where
> θ is defined as the angle between the central axis of a beamlet and the central axis of a broad beam
> d is the radiological depth

The patient coordinate system and pencil beam coordinate system used in this work are defined in Figure 14.1. For the beamlet central axis $\overline{SAB}$, d is the length of AO' and t is the corresponding radiological depth. $A(t, \theta)$ is a factor that accounts for off-axis effects. $F(*)$ is a beamlet profile at

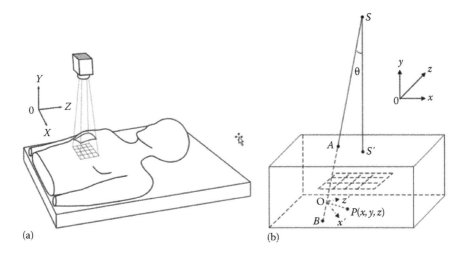

(a)   (b)

FIGURE 14.1 (a) An illustration of the FSPB model and the patient coordinate system. (b) An illustration of the pencil beam coordinate system for calculating the dose from a beamlet $\overline{SAB}$ to point P. We define the smallest rectangular prism that encases the patient CT images. θ is the angle between the central axes of a broad beam (SS') and a beamlet $\overline{SAB}$. A is the entrance point and B is the exit point of the beamlet central axis passing through the rectangular prism. O' is the perpendicular projection of point P onto $\overline{SAB}$. x' and z' are projections of x and z on the plane perpendicular to $\overline{SAB}$. (Reproduced from Gu, X.J. et al., *Phys. Med. Biol.*, 54(20), 6287, 2009.)

depth d without considering off-axis effects. x' and y' are the coordinates in the beamlet system, defined on the plane perpendicular to the beamlet central axis. x_0 and y_0, defined at the isocenter plane, represent the beamlet size. SAD is the source to beam isocenter distance. θ is the angle between the beamlet and its corresponding beam central axis. ω's denote weighting factors and u's are the steepness parameters of the beam's penumbra. Jelen et al. [32] model the function F as the summation of a primary dose term and a secondary dose (scattering components) term:

$$F(x,y,\omega,u_x,u_y,x_0,y_0) = \sum_{i=1}^{2} \omega_i p(x,u_{ix},x_0)p(y,u_{iy},y_0). \qquad (14.5)$$

Each term is a product of two independent exponential functions, specifically $p(x, u_{ix}, x_0)$ is defined as

$$p(x,u_{ix},x_0) = \begin{cases} \sin h(u_{ix}\,x_0)\exp(u_{ix}x) & \text{for } x < -x_0 \\ 1-\cos h(u_{ix}x)\exp(-u_{ix}\,x_0) & \text{for } -x_0 \le x \le x_0. \\ \sin h(u_{ix}\,x_0)\exp(-u_{ix}x) & \text{for } x > x_0 \end{cases} \qquad (14.6)$$

The term for $p(y, u_{iy}, y_0)$ is similarly defined. For this model, the beam dose distribution is depended on the parameters ω, u_x, u_y, which are obtained through fitting a broad beam dose distribution in homogenous water media.

14.2.2 Density-Corrected FSPB

The accuracy of the aforementioned FSPB model was improved by Jelen and Alber [33] with a three-dimensional (3D) density correction approach. The improved FSPB model, termed as the *DC-FSPB* model in this chapter, provides both lateral and longitudinal density corrections:

$$D_i^p(x,y,z) = A(t,\theta)\cdot\left(\frac{SAD}{d}\right)^2 \cdot F(x',y',\omega(d,\rho),u_x(d,\rho),u_y(d,\rho),x_0,y_0).$$

$$(14.7)$$

Notice that $F(*)$ term in Equation 14.7 is a function of geometry depth and local media density. By adjusting the parameters in Equation 14.7, we are able to shape the beamlet dose distribution in 3D. Along the beamlet direction, $A(t, \theta)$ is a function of radiological depth and off-axis angle, taking care of heterogeneity correction along beamlet depth direction as well as the horn effect at various off-axis distances. Perpendicular to the beamlet direction, the beam's penumbra steepness is tuned according to local density $\rho : u_1(\rho, t) = f_{u_1}(\rho) \cdot u_1^w(t)$, and a smoothed density $\hat{\rho} : u_2(\rho, t) = f_{u_2}(\hat{\rho}) \cdot u_2^w(t)$, where the smoothed density $\hat{\rho}$ is obtained by convolving the local density ρ with a 3D Gaussian kernel. Parameters $u_1^w(t)$ and $u_2^w(t)$ are commissioned in a homogenous water phantom at a geometrical depth t and $f_{u_1}(\hat{\rho})$ and $f_{u_1}(\hat{\rho})$ are penumbra widen factors. The weighting factors ω_i adjust the proportions of primary and secondary dose according to the smoothed density $\hat{\rho}$ and the beamlet passing history using a formula $\omega_i(\rho, t) = f_{\omega_i}(\rho) \cdot (\omega_i^w(t) + \omega_i^{corr}(t))$, where $f_{\omega_i}(\rho)$ adjusts weighting factors locally according to a smoothed density $\hat{\rho}$. $\omega_i^w(t)$ is the commissioned weighting factor in a homogenous water phantom at a depth t. $\omega_i^{corr}(t) = \int_0^t b(\rho(t)) dt'$, where $b(\rho(t'))$ is a parameter describing the changing of $\omega_i(\rho, t)$ values with the existence of heterogeneities. We commissioned the DC-FSPB parameters with MCSIM [34] simulation results in a homogenous water phantom and in heterogeneous slab geometry. The details of the DC-FSPB model can be found in the References 32 and 34.

14.3 PHYSICS MODEL OF PROTON PENCIL BEAM

The physics model of proton pencil beam dose calculation is very similar to photon. Equation 14.2 is valid for proton dose calculation, and only difference is the kernel K. This proton pencil beam kernel has been studied for many years [36–41]. One of the commonly used kernels is developed by Hong et al. [38], which follows Hogstrom's electron pencil beam model [42], a central elementary beam spreading through a Gaussian function. The dose distribution of a proton pencil beam can be formulated as

$$D_i^p(x, y, z) = C(z)O(x, y, z),$$

$$\text{with } C(z) = DD(d_{eff}) \left(\frac{SSD + d_{eff}}{z} \right)^2 ; \qquad (14.8)$$

$$O(x, y, z) = \frac{1}{2\pi\sigma(z)^2} \exp\left(-\frac{x^2 + y^2}{2\pi\sigma(z)^2} \right).$$

where
   $C(z)$ is the central axis term of the dose
   $O(x, y, z)$ denotes the lateral distribution of the central elementary beam

The central axis term further involves two terms, the depth-dose distribution and inverse square correction. The depth distribution term, which can be obtained through a broad beam measurement, is a function of effective depth d_{eff} (the water-equivalent length or the radiological depth) from the beam source to a point on the central axis and depth z. The second term, inverse square correction term, accounts for the fluence loss when the point has a geometrical distance away from the source. The lateral beam distribution is expressed as a Gaussian function with a beam size (or spread) parameter $\sigma(z)$, which is an increasing function of depth z. Physically, this spread parameter is due to the lateral scattering during the proton propagation. Practically, an empirical function form is usually employed which combines the contribution of nozzle device scattering and patient scattering [38]. The total dose distribution is obtained by superposition of contributions from elementary pencil beams, expressed as

$$D(x,y,z) = \sum D_i^p(x,y,z)T_i, \qquad (14.9)$$

where T_i refers to the intensity of ith proton elementary beam.

14.4 GPU IMPLEMENTATIONS

Both photon and proton pencil beam dose calculations include two steps: the first step is to calculate beamlet dose, following Equations 14.4, 14.7, or 14.8; the second step is to calculate voxel's dose by summarizing the contribution from all the pencil beams as shown in Equations 14.3 or 14.9. Algorithm 14.1 shows the basic implementation steps:

Algorithm 14.1: Basic steps for pencil beam dose calculation on GPU

Copy patient CT data from CPU to GPU:
1. *Kernel 1*: parallelized on beamlets, calculating beamlet-related parameters such as beamlet angels for photon dose calculation, radiological depth for both photon and proton dose calculation
2. *Kernel 2*: parallelized on voxels, calculating each voxel's dose contributed from each beamlet
3. Copy dose distribution matrix from GPU to CPU

As shown in Algorithm 14.1, GPU computation has to be conjunctive with a CPU. The CPU serves as the *host* while the GPU is called the *device.* The CUDA platform extends the concept of C functions to *kernels.* A kernel, invoked from the CPU, can be executed N times in parallel on the GPU by N different CUDA *threads.* For convenience, CUDA threads are grouped to form thread *blocks,* and blocks are grouped to comprise *grids.* The number of blocks and threads has to be explicitly defined when executing a kernel. The threads of a block are grouped into *warps* (32 threads per warp). A warp executes a single instruction at a time. On the CUDA platform, the main code runs on the host (CPU), calling kernels that are executed on a physically separate device (GPU). Due to the physical separation of the device and the host, communication between the two cannot be avoided and has to be carefully addressed in CUDA programming.

The beamlet dose calculation includes one term of depth dose as a function of water-equivalent depth and one term of dose spread function. The radiological depth calculation requires the integration of the density functions along the beamlet direction, which is a computational intensive ray tracing problem. On CPU platforms, Siddon's algorithm is often used [43]. However, since the segment length of the beamlet central axis intersecting each voxel is not a constant with Siddon's algorithm, the lookup table of the radiological depth (or the weighting factor correction term) for each beamlet has to include two arrays: one storing the radiological depth (or the weighting factor correction term) while the other auxiliary array listing the corresponding geometrical depth. With the established lookup table, the radiological depth can be calculated through searching the geometrical depth array. In order to reduce the memory accessing time and improve the computation efficiency, Gu et al. [35] adopt a novel approach which computes the radiological depth and the weighting factor correction term at the sampling points uniformly distributed along the beamlet central axis. The sampling step size $d = (1/2)\min(\delta_x, \delta_y, \delta_z)$, where $\delta_x, \delta_y, \delta_z$ represent the voxel size in x, y, and z dimensions. With this approach, the storing and searching of the geometrical depth array becomes unnecessary. The involved interpolation procedures can be conducted with high efficiency using the fast on-chip linear interpolation function. The dose spread function expressed with hyperbolic and exponential functions as shown in Equation 14.6. The computation of these functions could be realized with the use of an intrinsic function __expf(z), which is about an order of magnitude faster than the standard math function expf(z).

Another key to improve computational efficiency is to wisely utilize GPU memory. Available memory on GPU consists of constant memory, global memory, shared memory, and texture memory. The constant memory is cached and requires only one memory instruction (four clock cycles) to access, but the size is limited to 64 kB on a typical GPU card (such as NVIDIA Tesla C1060). The global memory is not cached and requires coalesced memory access to achieve an optimal usage, but it has a large capacity (4 GB on one Tesla C1060 card) and is writable. The texture memory is read-only memory, but it is cached and is not restricted by the coalescing memory access pattern to achieve high performance. Gu et al. [35] described the strategy for the GPU memory management in their g-DC-FSPB implementation. Due to the limited size of constant, they stored only those frequently accessed arrays with constant values in the constant memory, such as the beam setup parameters and the commissioned model parameters. Those arrays requiring memory writing, such as the dose distribution array, were stored in the global memory. The radiological depth array was copied to the texture memory to improve memory accessing speed by utilizing texture fetching.

14.5 EXPERIMENTAL RESULTS

14.5.1 Photon Dose Calculation Results

The performance of GPU-based FSPB (g-FSPB) and DC-FSPB (g-DC-FSPB) algorithm were evaluated with five head-and-neck patients' cases and five lung patients' cases [44]. All treatment plans were initially generated on the Eclipse planning system (Eclipse, Varian Medical Systems, Inc., Palo Alto, CA) and used to treat patients. The original CT images were downsampled to the resolution of $0.4 \times 0.4 \times 0.25$ cm^3 images. The accuracy of dose calculation was assessed with MCSIM dose calculation [34]. The resolution of the fluence maps (or the beamlet size) was selected as 0.2×0.5 cm^2 with 0.2 cm along the leaf motion direction of multi-leaf collimator (MLC). The dose distributions calculated with MCSIM were used as the ground truth, with the maximum relative uncertainty less than 0.1%. The 3D γ-index distributions were computed using a GPU-based algorithm. Dose distributions were normalized to the maximum MCSIM dose value (D_{max}) for each case and 3%/3 mm was used as the evaluation criteria. The following statistical parameters were calculated and used as metrics to evaluate the dose calculation accuracy: (1) γ^{max}: the maximum γ value of the entire dose distribution; (2) γ_{50}^{avg}: the average γ

values inside 50% isodose lines; and (3) P_{50}: the percentage of voxels inside 50% isodose lines with $\gamma < 1.0$. The efficiency evaluation was conducted on one single NVIDIA Tesla C1060 card. The data transferring time and the GPU computation time were recorded separately.

14.5.1.1 Accuracy

Table 14.1 summarizes the γ-index evaluation results of five head-and-neck cases and five lung cases. For all five cases, γ^{max} and γ_{50}^{avg} values are smaller and P_{50} values are larger for the g-DC-FSPB algorithm, indicating that g-DC-FSPB constantly outperforms g-FSPB algorithm. Specifically, we can put these five head-neck cases into three scenarios: *Scenario 1 (Case H1 and Case H2)—both g-FSPB and g-DC-FSPB algorithms are accurate.* For these two cases, the average γ-index values are low (~0.3) and the passing rates are high (>97%) for both g-FSPB and g-DC-FSPB algorithms. With minor inhomogeneities on beams' paths, the dose distributions of these two cases are quite accurately calculated with both g-FSPB and g-DC-FSPB. *Scenario 2 (Case H3)—g-FSPB algorithm is less accurate but g-DC-FSPB algorithm can greatly improve the accuracy.* Figure 14.2a through c shows the dose distributions for Case H3 calculated with the MCSIM, g-FSPB, and g-DC-FSPB algorithms in the XY plane through isocenter, respectively. The γ-index distributions in the same plane are presented in Figure 14.2d and e. The γ-index values decrease significantly at the nasal cavity region when the 3D density correction is applied. The statistical analysis of the γ-index also shows that

TABLE 14.1 Gamma Index Evaluation Results for Five Head-and-Neck Cases and Five Lung Cases Using the g-DC-FSPB Algorithm. The corresponding g-FSPB results are given in parenthesis for comparison purpose.

Case #	γ^{max}	γ_{50}^{avg}	P_{50} (%)
H1	2.12 (2.16)	0.30 (0.31)	97.53 (97.32)
H2	3.44 (4.11)	0.28 (0.28)	97.80 (97.01)
H3	2.27 (2.36)	0.46 (0.52)	92.29 (86.39)
H4	3.08 (3.11)	0.61 (0.63)	82.96 (81.56)
H5	3.33 (3.37)	0.61 (0.61)	86.19 (86.09)
L1	1.53 (1.92)	0.24 (0.45)	99.35 (94.81)
L2	2.35 (3.30)	0.36 (0.71)	96.64 (76.38)
L3	1.68 (3.07)	0.32 (0.75)	99.16 (76.60)
L4	2.70 (4.59)	0.63 (1.53)	81.33 (28.55)
L5	2.19 (4.34)	0.49 (1.13)	90.24 (57.03)

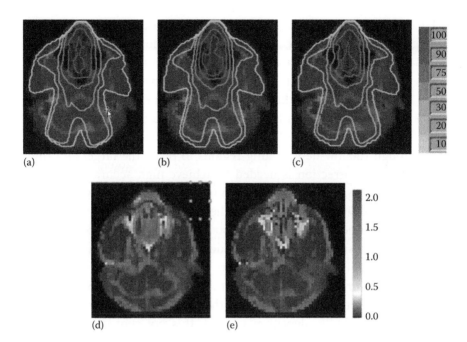

FIGURE 14.2 **(See color insert.)** Dose distributions for Case H3 calculated with the MCSIM (a), g-FSPB (b), and g-DC-FSPB (c) algorithms in the X-Y plane through the isocenter. The γ-index distributions are shown in (d) for g-FSPB and (e) for g-DC-FSPB dose distributions in the same plane. (Reproduced from Gu, X. et al., *Phys. Med. Biol.*, 56(11), 3337, 2011.)

the g-DC-FSPB dose distribution has a lower average γ-index value and a higher passing rate compared to the g-FSPB result. These results indicate that g-DC-FSPB is capable of calculating dose more accurately in a low-density region (e.g., nasal cavity) than g-FSPB. *Scenario 3 (Case H4 and Case H5)—both g-FSPB and g-DC-FSPB algorithms are less accurate.* For these two cases, the g-FSPB dose distribution has large average γ-index values $\left(\gamma_{50}^{avg} \sim 0.6\right)$ and low passing rates ($P_{50} \sim 86\%$). With 3D density correction, the accuracy of dose distributions has no much improvement. By carefully inspecting these two cases, we found that in both cases there are dental fillings of very high density (~ 4.0 g/cm³). The beam passing through the high density dental filling region before hitting the target causes a dose discrepancy between the g-DC-FSPB and MCSIM results up to 8% of D_{max}. This is because the density values near 4.0 g/cm³ are far beyond our commissioned density range and thus g-DC-FSPB cannot find proper parameters to accurately calculate the dose.

For five lung cases, the accuracy of calculated dose distributions has been greatly improved with the 3D density method. The γ^{max} and γ_{50}^{avg} values for all five testing cases have been considerably reduced and P_{50} values have been increased up to 53%. In Case L1, the tumor site is located in the lower lobe of the left lung, close to the pleura. Five out of total six beams do not pass through low-density lung regions before hitting the target, in which cases g-FSPB gives sufficient accuracy. The last beam goes through the low-density lung regions to reach the target and thus the 3D density correction is needed to achieve high accuracy. Overall, using g-DC-FSPB, γ_{50}^{avg} is reduced from 0.45 to 0.24 and P_{50} is increased from 94.81% to 99.35%. In Case L2, the tumor site is close to the vertebral body. Three out of six beams strike the targets without passing through low-density lung regions. The dose distributions calculated with the MCSIM, g-FSPB, and g-DC-FSPB are plotted in the XY plane through isocenter in Figure 14.3a through c. In Figure 14.3b, the g-FSPB dose is much higher than the MCSIM dose in the region indicated by the arrow. When the g-DC-FSPB algorithm is used, this hot spot disappears, as shown in Figure 14.3c. Figures 14.3d and e plot the γ-index distributions

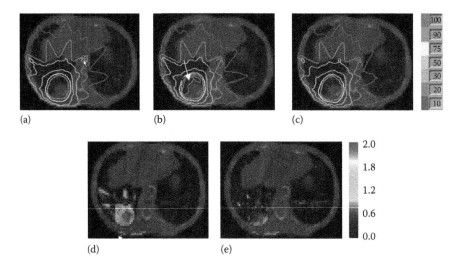

(a)   (b)   (c)

(d)   (e)

FIGURE 14.3   **(See color insert.)** The dose distribution of Case L2 calculated with MCSIM (a), g-FSPB (b), and g-DC-FSPB (c) in the X-Y plane through the isocenter. The γ-index distributions in the same plane are illustrated in (d) for g-FSPB and (e) for g-DC-FSPB. (Reproduced from Gu, X. et al., *Phys. Med. Biol.*, 56(11), 3337, 2011.)

TABLE 14.2   Dose Calculation Time Using the g-FSPB
(in parentheses) and g-DC-FSPB Algorithms for 10 Testing Cases

Case #	T_{tr} (s)	T_{gpu} (s)	T_{tot} (s)
H1	0.20	0.64 (0.55)	0.84 (0.75)
H2	0.20	0.40 (0.35)	0.60 (0.55)
H3	0.20	0.38 (0.34)	0.58 (0.54)
H4	0.19	0.35 (0.32)	0.54 (0.51)
H5	0.20	1.31 (1.10)	1.51 (1.30)
L1	0.21	0.22 (0.20)	0.43 (0.41)
L2	0.22	0.40 (0.36)	0.62 (0.58)
L3	0.21	0.30 (0.25)	0.51 (0.46)
L4	0.18	0.25 (0.23)	0.43 (0.41)
L5	0.21	0.33 (0.29)	0.54 (0.50)
Median	0.20	0.37 (0.33)	0.56 (0.53)

T_{tr} is the data transfer time between CPU and GPU; T_{gpu} is the
   GPU computation time; and $T_{tot} = T_{tr} + T_{gpu}$.

in the same plane calculated with the g-FSPB and g-DC-FSPB algorithms, respectively. The statistical values shown in Table 14.2 also indicate a significant improvement of γ_{50}^{avg} and P_{50} values using the 3D density correction method, where, γ_{50}^{avg} values decrease from 0.71 to 0.36 and P_{50} values increase from 76.38% to 96.64%. The situation for Case L3 is very similar to that of Case L2. The tumor in Case L4 is in the middle of the lung, indicating that all beams have to pass through the low-density lung regions before hitting the target. The dose distributions in the X-Y plane through isocenter calculated with the MCSIM, g-FSPB, and g-DC-FSPB algorithms are shown in Figure 14.4a through c. The γ-index distributions in the same plane calculated with the g-FSPB and g-DC-FSPB algorithms are plotted in Figure 14.4d and e. Figure 14.4 shows that in the high dose region, the g-FSPB algorithm heavily overestimates the calculated dose and the g-DC-FSPB algorithm can correct the overestimation of the g-FSPB algorithm and greatly improve the agreement with MCSIM, especially inside the target region. However, in lung regions outside the target, the density correction is overdone, resulting in an underestimated dose. Similarly, for Case L5, the improvement of dose distribution achieved by the 3D density correction method is dramatic. However, the γ-index passing rate in the 50% isodose line for the g-DC-FSPB algorithm is still less satisfactory due to the overcorrection issue.

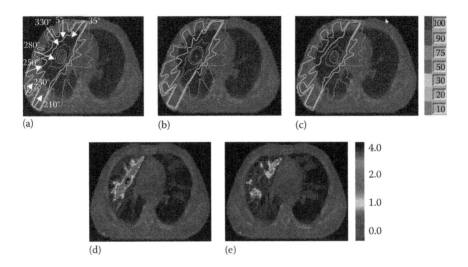

FIGURE 14.4 **(See color insert.)** Dose distributions for Case L4 in the X-Y plane through the isocenter calculated with the MCSIM (a), g-FSPB (b), and g-DC-FSPB (c) algorithms. γ-index distributions for the g-FSPB (d) and g-DC-FSPB (e) algorithms are given in the same plane. (Reproduced from Gu, X. et al., *Phys. Med. Biol.*, 56(11), 3337, 2011.)

14.5.1.2 Efficiency

Table 14.2 lists computation time for dose calculation using g-FSPB and g-DC-FSPB. The dose distribution of a realistic IMRT plan can be computed at a very high efficiency. For 9 out of 10 testing cases, the dose calculation can be completed within 1 s using either algorithm. For all 10 cases, the median data transfer time between CPU and GPU is 0.2 s, and the median GPU computation time is 0.37 s for the g-DC-FSPB algorithm and 0.33 s for the g-FSPB algorithm. Since the computation time is so short, the data transfer time takes a significant portion of the total computation time, up to 50% in Case L1. We can also see that, while the accuracy of the g-DC-FSPB algorithm is much higher than that of the g-FSPB algorithm, its efficiency sacrifice is quite mild (~5%–15% slower in terms of the total computation time).

14.5.2 Proton Dose Calculation Results

Fujimoto et al. [40] implemented a proton pencil beam dose calculation algorithm on GPU and compared to CPU version. The calculation was performed on a water phantom with an embedded cubic target. Table 14.2 lists the computation results, where S is the target size, $r_x = r_y$ are

TABLE 14.3   Conditions and Results for the Proton Dose Calculation Comparison Cases in Reference 40

S (mm²)	r_x, r_y (mm)	N_d (×10⁶)	N_b (×10³)	T_{CPU} (s)	T_{GPU} (s)	T_{CPU}/T_{GPU}
50 × 50	1.0	1.9	5	1.53	0.33	4.6
100 × 100	1.0	5.7	16	6.04	0.41	14.7
150 × 150	1.0	11.5	32	13.18	0.67	19.8
100 × 100	0.5	45.5	64	98.23	7.18	13.7
100 × 100	2.0	0.7	4	0.48	0.09	5.6

resolutions, N_d is the number of voxels where the deposited dose is calculated and N_b is the number of beamlets in each condition, and T_{CPU} and T_{GPU} are the computational times by CPUs and GPUs, respectively. T_{GPU} listed in the table includes data transfer time between CPUs and GPUs. As shown in Table 14.3, the GPU-based proton dose calculation algorithm accomplishes ~5–20 speedup compared to CPU algorithm. When comparing the dose accuracy, Fujimoto et al. [40] found the maximum difference is under 0.2%.

14.6 CONCLUSIONS AND DISCUSSIONS

In this chapter, we described both photon and proton pencil beam dose calculation models and detailed their GPU implementation. The dosimetric accuracy of g-FSPB and g-DC-FSPB were compared to MCSIM and both g-FSPB and g-DC-FSPB achieved reasonable accuracy in most evaluation cases, but g-DC-FSPB outperformed over g-FSPB. Regarding the efficiency, 9 out of 10 photon dose calculation cases could be completed well within 1 s with g-FSPB and g-DC-FSPB. Compared to the g-FSPB algorithm, the g-DC-FSPB algorithm slightly sacrificed the computation efficiency, about 5%–15% slower in terms of the total computation time. The accuracy of proton GPU algorithm was comparable to the corresponding CPU algorithm according to testing results. The GPU proton pencil beam dose calculation speeded up ~5–20 times compared to CPU algorithm, but the speedup factor depended on the size and resolution of the target.

In addition to the apparent advantages of computational efficiency, the pencil beam approach also provides dose distributions from each single pencil beam. Such information is of critical importance for many clinical applications such as intensity-modulated therapy, where the intensity of each pencil beam is optimized to yield a desired dose distribution. Yet, pencil beam dose calculation is only a temporary solution due to its questionable accuracy in some cases. Another limiting factor of this model is the associated difficulties in commissioning, where the empirical data must be determined and tedious parameters adjustment has to be conducted. GPU-based Monte Carlo will be the solution for accurate and efficient dose calculation.

REFERENCES

1. Wu, C. et al., Re-optimization in adaptive radiotherapy. *Physics in Medicine and Biology*, 2002. **47**(17): 3181–3195.
2. Wu, C. et al., Fast treatment plan modification with an over-relaxed Cimmino algorithm. *Medical Physics*, 2004. **31**(2): 191–200.
3. Court, L.E. et al., An automatic CT-guided adaptive radiation therapy technique by online modification of multileaf collimator leaf positions for prostate cancer. *International Journal of Radiation Oncology Biology Physics*, 2005. **62**(1): 154–163.
4. Mohan, R. et al., Use of deformed intensity distributions for on-line modification of image-guided IMRT to account for interfractional anatomic changes. *International Journal of Radiation Oncology Biology Physics*, 2005. **61**(4): 1258–1266.
5. Court, L.E. et al., Automatic online adaptive radiation therapy techniques for targets with significant shape change: A feasibility study. *Physics in Medicine and Biology*, 2006. **51**(10): 2493–2501.
6. Wu, Q.J. et al., On-line re-optimization of prostate IMRT plans for adaptive radiation therapy. *Physics in Medicine and Biology*, 2008. **53**(3): 673–691.
7. Lu, W.G. et al., Adaptive fractionation therapy: I. Basic concept and strategy. *Physics in Medicine and Biology*, 2008. **53**(19): 5495–5511.
8. Ahunbay, E.E. et al., An on-line replanning scheme for interfractional variations. *Medical Physics*, 2008. **35**(8): 3607–3615.
9. Fu, W.H. et al., A cone beam CT-guided online plan modification technique to correct interfractional anatomic changes for prostate cancer IMRT treatment. *Physics in Medicine and Biology*, 2009. **54**(6): 1691–1703.
10. Godley, A. et al., Automated registration of large deformations for adaptive radiation therapy of prostate cancer. *Medical Physics*, 2009. **36**(4): 1433–1441.
11. Men, C.H. et al., GPU-based ultrafast IMRT plan optimization. *Physics in Medicine and Biology*, 2009. **54**(21): 6565–6573.
12. Gu, X.J. et al., GPU-based ultra-fast dose calculation using a finite size pencil beam model. *Physics in Medicine and Biology*, 2009. **54**(20): 6287–6297.
13. Gu, X.J. et al., Implementation and evaluation of various demons deformable image registration algorithms on a GPU. *Physics in Medicine and Biology*, 2010. **55**(1): 207–219.
14. Men, C.H., X. Jia, and S.B. Jiang, GPU-based ultra-fast direct aperture optimization for online adaptive radiation therapy. *Physics in Medicine and Biology*, 2010. **55**(15): 4309–4319.
15. Men, C.H. et al., Ultrafast treatment plan optimization for volumetric modulated arc therapy (VMAT). *Medical Physics*, 2010. **37**(11): 5787–5791.

16. Ahunbay, E.E. et al., Online adaptive replanning method for prostate radiotherapy. *International Journal of Radiation Oncology Biology Physics*, 2010. **77**(5): 1561–1572.

17. Sharp, G.C. et al., GPU-based streaming architectures for fast cone-beam CT image reconstruction and demons deformable registration. *Physics in Medicine and Biology*, 2007. **52**: 5771–5783.

18. Zhao, Y. et al., Application of an M-line-based backprojected filtration algorithm to triple-cone-beam helical CT. *Physics in Medicine and Biology*, 2010. **55**(23): 7317.

19. Lu, W.G., A non-voxel-based broad-beam (NVBB) framework for IMRT treatment planning. *Physics in Medicine and Biology*, 2010. **55**(23): 7175–7210.

20. Samant, S.S. et al., High performance computing for deformable image registration: Towards a new paradigm in adaptive radiotherapy. *Medical Physics*, 2008. **35**(8): 3546–3553.

21. Lu, W.G. and M.L. Chen, Fluence-convolution broad-beam (FCBB) dose calculation. *Physics in Medicine and Biology*, 2010. **55**(23): 7211–7229.

22. Jia, X. et al., GPU-based fast cone beam CT reconstruction from undersampled and noisy projection data via total variation. *Medical Physics*, 2010. **37**(4): 1757–1760.

23. Hissoiny, S., B. Ozell, and P. Despres, Fast convolution-superposition dose calculation on graphics hardware. *Medical Physics*, 2009. **36**(6): 1998–2005.

24. Jia, X. et al., Development of a GPU-based Monte Carlo dose calculation code for coupled electron-photon transport. *Physics in Medicine and Biology*, 2010. **55**(11): 3077–3086.

25. Jacques, R. et al., Towards real-time radiation therapy: GPU accelerated superposition/convolution. *Computer Methods and Programs in Biomedicine*, 2010. **98**(3): 285–292.

26. Diot, Q. et al., Dosimetric effect of online image-guided anatomical interventions for postprorstatectomy cancer patients. *International Journal of Radiation Oncology Biology Physics*, 2011. **79**(2): 623–632.

27. Jia, X. et al., Proton therapy dose calculations on GPU: Advances and challenges. *Translational Cancer Research*, 2012. **1**(3): 207–216.

28. Gustafsson, A., B.K. Lind, and A. Brahme, A generalized pencil beam algorithm for optimization of radiation therapy. *Medical Physics*, 1994. **21**(3): 343–356.

29. Bourland, J.D. and E.L. Chaney, A finite-size pencil beam model for photon dose calculation in three dimensions. *Medical Physics*, 1992. **19**: 1401–1412.

30. Ostapiak, O.Z., Y. Zhu, and J. VanDyk, Refinements of the finite-size pencil beam model of three-dimensional photon dose calculation. *Medical Physics*, 1997. **24**(5): 743–750.

31. Jiang, S.B., *Development of a Compensator Based Intensity-Modulated Radiation Therapy System*, Toledo, OH: Medical College of Ohio, 1998.

32. Jelen, U., M. Sohn, and M. Alber, A finite size pencil beam for IMRT dose optimization. *Physics in Medicine and Biology*, 2005. **50**(8): 1747–1766.

33. Jelen, U. and M. Alber, A finite size pencil beam algorithm for IMRT dose optimization: Density corrections. *Physics in Medicine and Biology*, 2007. **52**(3): 617–633.

34. Ma, C.-M. et al., A Monte Carlo dose calculation tool for radiotherapy treatment planning. *Physics in Medicine and Biology*, 2002. **47**(10): 1671.

35. Gu, X. et al., A GPU-based finite-size pencil beam algorithm with 3D-density correction for radiotherapy dose calculation. *Physics in Medicine and Biology*, 2011. **56**(11): 3337–3350.

36. Petti, P.L., Differential-pencil-beam dose calculations for charged particles. *Medical Physics*, 1992. **19**(1): 137–149.

37. Lee, M., A.E. Nahum, and S. Webb, An empirical method to build up a model of proton dose distribution for a radiotherapy treatment-planning package. *Physics in Medicine and Biology*, 1993. **38**(7): 989.

38. Hong, L. et al., A pencil beam algorithm for proton dose calculations. *Physics in Medicine and Biology*, 1996. **41**(8): 1305–1330.

39. Soukup, M., M. Fippel, and M. Alber, A pencil beam algorithm for intensity modulated proton therapy derived from Monte Carlo simulations. *Physics in Medicine and Biology*, 2005. **50**(21): 5089–5104.

40. Fujimoto, R., T. Kurihara, and Y. Nagamine, GPU-based fast pencil beam algorithm for proton therapy. *Physics in Medicine and Biology*, 2011. **56**(5): 1319–1328.

41. Szymanowski, H. and U. Oelfke, Two-dimensional pencil beam scaling: An improved proton dose algorithm for heterogeneous media. *Physics in Medicine and Biology*, 2002. **47**(18): 3313–3330.

42. Hogstrom, K.R., M.D. Mills, and P.R. Almond, Electron beam dose calculations. *Physics in Medicine and Biology*, 1981. **26**(3): 445–459.

43. Siddon, R.L., Fast calculation of the exact radiological path for a 3-dimensional CT array. *Medical Physics*, 1985. **12**(2): 252–255.

44. Gu, X., X. Jia, and S.B. Jiang, GPU-based fast gamma index calculation. *Physics in Medicine and Biology*, 2011. **56**(5): 1431–1441.

CHAPTER 15

Photon Monte Carlo Dose Calculation

Sami Hissoiny

CONTENTS

15.1 INTRODUCTION

DOSE CALCULATION IS AT the center of radiation therapy planning. Its result serves as input to the optimization algorithm and is at the core of plan creation and evaluation. Dose calculation methods are diverse, going not only from fast to slow but also from inaccurate to accurate. Monte Carlo (MC) methods, now alongside the relatively recently introduced grid-based Boltzmann solvers, are at the "slow but accurate" extreme of the diversity spectrum of dose calculation methods used in radiation therapy [1]. The appearance of absolute parallelism, where the contributions of millions of individual particles are added to generate the final result, also made it a seemingly ideal candidate for parallelization schemes.

First attempts at parallelizing the MC dose calculation algorithm were targeted at the central processing unit (CPU). If the workload is large enough, in order to hide communication costs, close to ideal accelerations are observed on multicore CPUs and CPU clusters [2–4]. Both multicore CPUs and CPU clusters are ideally suited for the parallelization of MC calculations as they are able to perform task parallelism, a form of parallelization where two computing elements are able to perform two disjoints and independent tasks on a joint or disjoint set of data. Each particle, or batch of particles, is considered as an independent task and is distributed across the computing elements. A final round of accumulation is performed on the individual set of results generated by the individual computing elements, and the final, aggregated, result is then available.

Using similar strategies as for multicore CPUs and CPU clusters, cloud computing is a more recent method for leveraging the computational power of large quantities of CPUs. This method is based on large, offsite, arrays of computing elements made available by third-party companies on a pay-per-use basis. Users buy time for a specific hardware configuration offered by these third-party suppliers to run their parallel software. Again, if the workload is large enough to hide the communication costs, close to ideal speedups are observed [5,6].

Starting in 2008, the graphics processing unit (GPU) started to make waves as a hardware platform for dose calculation, not only for MC dose calculation but also for a number of other analytical methods such as the pencil beam method [7–9] and convolution-superposition-based methods [10–12]. Due to its key advantages of large single-precision computing power and wide memory bandwidth, the GPU is ideally situated to perform well for the task of dose calculation. This is especially true for analytical dose calculation methods, where the problem can usually be formatted such that the computation of dose for each individual voxel is a completely individual task that can be assigned to the different cores of the GPU.

The first appearance of GPU-based MC transport dates back to 2008 with the work of Alerstam et al. [13]. Their application is targeted at light propagation in random media. It is not strictly applicable to dose calculation, but it recognized that the GPU was an attractive, but certainly not ideal, platform for MC-based energy transport algorithms. It was the first to highlight the difficulties of mapping the MC algorithm directly to the GPU. The first application of GPU-based photon transport application to radiation therapy was done in 2009 [14]. They ported part of the photon transport code of PENELOPE [15] to the GPU for applications in radiographic image creation.

This chapter will cover the state of the use of GPUs for the task of dose calculation through the MC method. First, the relevant physics for the transport of photons and electrons in the keV and MeV energy range will be detailed. Then, some challenges of mapping the MC algorithm to the GPU architecture will be covered. Finally, a list of existing keV and MeV packages and their characteristics will be presented.

15.2 PHYSICS OF PHOTON AND ELECTRON TRANSPORTS IN KV/MV ENERGY RANGE

The current section aims to deal with the photon and electron physics involved in the keV and MeV energy range. However, as this chapter, and this book, is oriented toward software techniques in radiation therapy, a data management and workflow analysis view of the physics involved is presented. The interactions for a given type of particles are first introduced, and a more general view of the flow of a typical simulation is detailed. Readers more interested in the physics and mathematics aspects of the MC simulation for energy transport are kindly redirected to more general MC platform manuals (e.g., [15,16]), which serve as a reference for the content of this section.

15.2.1 Photon Physics

It is customary for all interactions encountered by the photon to be simulated discretely, interaction by interaction. From a task perspective, the flow of a photon can be detailed as follows. A photon continues to move forward, from interaction to interaction, as long as its energy is above a certain threshold given by the user and that it is within the geometric bounds of the simulation. A step between two interactions is characterized by (1) sampling the distance to the next interaction, (2) moving the particle to the point of interaction and verifying that it is still within the geometric bounds of the problem, (3) sampling the interaction type, and (4) simulating the interaction. Within steps (1) and (2), the Woodcock algorithm is often used in dose calculation orientation MC algorithms [17]. This method eliminates the need for explicit ray tracing while moving to the position of next interaction, which in turns makes it so that those per-voxel memory lookups are not necessary and that the particle can be moved forward by a geometrical distance. A schematic of the simulation of a photon is shown in Figure 15.1.

There are four main ways in which photons interact in the energy range for radiation therapy. They are as follows: Rayleigh scattering, Compton scattering, the photoelectric effect, and pair production.

In Rayleigh scattering, the direction of momentum of a photon changes while its magnitude remains the same. The scattering angle is sampled from the Rayleigh differential cross section, which is a function of the molecule form factor, which is itself a function of energy. Thus, for sampling the scattering angle, a memory lookup is necessary to get the form

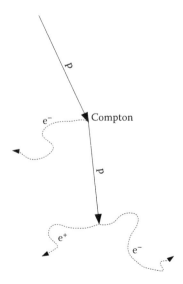

FIGURE 15.1   Flow of the simulation of a photon. The straight lines represent a photon and the dashed lines represent an electron or a positron.

factor for that molecule and that energy. Rayleigh scattering produces no secondary particles.

Compton scattering is the inelastic scattering of a photon impinging on a charged particle. Both the magnitude and direction of the momentum of the incoming photons are changed and a recoil electron is put in movement. If the target electrons are assumed to be free, which is an often used approximation for MeV dose calculation, the differential cross section for the scattering angle in Compton scattering on free electrons is the Klein–Nishina formula [18], and it can be sampled analytically [19–21], thus no data lookup is necessary. The energy of the outgoing particles (both the recoil electron and scattered photon) is fixed by the kinematics once the scattering angle has been sampled. The approximation that the photon is interacting with free electrons is a valid approximation for energies above 1 MeV [22]. For applications at lower energies, a sampling that takes into account binding effects is recommended.

In the photoelectric effect, the incoming photon gets absorbed by an electron of an atom. The electron then escapes the atom and is put into movement. In the simple form described here, the result of the photoelectric effect is then the absorption of the incoming photon and a new electron, where all of the energy of the photon is transferred to the electron. The only quantity needing sampling is the scattering angle of the electron with regard to the incoming direction of the photon, which is usually sampled from the Sauter distribution [23], which can be sampled analytically. In a more complete description of the photoelectric effect, a number of low energy photons are emitted as the atom gets back to its ground state. This modeling is complex and relies on a large quantity of external data describing all the possible transitions as well as their possibilities and the energy of the emitted photon. As GPU implementations of the MC algorithm are typically targeted at fast dose calculation algorithms, such modeling is usually not taken into consideration.

In the pair production interaction in the context of dose calculation, a photon interacts with a nuclear field, annihilates, and produces a pair of electron-positron. Occasionally, the photon can interact with an electron of the atomic cloud and is then called triplet production. A usual approximation is for the total cross section for pair production to be the sum of the pair production in the nuclear field and electron field but for the interaction to be modeled as always being in the nuclear field [15]. The effect of the interaction is then the absorption of the incoming photon and the creation of two new particles: an electron and a positron.

The mathematics for the correct sampling of the energy and direction of the outgoing particles is complex, and it is often simplified in fast dose algorithms by trivially splitting the energy of the incoming photon and following the initial direction of the incoming photon [22].

15.2.2 Electron Physics

In this section, the relevant physics process for electrons is discussed. The handling of positrons is the same insofar as the data workflow is concerned, with the exception that the positron will emit a pair of annihilation photon at the end of its track. The possible interactions for electrons are elastic scattering, inelastic (Møller) scattering, and bremsstrahlung. All fast MC codes for dose calculation employ a condensed history technique. A detailed description is beyond the scope of this book, but let us resume it by mentioning that only certain inelastic interactions and certain bremsstrahlung are modeled explicitly and that the rest of the interactions are grouped together.

From a task perspective, the flow of an electron is as follows. An electron moves forward as long as it has more energy than an energy threshold set by the user. The process of moving forward is modeled by (1) sampling a distance to the next hard interaction, (2) moving the electron through a series of sub-steps until it reaches the point of interaction, (3) sampling the interaction type, and (4) simulating the interaction.

Each sub-step mentioned in (2) is itself divided into three steps: (2.1) establishing the distance to travel in that sub-step (as a function of maximum energy loss, maximum distance allowed, remaining energy, remaining distance to the next hard interaction, etc.); (2.2) through an implementation-specific method, computing the position and direction of the particle after the step, accounting for all the soft interactions grouped together; and (2.3) through an implementation-specific method, computing the energy loss during that sub-step and deposit it in the results grid. An example of the flow of an electron is detailed in Figure 15.2.

In the Møller scattering interaction, an incident electron collides with an atomic electron (which is, again, approximated to be free). The result of the interaction is that the incoming electron changes momentum in both magnitude and direction, and a new electron is set in motion. The Møller cross section can be sampled analytically to establish the ratio of the scattered energy to the incoming energy, after which the kinematics set the directions of the resulting electrons.

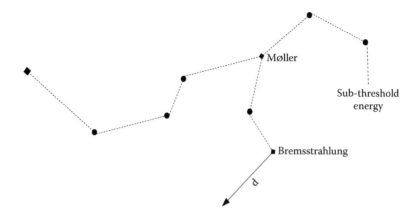

FIGURE 15.2   Flow of the simulation of an electron. The dashed lines represent electrons and the straight line represents photons. The circles represent multiple scattering and the squares a hard interaction. All along the dashed lines, energy loss is applied.

In the bremsstrahlung interaction, a photon is created in the field of an atom. The result of the interaction is that the incoming electron changes momentum in both magnitude and direction, and a new photon is set in motion. The energy of the created photon can be sampled analytically but the sampling depends on a number of energy-independent values that characterize the interacting medium. The scattering angles for both the scattered electron and created photon can be sampled analytically. A typical approximation is to leave the direction of the electron untouched and deterministically set the direction of the photon based on the energy of the incoming electron [22].

15.3 CHALLENGES OF GPU-BASED MC SIMULATIONS AND CURRENT SOLUTIONS

The GPU offers an attractive solution to the acceleration of MC simulations, but it is certainly not ideal. This section concentrates on the challenges of mapping the MC algorithm on the GPU and offers some solutions.

15.3.1 Thread Divergence

Thread divergence is the most common source of issues when mapping the MC algorithm to the GPU architecture. It refers to the situation where threads within a warp have to execute difference lines of code and follow different code paths. This situation occurs frequently as the statistical trials that constitute the MC algorithm are all independent and thus free to evolve in independent and likely different ways. Thread divergence is also present in many levels of the MC algorithm. The following paragraphs will elaborate on these different levels and offer possible solutions.

The highest, and most detrimental, source of divergence happens at the particle-type level. Many types of simulations require more than one particle type to be simulated. For example, in external beam photon therapy there are electrons and photons, or in proton therapy there are protons and electrons. The way most traditional MC packages work is through a particle stack. This stack is used to place secondary particles as they are created. It is then very likely that, at some point in the simulation, there are both, for example, photons and electrons to simulate. Based on the details of the previous section, it has been shown that electrons can produce secondary photons, and photons can produce secondary electrons. If this situation is taken as-is and ported to the GPU, it is then likely that at some point two threads of the same warp would be assigned to simulate two particles of different types, because the primary particles attached to those threads interacted in different ways and created different types of secondary particles. Since the threads within a warp can only follow one code path at the same time and the simulation of photons and electrons likely follows two completely distinct code paths, only one type of particle can be simulated at a time while all the threads assigned to particles of another type would be idle. The solution to that issue is to never try to simulate more than one particle type at a time [24]. This can be accomplished by launching specific kernels that only deal with one particle type. Each launch of a kernel for a given particle type will likely generate secondary particle of the other types. These secondary particles will be simulated at a later time, rather than on the fly as with traditional MC packages. The disadvantage of this method is that large particle buffers need to be allocated a priori to house the created secondary particles. Jia et al. have shown that this method is at the heart of GPU optimization by not using the method at first [25] and then using it [26]. A difference of about 16× in acceleration can be seen between the two publications.

The second level of divergence is particle lifetime. As noted in the previous section, each particle is simulated "as long as its energy is above a user-defined energy threshold." The amount of iteration of the loop that simulates one particle step is the result of a stochastic process, and thus, each thread within the warp is susceptible of requiring a different number of iterations to complete its particle. Following the GPU architecture requirements, a warp will not be considered as terminated before all of the threads within the warp have exited. It is then likely that, if one particle is assigned to one thread, some threads will finish before others and the warp will only exit when the "slowest" thread, the one requiring the most loop iterations, will be done. One solution to that problem is to simulate a fixed number of loop iteration per kernel launch [24]. After that number is reached, all particles are written back to global memory, the terminated particles are removed through a stream compaction operation, and the simulation kernel is launched again with the advantage that it contains only live particles. A similar result can be achieved by double buffering the particle vectors. The disadvantage of these methods is that it comes with constant reading and writing of particles from and to global memory and that the stream compaction operation is not free. Another method for reducing that source of divergence is to not use a thread per particle but rather use a fixed number of threads to handle all particles [27]. When one thread is done with its current particle, it is free to fetch a new particle from global memory and continue working.

The third level of thread divergence is at the interaction level. The sampling of the interaction type for a given particle in a given material is a stochastic process. It is thus possible that many different interaction types be sampled for the particles within a warp. On the GPU, this will result in all interaction simulations being serialized where, for example, all the threads handling a photon that undergo Compton interaction are active but all other threads are idle. A solution to that issue is to employ the same mechanism as the first level of thread divergence but to further differentiate by interaction type [28]. Using this method, one kernel is responsible for, for example, moving the photons to their position of interaction and sampling the interaction type. Then, based on the sampled interaction type, particles will be outputted to different buffers. One kernel per interaction type will then be launched on its respective buffer, ensuring that all threads only deal with, for example, the Compton interaction code. The disadvantage of this method is the added burden on global memory bandwidth since particles need to be read and written often.

15.3.2 Other Typical Parallel Monte Carlo Considerations

There are other issues associated with the use of multiple threads to carry out an MC simulation, which are affecting the use of not only a GPU but also multiple CPU threads.

A first example is the usage of the pseudo-random number generator (PRNG). It is necessary for a simulation running on multiple threads to have multiple PRNG streams. If all threads were to share one PRNG, its access would have to be guarded so that only one thread can access it at a time, which would essentially serialize the simulation. The user of multiple PRNG streams must make sure that each stream at least has the appearance of being independent, meaning that each stream does not go through the same sequence of random numbers. This easiest but most fragile way to accomplish this is to seed each stream differently. The fragility comes from the fact that the seed combination for two identically parameterized PRNG could result in both streams that are simply lagged by a few random numbers, that is, stream i goes through the sequence rn_1-rn_2-rn_3-rn_4 while stream j goes through the sequence rn_2-rn_3-rn_4-rn_5. If the properties of the PRNG are amenable to it, the leapfrogging method can be used [29]. Using this method, stream i generates random numbers rn_1-rn_2-rn_3-rn_4 while stream j goes through the sequence rn_{1+l}-rn_{2+l}-rn_{3+l}-rn_{4+l}, where l is a large enough number so that the two streams will not overlap during their use. Finally, another method is to use a difference parameterization of the PRNGs. For example, for a multiple-with-carry [30], one could use a different multiplier for each stream which essentially means that the two streams are different PRNGs.

A second example is the use of atomic operations. If there is to be only one results grid for a given type of tally, its access must be protected so that only one thread can write to it at a time. This can be achieved through atomic operations that will guarantee that the operation of read–update–write to memory cannot be interrupted by another thread or by using mutexes that essentially do the same thing through the acquisition of a "right-to-access" token by a given thread. Atomic operations are available on the GPU.

15.4 AVAILABLE KV PACKAGES

15.4.1 MC-GPU

The MC-GPU platform has its roots in one of the first GPU MC platforms to be published on Reference 14. The latest code to have been released is MC-GPU 1.3 in 2012 [31]. The code implements the photon physics of PENELOPE and electrons are not tracked. The primary use of the code is to transport x-rays in voxelized geometries and tally the response at the detector, but the code can also tally doses within the voxelized phantom itself to compute organ doses due to imaging [32].

15.4.2 gCTD

gCTD [33] has been developed for organ dose calculations in the context of imaging. Electrons are not tracked at all and photons can interact using Rayleigh scattering, Compton scattering, and the photoelectric effect. The atomic form factors used for Rayleigh scattering are taken from PENELOPE, and a direct sampling method has been developed as it is better suited for a GPU implementation. A fluence map derived from an air scan as well as spectrum is used to model the source. The authors report that results with sufficient precision can be calculated in tens of seconds.

15.4.3 gDRR

gDRR [34] shares its roots with the previously mentioned gCTD package, but has been specifically designed for the simulation of x-ray projections. The same fluence map and spectrum method is used to characterize the source. A powerful noise removal algorithm is part of the software to reduce the required number of simulated photons for a given level of statistical uncertainty. Simulation times of 30–95 s per projection are reported.

15.4.4 GPUMCD

The kV expansion of the GPUMCD platform [35,36] was accomplished in the context of brachytherapy dose calculations. A non-voxelized framework for the modeling of low dose rate (LDR) brachytherapy seeds was added, and high dose rate (HDR) seeds were modeled using phase-space files. The authors reported that HDR brachytherapy plans could be computed in a small number of seconds.

15.5 AVAILABLE MV PACKAGES

15.5.1 gDPM

The gDPM platform [25,26] implements a fully coupled electron–photon simulation. The physics is taken from the DPM MC platform [22] for both electron/positron and photon transport. Test cases were run against the original single-threaded CPU DPM implementation and accelerations of up to ~80× were found, with good accuracy results.

15.5.2 GPUMCD

The GPUMCD platform [24,37] was developed around completely new code but existing physics taken from existing MC packages. On top of the fully coupled photon–electron simulation, it supports the transport of charged particles in a geometry under the influence of a magnetic field. Acceleration factors of 200×–500× were reported against the single-threaded DPM CPU executable with good dose agreement.

15.5.3 GMC

GMC [28] is an implementation of the low-energy (<100 GeV) electromagnetic physics part of the Geant4 toolkit, where only voxelized geometries are supported. Accelerations of three orders of magnitudes, in terms of histories per seconds, are reported for a given simulation compared to the CPU version of Geant4 to its GPU reimplementation. The results, however, show that the two simulations are not strictly equivalent as a relaxed 3%-3 mm gamma criterion is not passed by all voxels in the resulting volumes.

15.6 CONCLUSION AND DISCUSSIONS

The mapping of the MC algorithm for energy transport for photons/electrons on the GPU is still an active field of research with much room for improvement. By its nature, the algorithm is not well suited for a GPU implementation. However, with the appreciable advantage the GPU has, in terms of raw computational power, over the CPU, good acceleration factors are found in the literature. Care has to be taken in analyzing these acceleration factors as they are mostly comparing a fine-tuned application-specific GPU MC code to a single-threaded general-purpose CPU counterpart. With that in mind, when publications compare to the DPM platform and still report 100×–500× accelerations, even if that number is divided by 15, representing (as of the writing of these lines) the number of cores in a top-of-the-line Intel Xeon processor, the acceleration factors are still well over 1.0.

MC-based dose calculations for photons and electrons are poised to be used more and more for certain types of calculations and modalities. For example, several groups, and vendors, are working on the integration of the magnetic resonance imaging (MRI) and linear accelerator [38–41] where dose has to be computed within a magnetic field. MC is currently the only reliable method to compute dose in such a situation. In another context, the AAPM task group 187 report [42] covers recommendations for early adopters of model-based dose calculation methods in brachytherapy, of which MC is an example.

REFERENCES

1. I. J. Chetty, B. Curran, J. E. Cygler, J. J. DeMarco, G. Ezzell, B. A. Faddegon, I. Kawrakow et al., Report of the AAPM Task Group No. 105: Issues associated with clinical implementation of Monte Carlo-based photon and electron external beam treatment planning, *Medical Physics*, 34, 4818–4853, 2007.
2. L. Deng and Z.-S. Xie, Parallelization of MCNP Monte Carlo neutron and photon transport code in parallel virtual machine and message passing interface, *Journal of Nuclear Science and Technology*, 36, 626–629, 1999.
3. K. Sutherland, S. Miyajima, and H. Date, A simple parallelization of GEANT4 on a PC cluster with static scheduling for dose calculations, *Journal of Physics: Conference Series*, 74, 021020, 2007.
4. N. Tyagi, A. Bose, and I. J. Chetty, Implementation of the DPM Monte Carlo code on a parallel architecture for treatment planning applications, *Medical Physics*, 31, 2721–2725, 2004.
5. G. Pratx and L. Xing, Monte Carlo simulation of photon migration in a cloud computing environment with MapReduce, *Journal of Biomedical Optics*, 16, 125003–125003-9, 2011.
6. H. Wang, Y. Ma, G. Pratx, and L. Xing, Toward real-time Monte Carlo simulation using a commercial cloud computing infrastructure, *Physics in Medicine and Biology*, 56, N175, 2011.
7. X. Gu, D. Choi, C. Men, H. Pan, A. Majumdar, and S. B. Jiang, GPU-based ultra-fast dose calculation using a finite size pencil beam model, *Physics in Medicine and Biology*, 54, 6287, 2009.
8. X. Gu, U. Jelen, J. Li, X. Jia, and S. B. Jiang, A GPU-based finite-size pencil beam algorithm with 3D-density correction for radiotherapy dose calculation, *Physics in Medicine and Biology*, 56, 3337, 2011.
9. R. Fujimoto, T. Kurihara, and Y. Nagamine, GPU-based fast pencil beam algorithm for proton therapy, *Physics in Medicine and Biology*, 56, 1319, 2011.
10. S. Hissoiny, B. Ozell, and P. Després, Fast convolution-superposition dose calculation on graphics hardware, *Medical Physics*, 36, 1998, 2009.
11. R. Jacques, R. Taylor, J. Wong, and T. McNutt, Towards real-time radiation therapy: GPU accelerated superposition/convolution, *Computer Methods and Programs in Biomedicine*, 98, 285–292, 2010.
12. Q. Chen, M. Chen, and W. Lu, Ultrafast convolution/superposition using tabulated and exponential kernels on GPU, *Medical Physics*, 38, 1150–1161, 2011.
13. E. Alerstam, T. Svensson, and S. Andersson-Engels, Parallel computing with graphics processing units for high-speed Monte Carlo simulation of photon migration, *Journal of Biomedical Optics*, 13, 060504–060504-3, 2008.
14. A. Badal and A. Badano, Accelerating Monte Carlo simulations of photon transport in a voxelized geometry using a massively parallel graphics processing unit, *Medical Physics*, 36, 4878, 2009.

15. J. Baro, J. Sempau, J. Fernandez-Varea, and F. Salvat, PENELOPE: An algorithm for Monte Carlo simulation of the penetration and energy loss of electrons and positrons in matter, *Nuclear Instruments and Methods in Physics Research Section B: Beam Interactions with Materials and Atoms,* 100, 31–46, 1995.

16. I. Kawrakow and D. Rogers, Technical report PIRS-701, National Research Council of Canada, Ottawa, Ontario, Canada, 2000.

17. I. Lux and L. Koblinger, *Monte Carlo Particle Transport Methods: Neutron and Photon Calculations,* CRC Press, Boca Raton, FL, U.S., 1991.

18. O. Klein and Y. Nishina, Über die Streuung von Strahlung durch freie Elektronen nach der neuen relativistischen Quantendynamik von Dirac, *Zeitschrift für Physik,* 52, 853–868, 1929.

19. U. Fano, H. Hurwitz, and L. V. Spencer, Penetration and diffusion of X-rays. V. Effect of small_deflections upon the asymptotic behavior, *Physical Review,* 77(4), 425–426, February 1950.

20. C. J. Everett and E. D. Cashwell, *A New Method of Sampling the Klein-Nishina Probability Distribution for All Incident Photon Energies above 1 keV,* Los Alamos, NM, 1978.

21. H. Kahn, *Applications of Monte Carlo,* U.S. Atomic Energy Commission, Washington, DC, 1954.

22. J. Sempau, S. J. Wilderman, and A. F. Bielajew, DPM, a fast, accurate Monte Carlo code optimized for photon and electron radiotherapy treatment planning dose calculations, *Physics in Medicine and Biology,* 45, 2263, 2000.

23. F. Sauter, Über den atomaren Photoeffekt in der K-Schale nach der relativistischen Wellenmechanik Diracs, *Annalen der Physik,* 403, 454–488, 1931.

24. S. Hissoiny, B. Ozell, H. Bouchard, and P. Després, GPUMCD: A new GPU-oriented Monte Carlo dose calculation platform, *Medical Physics,* 38, 754–764, 2011.

25. X. Jia, X. Gu, J. Sempau, D. Choi, A. Majumdar, and S. B. Jiang, Development of a GPU-based Monte Carlo dose calculation code for coupled electron/photon transport, *Physics in Medicine and Biology,* 55, 3077, 2010.

26. X. Jia, X. Gu, Y. J. Graves, M. Folkerts, and S. B. Jiang, GPU-based fast Monte Carlo simulation for radiotherapy dose calculation, *Physics in Medicine and Biology,* 56, 7017, 2011.

27. X. Jia, X. Gu, T. McNutt, and S. Hissoiny, TH-A-108-01: Radiation dose calculations on graphics processing units (GPUs): Advances and challenges, *Medical Physics,* 40, 519, 2013.

28. L. Jahnke, J. Fleckenstein, F. Wenz, and J. Hesser, GMC: A GPU implementation of a Monte Carlo dose calculation based on Geant4, *Physics in Medicine and Biology,* 57, 1217, 2012.

29. A. Badal and J. Sempau, A package of Linux scripts for the parallelization of Monte Carlo simulations, *Computer Physics Communications,* 175, 440–450, 2006.

30. P. L'Ecuyer, *Uniform Random Number Generators: A Review,* IEEE Computer Society, Washington, DC, U.S., 1997.

31. mcgpu, Monte Carlo simulation of x-ray transport in a GPU with CUDA. https://code.google.com/p/mcgpu/ (accessed on May 11, 2015).

32. A. Badal and A. Badano, SU-E-I-68: Fast and accurate estimation of organ doses in medical imaging using a GPU-accelerated Monte Carlo simulation code, *Medical Physics*, 38, 3411, 2011.

33. X. Jia, H. Yan, X. Gu, and S. B. Jiang, Fast Monte Carlo simulation for patient-specific CT/CBCT imaging dose calculation, *Physics in Medicine and Biology*, 57, 577, 2012.

34. X. Jia, H. Yan, L. Cervino, M. Folkerts, and S. B. Jiang, A GPU tool for efficient, accurate, and realistic simulation of cone beam CT projections, *Medical Physics*, 39, 7368–7378, 2012.

35. S. Hissoiny, B. Ozell, P. Després, and J.-F. Carrier, Validation of GPUMCD for low-energy brachytherapy seed dosimetry, *Medical Physics*, 38, 4101–4107, 2011.

36. S. Hissoiny, M. D'Amours, B. Ozell, P. Després, and L. Beaulieu, Subsecond high dose rate brachytherapy Monte Carlo dose calculations with bGPUMCD, *Medical Physics*, 39, 4559–4567, 2012.

37. S. Hissoiny, B. Ozell, P. Després, and B. W. Raaymakers, Fast dose calculation in magnetic fields with GPUMCD, *Physics in Medicine and Biology*, 56, 5119, 2011.

38. J. J. Lagendijk, B. W. Raaymakers, A. J. Raaijmakers, J. Overweg, K. J. Brown, E. M. Kerkhof, and B. Härdemark, MRI/linac integration, *Radiotherapy and Oncology*, 86, 25–29, 2008.

39. B. G. Fallone, B. Murray, S. Rathee, T. Stanescu, S. Steciw, S. Vidakovic, E. Blosser, and D. Tymofichuk, First MR images obtained during megavoltage photon irradiation from a prototype integrated linac-MR system, *Medical Physics*, 36, 2084–2088, 2009.

40. P. J. Keall, M. Barton, and S. Crozier, The Australian magnetic resonance imaging-linac program, *Seminars in Radiation Oncology*, 24, 203–206, 2014.

41. S. Mutic and J. F. Dempsey, The viewray system: Magnetic resonance-guided and controlled radiotherapy, *Seminars in Radiation Oncology*, 24, 196–199, 2014.

42. L. Beaulieu, A. Carlsson Tedgren, J.-F. Carrier, S. D. Davis, F. Mourtada, M. J. Rivard, R. M. Thomson, F. Verhaegen, T. A. Wareing, and J. F. Williamson, Report of the Task Group 186 on model-based dose calculation methods in brachytherapy beyond the TG-43 formalism: Current status and recommendations for clinical implementation, *Medical Physics*, 39, 6208–6236, 2012.

CHAPTER 16

Monte Carlo Dose Calculations for Proton Therapy

Xun Jia

CONTENTS

16.1 INTRODUCTION

Proton therapy is an advanced form of radiation therapy. Because of the unique physics of proton beam interaction with matters, proton therapy can achieve higher dose conformality compared to conventional photon-based radiation therapy. However, because of the sharper dose falloff, proton therapy is more susceptible to uncertainties. For instance, a small shift of proton range due to dose calculation in the treatment planning stage can result in a significant underdosage of the tumor and/ or overdosage to the critical structures nearby. Hence, an accurate and efficient dose calculation method is of vital importance for the success of a treatment [1].

At present, pencil beam–based algorithms are widely employed in routine clinical practice [2,3] mainly because of their high computational efficiency and acceptable calculation accuracy in most of the cases. Yet, in some cases, especially for those with a large degree of tissue heterogeneity, the accuracy of pencil beam algorithms is not satisfactory. Although Monte Carlo (MC) dose calculation method is highly desirable due to its well-demonstrated accuracy and its capability of significantly reducing treatment planning margins [1], this novel method still largely remains in research. For instance, it has been used to recalculate existing treatment plans for research studies. The routine clinical application of MC dose calculation is still prohibited by the extremely low computational efficiency [4]. Over the years, despite the great efforts devoted to accelerating the MC dose calculation process [5–8], the achieved efficiency still cannot meet clinical requirement. The low efficiency issue has also hindered the developments of advanced treatment techniques for proton therapy, for example, MC-based treatment planning, adaptive proton radiotherapy, and biological treatment optimization, to name a few.

Recently, with the introduction of GPU in medical physics problems, MC dose calculations in radiotherapy have been substantially accelerated. A number of MC packages have been successfully developed, ranging from MV photon beam dose calculations [9–12] to KV photon transport simulations for imaging purposes [13,14]. In the proton dose calculation regime, a track-repeating algorithm [15] and simplified MC simulation [16] have been developed. Full MC simulations of proton transport have also become available [17,18]. These successes have substantially shortened the computation time, and dose calculation for a real treatment plan can be accomplished in a sub-minute timescale, which greatly facilitates the clinical introduction of MC method in proton therapy practice.

This chapter aims at providing a review of the available GPU-based MC dose calculation packages. In fact, it is quite challenging to develop a proton MC dose calculation package on GPU to achieve both satisfactory accuracy and efficiency. In addition to the complicated physical process of proton transport, the inherent conflict between the GPU's SIMD (single instruction multiple data) processing scheme and the stochastic nature of an MC process also impedes the computation process. Moreover, GPU-based proton MC dose calculation encounters its own difficulties such as memory writing conflict, also known as race condition in parallel computation field. While this problem also occurs in other GPU-based MC dose calculations, for example, for the photon beam, proton beam's unique physics exacerbates this issue. Hence, in this chapter, in addition to summarizing techniques employed in each of the available packages, GPU-specific technical issues will also be discussed.

16.2 GPU-BASED PROTON MC DOSE CALCULATION PACKAGES

16.2.1 Track-Repeating MC

The first attempt to use GPU to accelerate MC-based proton dose calculations was in 2010, where a track-repeating method was implemented [15]. Track repeating is actually a variance reduction technique in MC particle transport simulations. Specific for the proton dose calculation problem it was first utilized by Li et al. [5]. In this method, a database of proton transport histories is needed, which is typically generated in homogeneous phantoms. Each of the particle history stored in the database records information for each proton step, the direction, step length, and energy loss. Because this database is prepared only once before actual dose calculation tasks, the computational load for it is not a concern. One typically performs careful simulations using a package that have well-demonstrated accuracy, for example, Geant4 [19] for the database generation.

Once the database is prepared, dose calculation in a patient case can be achieved by repeating appropriate proton tracks in the database but in the patient's body. Specifically, it first selects a track in the database corresponding to a source proton. The proton is then propagated following the exact path specified in the track database. The underlying assumption is that the random numbers generated while transporting this proton in the patient are identical to what occurred when generating the track in the database, and hence yielding an identical trajectory. To account for the different material properties between the patient and those in which the tracks were generated, scaling of step length and scattering angle is needed.

This track-repeating method is computationally efficient and GPU-friendly. First, it eliminates the needs of sampling of physical interactions on the fly, the majority of the computational burden in particle transport simulations. Second, in the context of GPU, the track-repeating method is easily parallelizable, as each GPU thread essentially performs the same operations all the time, namely, fetching a step from the database and repeating it in the patient's body. Hence, a high computational efficiency can be achieved with the full GPU power employed. In the validation studies conducted in Reference 15, it was reported that dose calculation can be achieved in less than 1 min with a dual-GPU system equipped with Geforce GTX 295 GPUs, corresponding to a speedup factor of 75.5 with respect to the implementation on CPU. In terms of accuracy, the dosimetric results of this method agree with those from a full MC simulation using Geant4 within 1% discrepancy.

16.2.2 Simplified MC Simulations

To speed up dose calculations, efforts have also been devoted to the development of MC simulations for proton transport with simplified physics [6]. The package, named SMC, has been developed on a GPU platform [16]. The SMC package transports a proton by first setting its location, velocity direction, and residual range in water. The proton is then set to move and travel through a voxelized geometry defined by the patient's CT. Two effects are modeled at each voxel. First, the proton's residual range in water is decreased according to the local material property. Correspondingly, a certain amount of energy deposition is calculated through a water equivalent model [20] based on the measured depth-dose distribution in water, which is deposited to the voxel. Second, multiple Coulomb scattering of the proton is also modeled. The scattering angle is sampled from a normal distribution with a standard deviation given by Highland's formula [21], and the proton direction is changed. This is essentially an effective model that phenomenologically describes the proton transport and dose deposition process. For instance, as opposed to calculating dose deposition according to the actual physical interactions, SMC determines this quantity using measured depth-dose distribution in water.

Similar to the track-repeating method, this SMC algorithm is also very suitable for the GPU's SIMD structure. In fact, the aforementioned proton transport process can be implemented easily by setting each GPU for one proton independently. All of the threads essentially perform the same instructions but using different data according to the current proton status. High-speed shared memory has also been employed in SMC implementation to further reduce the burden of repeatedly accessing the relatively slow global memory of a GPU card. A speedup factor of 12–16 compared to CPU implementation has been observed in real clinical cases. The absolute dose calculation time was found to be 9–67 s with a clinically acceptable uncertainty level on an NVIDIA Tesla C2050 GPU.

16.2.3 Full MC Simulations

Although the accuracies for the two aforementioned MC codes are sufficient for most clinical applications, in some difficult cases, for example, those with a high level of heterogeneity such as head and neck area, a full MC simulation with detailed considerations of proton transport and physical interaction process is still desired. Aiming at realizing a full MC simulation on GPU, Jia et al. first developed a package called gPMC [17].

Proton transport is a complicated process. Generally speaking, interactions can be divided into electromagnetic (EM) interaction channel and nuclear interaction channel, with the former one dominating dose depositions. In gPMC, the simulations of EM channel are modeled by a Class II condensed history simulation scheme using the continuous slowing down approximation. Specifically, the proton is transported in a step-by-step fashion until its energy is below a user-defined cutoff energy or it exits the phantom region. Each step terminates at an interaction point, a voxel boundary, or an upper bound set by the user, whichever is the closest to the current position. Once a step length d in the medium is determined, its equivalent length in water is calculated based on the mass stopping power ratio as

$$d_w = df_s(E,i) \frac{\rho}{\rho_w} \tag{16.1}$$

where ρ_w is the water density. The mean energy decrease in this step $\overline{\Delta E}$ is calculated by numerically solving an equation

$$d_w = -\int_E^{E-\overline{\Delta E}} \frac{dE'}{L_w(E')}, \tag{16.2}$$

using the scheme developed by Kawrakow [22]. The actual energy decrease is further calculated as $\Delta E = \overline{\Delta E} + \zeta$, with ζ being a Gaussian random number with a zero mean and a certain variance calculated for this step [23], to account for the energy straggling. After moving the electron for a distance d, its direction is changed by an angle sampled from a Gaussian distribution [8,24] to model multiple scattering due to elastic Coulomb interactions. For simplicity, lateral deflection within each step is not modeled in gPMC. This has been shown to be accurate enough for the heavy proton particle [8].

At the end of each step, secondary particles may be produced. If a δ-electron from an ionization event is generated, it is simply terminated and the energy is locally deposited. Neglecting the electron transport will lead to negligible error in most clinical cases due to the small electron range in human tissue in this energy range. As for nuclear interactions, gPMC follows an empirical strategy developed previously by Fippel and Soukup [8]. It only considers three possible channels that are most relevant to human body: proton–proton elastic interactions and proton–oxygen elastic and inelastic interactions. The secondary protons generated in these interactions are put in a stack and are transported by the same proton transport physics mentioned earlier. All other heavy ions are not followed and their energies are locally deposited. Charge-neutral particles produced in the proton–oxygen inelastic events are neglected.

Regarding implementations on GPU, gPMC uses a kernel function to simultaneously launch a set of GPU threads, each transporting a proton till it is absorbed or is out of the regions of interest. Secondary protons during this step are put in a stack located in the global memory of the GPU card. This kernel is repeatedly executed, till all the source protons are simulated and all the stacks are empty. In addition, gPMC performs the simulation in a batched fashion. The results from different batches are used for statistical analysis to obtain the average dose to each voxel and the corresponding uncertainties. To further ensure the computational efficiency, a pseudo-random number generator CURAND [25] developed by NVIDIA is utilized, which offers simple and efficient generation of high-quality pseudo-random numbers using the XORWOW algorithm. GPU texture memory is also used to support hardware-based interpolation on data, such as cross section and stopping power data.

The calculation accuracy of gPMC has been established by comparing the dose calculation results with those from TOPAS/Geant4 [26], a golden standard MC simulation package for proton therapy. For a set of cases with different patient sizes and heterogeneity levels, satisfactory agreements between gPMC and TOPAS/Geant4 have been observed. Gamma test with 2%/2 mm criteria has been employed to quantify the dose agreement, and typical passing rate is over 98% in the region with dose greater than 10% maximum dose. Figure 16.1 shows a dose distribution of a typical treatment plan overlaid on the patient's CT images. Comparisons of the dose profiles along two lines are also shown in Figure 16.1. With respect to computational efficiency, it takes only 6–22 s to simulate 10 million source protons to yield ~1% relative statistical uncertainty on an NVIDIA C2050

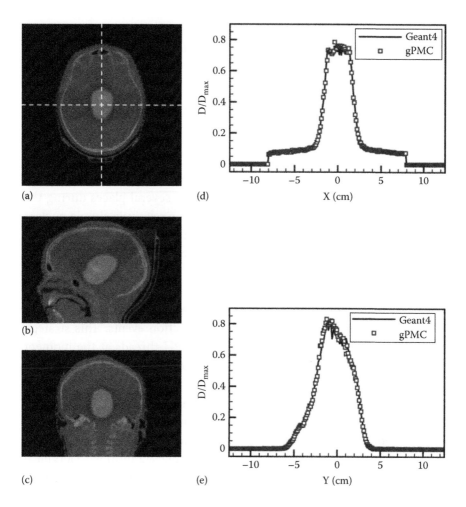

FIGURE 16.1 (a through c) Dose deposition of a typical proton therapy plan. (d through e) Dose profiles along the horizontal and the vertical dash lines in (a).

GPU card. The variation of time is mainly due to phantom voxel size and beam energy. Compared to hours of computation time for TOPAS/Geant4, gPMC has achieved extremely high efficiency.

Recently, another GPU-based MC package for proton transport has been reported [18]. While the proton transport via condensed history technique is a standard method, this code has several improvements over gPMC both on the physics side and on the implementation side. First of all, a detailed treatment of nuclear interaction process is implemented, instead of the simple approach employed in gPMC [8]. This is important in certain situations, for example, in patients with metallic implants.

In their work, a nonelastic proton–nucleus interaction process is simulated via a two-step model, a fast intranuclear (INC) cascade that leaves the nucleus in an excited state, followed by an evaporation stage. With a careful GPU-based implementation, a speedup factor of ~50 for this nuclear interaction part has been reached compared to a CPU implementation in Geant4 [27].

Second, the simulation workflow of proton transport is designed to effectively incorporate this nuclear interaction module. In gPMC, a thread transports a proton all the way to its end (being absorbed or exit from the phantom). Nuclear interactions could occur at several places during the thread execution. Yet, if one were to replace the relatively simple nuclear interaction simulations in gPMC with this complicated new model, efficiency would be a concern due to the associated thread branching issue. To overcome this difficulty, a new workflow is employed, where each GPU thread transports a proton to a place at which a nuclear interaction occurs. After that, GPU kernels that simulate nuclear interaction processes are invoked to simultaneously treat all the interaction events. This strategy manually imposes synchronization among GPU threads at the moment where nuclear interactions are needed, hence ensuring computational efficiency.

With the improved nuclear interaction module, energy and angular distributions of secondary particles from nonelastic collisions are found to be in good agreement with the Geant4 calculation results. The resulting dose distributions also pass 3D gamma test (2%/2 mm) with a typical passing rate of 98%. It takes ~20 s to simulate 1×10^7 proton histories on an NVIDIA GTX680 card, including all CPU–GPU data transfers.

16.3 GPU MEMORY WRITING CONFLICT

16.3.1 Memory Writing Conflict Problem

One issue in GPU-based proton MC simulations is memory writing conflict, which was reported by Jia et al. in gPMC [17]. This problem occurs when two GPU threads happen to update the same dose counter. At this moment, when two threads attempt to deposit dose to the same voxel, the writing process has to be serialized, as otherwise the result would be unpredictable depending on the actual sequence of visiting this memory address, known as racing condition. This serialization apparently reduces the overall parallel processing capability of a GPU, prolonging computation time. A higher frequency of conflict causes a lower computational efficiency.

However, this memory writing conflict is a random event. Proton transport processes in different GPU threads are independent and it is not possible to predict when two threads would deposit dose to the same voxel at the same time. Yet, this problem occurs more often in proton beam dose calculations than in photon beam dose calculations. This is because protons travel almost in a straight line with little lateral deflection. A parading column of protons in a beam, especially in a small-size beam, marches in almost synchronized steps, which leads to a high frequency of memory writing conflicts.

The impacts of this memory writing conflict problem on computational efficiency have been demonstrated in Reference 17 in a simple test case where a fixed proton beam energy of 200 MeV impinges to a water phantom. MC simulation experiments were repeatedly conducted with different field sizes f and the deposition time was recorded. It was found that, as the field size decreased, the deposition time monotonically increased. In particular, when it comes to the cases with a field size of $f=1$ cm or less, the dose deposition time is comparable to or even longer than the particle transport time.

A theoretical investigation was also proposed to further understand this effect. Suppose the dose deposition time per event is Δt and there are N events occurring in an MC simulation. Among them, a fraction of $p < 1$ events gets the memory conflict problem and the depositions are serialized, which leads to a time of $Np\Delta t$. The rest of $(1-p)$ are deposited in parallel by all GPU threads, and hence, the time is $N(1-p)\Delta t/N_{thread}$, where N_{thread} is the number of GPU threads during the simulations. Assuming that the probability of this memory conflict is inversely proportional to the field size, $p = \alpha/f^2$. Hence, the total dose deposition time is

$$t = \frac{\alpha N \Delta t}{f^2} + \frac{N \Delta t}{N_{thread}} \left(1 - \frac{\alpha}{f^2}\right). \tag{16.3}$$

The data for dose deposition time as a function of field size can be nicely fitted by this function form, as shown in Figure 16.2.

16.3.2 Solutions to the Problem

This memory writing issue is indeed a severe problem, as in many situations, calculations of dose for small proton beams are necessary. For instance, in the emerging intensity-modulated proton therapy (IMPT)

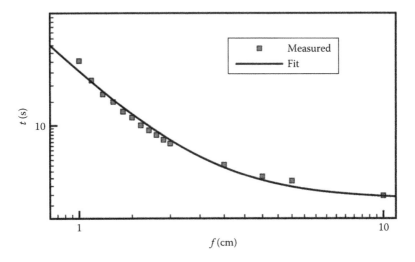

FIGURE 16.2   Dose deposition time as a function of the field size. (Reproduced from Jia, X. et al., *Phys. Med. Biol.*, 57(23), 7783, 2012.)

technology, dose of each proton pencil beam is the prerequisite for the subsequent treatment plan optimization stage. The key to this problem is to reduce the chances of memory writing conflict. There have been two solutions at present depending on the context.

When dose distribution for a small-sized proton beam is necessary, for example, in the case of a small tumor, the memory writing conflict problem can be alleviated via a multi-counter technique. Specifically, multiple dose counters, as opposed only one in gPMC, are allocated in the GPU's global memory. During dose calculations, each dose deposition event can be randomly directed to one of the counters with equal probability. Only at the end of dose calculation, doses from these counters are added to yield the total dose. This strategy apparently reduces the chances of memory writing conflicts, at the price of using more GPU space and one additional random number generation at each dose deposition event to determine the counter index. Figure 16.3 demonstrates the effectiveness of this strategy. Although the dose deposition time increases as the field size is reduced, the speed of increase is much lower when 10 dose counters are utilized, compared to the single-counter situation. Another advantage of this multi-counter technique is the feasibility of estimating statistical uncertainties using doses in them. Statistically speaking, doses recorded in these counters are independent due to the randomness when selecting

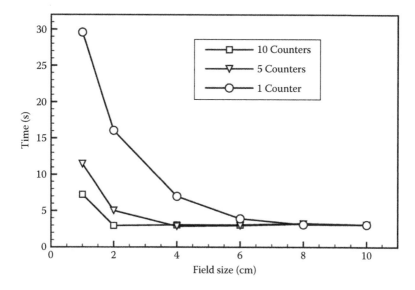

FIGURE 16.3 Dose deposition time as a function of the field size with different numbers of dose counters.

a counter. Results in these counters can be used to estimate uncertainties, which hence eliminates the cumbersome batched method employed in gPMC for the same purpose.

When dose distributions for a number of beams with small sizes are needed, for example, in the case of IMPT for pencil beam dose calculations, a particle labeling method can be used. In this method, as opposed to computing doses for each beamlet sequentially, those for a group of beamlets are commutated simultaneously. As such, source protons are sampled from these beamlets of interest and transport by the MC algorithm. Each proton is tagged by an index that specifies which beamlet it is from. The index is also passed to all the secondary particles generated. Meanwhile, a set of dose counters are allocated to score doses from each beamlet. Each dose deposition event is directed to a counter corresponding to the beamlet index. In essence, this particle labeling method also alleviates the memory writing conflict problem by directing dose depositions to multiple memory addresses. Because in the context of beamlet dose calculations for IMPT each individual beamlet dose is needed, the separation of dose deposition is naturally realized. It is the particle labeling that enables one to achieve this separation. In practice, it is found that this strategy enables the computations of beamlet doses in an efficient fashion, such that the time it takes to compute doses for a group of beamlets that reside in a broad beam is slightly longer than the time to compute the dose for the broad beam [28]. The additional time is ascribed to the required operations of computing memory address for dose depositions according to the beamlet tag.

16.4 CONCLUSION AND DISCUSSIONS

In this chapter, we have reviewed a set of GPU-based MC dose calculation packages for proton therapy. Simplified MC and track-repeating MC simulations were first developed on the GPU platform. Recently, with the increasing power of GPU cards, full MC simulation is not burdened by the low efficiency anymore. MC packages with detailed proton transport process have become available. Substantially improved computational speed compared to that on the conventional CPU platform has been achieved. We believe these developments will greatly facilitate clinical introduction of GPU-based MC into proton therapy clinical practice. For instance, Mayo clinic has implemented their MC package to perform routinely quality assurance tests on treatment plans generated by a commercial treatment planning system that uses pencil beam–type dose calculation algorithm [18].

Despite the success so far, we should also be aware that a few challenges still exist. Overcoming these challenges will not only help to bring MC simulations into routine clinic, it will also hold the potential of advancing the proton therapy with novel technologies. First, one would probably never stop pursuing higher and higher computational speed. While it is always possible to gain speed through brute-force ways, for example, using multiple GPU cards or GPUs with higher performance, it is desired to develop algorithms to accelerate the computations. Possible directions include employing variance reduction techniques. In addition, designing simulation schemes with controlled synchronization among threads would also help to relieve the inherent conflict between the GPU SIMD structure and the MC randomness. Another example is the memory writing conflict issue discussed in this chapter. Developing novel methods to maximally reduce this conflict is an interesting and important topic in GPU-based proton MC simulations. Second, there is also a desire to model more detailed physics in the MC simulation process. This is probably task dependent. At present, we can achieve satisfactory level of accuracy for physical dose calculations, as having been demonstrated in many studies summarized in this chapter. However, with the advancements of proton radiotherapy, one will demand other information derived from MC simulations to facilitate clinical investigations. One example is biological dose calculations, where linear energy transfer for all different particle types is needed. Although MC is potentially capable of computing this quantity, it certainly requires modeling the transport of all secondary particles, which is currently not yet available. It is certainly an open question regarding to what degree of details one should model the transport process for a satisfactory level of accuracy. The associated increase in computation load along this way will also pose a problem.

REFERENCES

1. Paganetti, H., Range uncertainties in proton therapy and the role of Monte Carlo simulations. *Physics in Medicine and Biology*, 2012. **57**(11): R99–R117.
2. Hong, L. et al., A pencil beam algorithm for proton dose calculations. *Physics in Medicine and Biology*, 1996. **41**(8): 1305–1330.
3. Schaffner, B., E. Pedroni, and A. Lomax, Dose calculation models for proton treatment planning using a dynamic beam delivery system: An attempt to include density heterogeneity effects in the analytical dose calculation. *Physics in Medicine and Biology*, 1999. **44**(1): 27–41.
4. Paganetti, H. et al., Clinical implementation of full Monte Carlo dose calculation in proton beam therapy. *Physics in Medicine and Biology*, 2008. **53**(17): 4825–4853.
5. Li, J.S. et al., A particle track-repeating algorithm for proton beam dose calculation. *Physics in Medicine and Biology*, 2005. **50**(5): 1001–1010.
6. Kohno, R. et al., Experimental evaluation of validity of simplified Monte Carlo method in proton dose calculations. *Physics in Medicine and Biology*, 2003. **48**(10): 1277–1288.
7. Yepes, P. et al., Monte Carlo fast dose calculator for proton radiotherapy: Application to a voxelized geometry representing a patient with prostate cancer. *Physics in Medicine and Biology*, 2009. **54**(1): N21–N28.
8. Fippel, M. and M. Soukup, A Monte Carlo dose calculation algorithm for proton therapy. *Medical Physics*, 2004. **31**(8): 2263–2273.
9. Jia, X. et al., Development of a GPU-based Monte Carlo dose calculation code for coupled electron-photon transport. *Physics in Medicine and Biology*, 2010. **55**: 3077.
10. Hissoiny, S. et al., GPUMCD: A new GPU-oriented Monte Carlo dose calculation platform. *Medical Physics*, 2011. **38**(2): 754–764.
11. Jia, X. et al., GPU-based fast Monte Carlo simulation for radiotherapy dose calculation. *Physics in Medicine and Biology*, 2011. **56**(22): 7017–7031.
12. Jahnke, L. et al., GMC: A GPU implementation of a Monte Carlo dose calculation based on Geant4. *Physics in Medicine and Biology*, 2012. **57**(5): 1217–1229.

13. Badal, A. and A. Badano, Accelerating Monte Carlo simulations of photon transport in a voxelized geometry using a massively parallel graphics processing unit. *Medical Physics*, 2009. **36**(11): 4878–4880.

14. Jia, X. et al., Fast Monte Carlo simulation for patient-specific CT/CBCT imaging dose calculation. *Physics in Medicine and Biology*, 2012. **57**(3): 577–590.

15. Yepes, P.P., D. Mirkovic, and P.J. Taddei, A GPU implementation of a track-repeating algorithm for proton radiotherapy dose calculations. *Physics in Medicine and Biology*, 2010. **55**(23): 7107–7120.

16. Kohno, R. et al., Clinical implementation of a GPU-based simplified Monte Carlo method for a treatment planning system of proton beam therapy. *Physics in Medicine and Biology*, 2011. **56**(22): N287.

17. Jia, X. et al., GPU-based fast Monte Carlo dose calculation for proton therapy. *Physics in Medicine and Biology*, 2012. **57**(23): 7783–7797.

18. Tseung, H.W.C., J. Ma, and C. Beltran, *Medical Physics*, 42, 2967–2978, (2015). doi:http://dx.doi.org/10.1118/1.4921046.

19. Agostinelli, S. et al., GEANT4–A simulation toolkit. *Nuclear Instruments & Methods in Physics Research Section A-Accelerators Spectrometers Detectors and Associated Equipment*, 2003. **506**(3): 250–303.

20. Chen, G.T.Y. et al., Treatment planning for heavy ion radiotherapy. *International Journal of Radiation Oncology Biology Physics*, 1979. **5**(10): 1809–1819.

21. Highland, V.L., Some practical remarks on multiple scattering. *Nuclear Instruments & Methods*, 1975. **129**(2): 497–499.

22. Kawrakow, I., Accurate condensed history Monte Carlo simulation of electron transport. I. EGSnrc, the new EGS4 version. *Medical Physics*, 2000. **27**(3): 485–498.

23. Geant4 Collaboration. *Geant4 Physics Reference Manual*. 2011. Available from: http://geant4.web.cern.ch/geant4/UserDocumentation/UsersGuides/PhysicsReferenceManual/fo/PhysicsReferenceManual.pdf.

24. Hagiwara, K. et al., Review of particle physics. *Physical Review D*, 2002. **66**(1), 010001.

25. NVIDIA, *CUDA CURAND Library*. 2010. http://docs.nvidia.com/cuda/curand/#axzz3cRLZ4KqT.

26. Perl, J. et al., TOPAS: An innovative proton Monte Carlo platform for research and clinical applications. *Medical Physics*, 2012. **39**(11): 6818–6837.

27. Tseung, H.W.C. and C. Beltran, A graphics processor-based intranuclear cascade and evaporation simulation. *Computer Physics Communications*, 2014. **185**(7): 2029–2033.

28. Li, Y. et al., Development of a high performance dose calculation and plan optimization system for intensity modulated proton therapy. *Presentation at the 54 Annual Conference of the Particle Therapy Co-Operative Group*, San Diego, CA, 2015.

CHAPTER 17

Treatment Plan Optimization for Intensity-Modulated Radiation Therapy (IMRT)

Chunhua Men

CONTENTS

17.1 INTRODUCTION

INTENSITY-MODULATED RADIATION THERAPY (IMRT) is one of the most used techniques to treat many forms of cancer. In IMRT, the linear accelerator is equipped with a multileaf collimator (MLC) system. By dynamically moving the leaves, very complex dose distributions can be delivered to the patient; therefore, sufficient dose can be delivered to the tumor, while nearby organs and tissues can be spared.

There are different kinds of optimization problems associated with IMRT treatment planning. In a typical IMRT treatment setup, 3–9 beams are usually used. Choosing the right number of beams with associated beam angles is called the *beam angle optimization* (BAO) problem. Furthermore, IMRT treatment plans are traditionally developed using a two-stage process. In the first stage, each beam is modeled as a collection of small beamlets, and the intensities of each of these beamlets are assumed to be controllable. The problem of finding an optimal intensity profile or fluence map for each beam is called the *fluence map optimization* (FMO) problem. The FMO problem must then be followed by a leaf sequencing stage in which the fluence maps are decomposed into a manageable number of apertures that are deliverable using an MLC system, which is called the *leaf sequencing optimization* (LSO) problem. A major drawback of decoupling of the treatment planning problem into a beamlet-based FMO problem and an LSO problem is that there is a potential loss of treatment quality. This has led to the development of approaches that either further improve the treatment quality by modifying the existing apertures from the LSO stage, which are referred to as *aperture shape optimization* (ASO) approaches, or integrate the beamlet-based FMO and leaf sequencing problems into a single optimization model, which are usually referred to as *direct aperture optimization* (DAO) approaches.

IMRT optimization problems are usually large scale, and it can be very time consuming to find the optimal solution. The current state-of-the-art optimization models implemented on a single central processing unit (CPU) often take minutes or even hours to optimize a treatment plan. Speedup of the IMRT treatment plan optimization process is highly desirable and can help (1) save planners' time, (2) enable online adaptive planning, and (3) make interactive planning possible. Parallel computing techniques offer great potential to improve computational efficiency. One way to take advantage of parallel computing is to use computer clusters or traditional supercomputers. However, they are not readily available to most clinical users. In addition, communications among CPUs can be expensive, leading to marginal efficiency gains for solving IMRT optimization problems. Multi-core processor PCs make parallel computing affordable for most users but the level of parallelism is still limited—in the best case, the speedup factor gained by the use of a multi-core processor is equal to the limited number of cores. Modern graphics processor units (GPUs), on the other hand, offer hundreds of processing cores, can be effectively used for parallel computing, and are affordable to most users. While GPUs were traditionally used primarily for graphics applications, they have recently been widely adopted for general-purpose scientific computing, especially since the introduction of the CUDA® (Compute Unified Device Architecture) platform by NVIDIA in 2007.

In this chapter, we describe the design of suitable optimization algorithms to solve FMO and DAO problems in IMRT and their parallelized implementation on CUDA-enabled GPUs [1,2].

17.2 FLUENCE MAP OPTIMIZATION

17.2.1 Model of the Problem

For fluence map optimization in IMRT, each beam is decomposed into a set of beamlets. Let us denote the number of beamlets by N and the set of voxels in a patient's planning image by V. In addition, we denote the dose to voxel $j \in V$ by z_j and the intensity of a beamlet by x_i ($x_i \geq 0$) for $i \in N$. Without loss of generality, the FMO model can be formulated as follows:

$$\text{min } F(\boldsymbol{d}),$$
$$\text{Subject to} : G(\boldsymbol{d}) \leq 0. \tag{17.1}$$

In this equation, $\boldsymbol{d}$ represents the voxel dose distribution. The dose for each voxel is calculated as $d_j = \sum_{i \in N} D_{ij} x_i$, where D_{ij} is called the *dose deposition coefficient*, which represents the dose received by voxel j from beamlet i of unit intensity. $F(\boldsymbol{d})$ denotes the collection of treatment plan evaluation criterion objective functions, and $G(\boldsymbol{d})$ is the collection of treatment plan evaluation criterion constraints. Each of these criteria is usually a function of the dose distribution for a particular structure only. Criteria in the objective functions are expressed in such a way that smaller values are preferred to larger values. The FMO model can be assumed to be convex. This is justified by the fact that most criteria proposed in the literature to date are indeed convex or they can be replaced by a convex equivalent.

In this work, we consider an FMO model that employs treatment plan evaluation criteria based on quadratic, one-sided voxel-based penalties. If we denote the set of target voxels by V_T, we can write the criteria as

$$F_j^-(z_j) = \alpha_j \left(\max\{0, T_j - z_j\} \right)^2 \quad j \in V_T$$
$$F_j^+(z_j) = \beta_j \left(\max\{0, z_j - T_j\} \right)^2 \quad j \in V. \tag{17.2}$$

where

α_j and β_j represent the penalty weights for underdosing and overdosing, respectively

T_j represents the penalty threshold

If generating a plan from scratch, T_j can be set to be the prescription dose; when replanning or re-optimizing a plan as in adaptive radiotherapy, T_j can be the dose from the original plan. In this work, we let $F(\mathbf{d}) = \sum_{j \in V} \left[F_j^-(z_j) + F_j^+(z_j) \right]$. For $G(\mathbf{d})$, we consider the nonnegative constraints for beamlet intensities. There is no other constraint in the following implementation.

17.2.2 GPU Implementation

The objective function is convex and quadratic in the model described earlier; thus, a gradient descent method can be used to find its solution. In particular, the direction of steepest descent is that of the negative gradient of the objective function. However, since the decision variables, which are the intensities of the beamlets, need be nonnegative, we use the gradient projection method that projects the negative gradient in a way that improves the objective function while maintaining the feasibility of the solution.

In our implementation, the CPU serves as the host while the GPU is called the device. The main code runs on the host, invoking kernels that are executed in parallel on the device by using a large number of CUDA threads. Figure 17.1 depicts the flowchart of the GPU implementation of

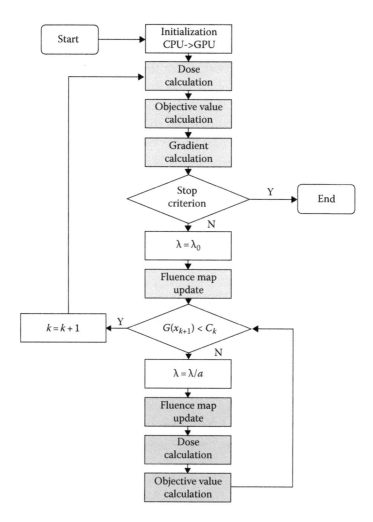

FIGURE 17.1 Flowchart of our GPU-based gradient projection algorithm. Gray boxes indicate CUDA kernels running on the device (GPU) and white boxes indicate C code running on the host (CPU). (Reprinted from Men, C. et al., *Phys. Med. Biol.*, 54, 6565, 2009.)

the gradient projection algorithm. The majority of the algorithm runs on the device as CUDA kernels (gray boxes), while the remaining simple arithmetic operations run on the host (white boxes). Four kernels are used in the CUDA implementation: dose calculation, objective function calculation, gradient calculation, and fluence map update. For the dose calculation, each thread calculates the dose received by a certain voxel. For the objective function calculation, each thread calculates the objective value contributed by a certain voxel, which is summarized to a total objective value using parallel reduction and loop unrolling operations. For the gradient calculation, each thread calculates the gradient value at a certain beamlet. Some global variables, such as x_i, may be accessed by various threads simultaneously, causing massive memory accesses that can significantly impair the efficiency. To avoid this, we store those variables in texture memory which can be accessed as cached data.

The dose deposition coefficient matrix is a sparse matrix due to the fact that a beamlet only contributes to voxels close to its path. We, therefore, store the D_{ij} using the compressed sparse row (CSR) format, which is the most popular general-purpose sparse matrix representation. To improve the performance of our CUDA code for calculating the voxel dose, which is actually a sparse matrix–vector multiplication, a CUDA/GPU specified method [3] was implemented.

17.2.3 Experimental Results

To test the implementation, one clinical case of prostate cancer is used with three scenarios of different beamlet/voxel sizes (Table 17.1). For each scenario, nine coplanar beams are evenly distributed around the patient and the prescription dose to planning target volume (PTV) is 73.8 Gy. For target and organs at risk (OARs), a voxel size of $4 \times 4 \times 4$ mm^3 is used in the first two scenarios and $2.5 \times 2.5 \times 2.5$ mm^3 in the last scenario. For unspecified tissues (i.e., tissues outside the target and OARs), the voxel size is increased in each dimension by a factor of 2 to reduce the optimization problem size. The full resolution is used when evaluating the treatment quality (dose–volume histograms [DVHs], dose color wash, isodose curves, etc.).

The CUDA implementation is tested on an NVIDIA Tesla C1060 GPU card, which has 30 multiprocessors (each with 8 SIMD processing cores) and 4 GB of memory. For comparison purposes, the algorithm is also implemented in sequential C code and runs on an Intel Xeon 2.27 GHz CPU. The same sparse matrix format to represent the sparse matrix D is used for both CPU and GPU implementations.

The accuracy of the results is first analyzed. The input data are stored in single floating point precision on both CPU and GPU. The difference between the final objective function values calculated by CPU and GPU is around $10^{-6} \sim 10^{-5}$ which is negligible in clinical practice. In fact, when checking the DVHs, dose color wash, and isodose curves, no differences can be observed between the CPU and the GPU results.

TABLE 17.1   Running Times for Plan Optimization for CPU and GPU Implementations Tested on a Clinical Case with Various Beamlet and Voxel Sizes

#	# Beams	Beamlet Size (mm²)	# Beamlets	Voxel Size (mm³)	# Voxels	# Nonzero D_{ij}'s	CPU (s)	GPU (s)	Speedup
1	9	10×10	2055	$4 \times 4 \times 4$	35,988	3,137,805	3.81	0.19	20.1
2	9	5×5	6453	$4 \times 4 \times 4$	35,988	10,612,611	16.4	0.49	33.5
3	9	5×5	6453	$2.5 \times 2.5 \times 2.5$	143,329	43,266,357	111.8	2.79	40.1

Source: Reprinted from Men, C. et al., *Phys. Med. Biol.*, 54, 6565, 2009.

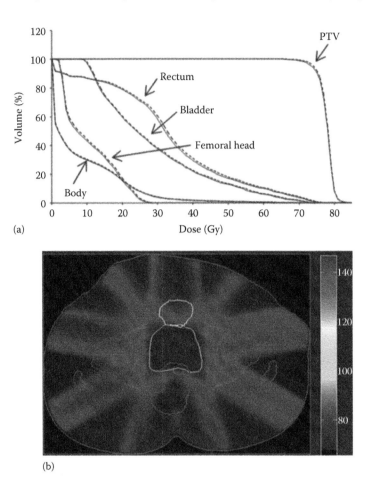

(a)

(b)

FIGURE 17.2 The optimal treatment plan obtained for Scenario 3 on GPU: (a) DVHs (the solid lines represent the new plan; the dot lines represent the original plan); (b) dose color wash on a representative computed tomography (CT) slice. (Reprinted from Men, C. et al., *Phys. Med. Biol.*, 54, 6565, 2009.)

Figure 17.2 shows the DVHs and dose wash superimposed on a representative CT slice of this clinical case, both corresponding to an optimal treatment plan for Scenario 3 on GPU, which corresponds to a replanning case. For comparison purposes, the DVHs from the original plan are also shown in this figure. The running time for plan optimization for the three scenarios on both CPU and GPU is shown in Table 17.1. The speedup achieved by the GPU implementation over the CPU implementation improves from 20.1 to 40.1 as the problem size increased from Scenario 1 to Scenario 3.

17.3 DIRECT APERTURE OPTIMIZATION

17.3.1 Model of the Problem

Even though the DAO problems are well studied in the literature, most of the works focus on heuristic search methods, such as simulated annealing and genetic method. These search algorithms in general are inefficient. Another method has been developed to solve DAO problem in a deterministic way [4,5], which uses a column generation method to handle its large dimensionality and high-quality treatment plans can be obtained in ~2 min. To further improve its efficiency for treatment planning, the method can be implemented on GPU. In the following, we describe the implementation of the column generation method on GPU to solve IMRT DAO problems.

The notation for FMO in previous section is still applicable here. In addition, we will denote the set of deliverable apertures by K, and a particular set of beamlets $A_k \in N$ forms each aperture $k \in K$. With each aperture k, we also associate a decision variable y_k that indicates the intensity of the aperture. The voxel dose d_j ($j \in V$) is calculated using a linear function of the intensities of the apertures through the dose deposition coefficients D_{kj}, the dose received by voxel $j \in V$ from aperture $k \in K$ at unit intensity:

$$d_j = \sum_{k \in K} D_{kj} y_k \qquad (17.3)$$

D_{kj} is obtained from the following equation:

$$D_{kj} = \sum_{i \in A_k} D_{ij} \qquad (17.4)$$

The DAO model employs the same treatment plan evaluation criteria as our FMO model and it can be written as

$$\min F(\boldsymbol{d}) \qquad (17.5)$$

subject to:

$$d_j = \sum_{k \in K} D_{kj} y_k \qquad j \in V$$

$$y_k \geq 0 \qquad k \in K.$$

In addition, the following relationship between the sets of decision variables has to hold:

$$x_i = \sum_{k \in K_i} y_k \quad i \in N. \tag{17.6}$$

17.3.2 The Optimization Algorithm

The model is convex and the decision variables are the intensities of all MLC deliverable apertures. Mathematically, the DAO model is the same as the FMO model developed in the previous work [1] in which the decision variables are the intensities of all beamlets. While the same gradient projection method may be used to solve DAO problem, it is neither practical nor necessary. It is clear that the number of deliverable apertures is huge. Since the FMO model is already a challenging large-scale optimization problem (number of decision variables is around 10^3–10^4), the DAO model is intractable due to the extremely large solution space (number of decision variables is more than 10^{18}). On the other hand, we only need to select 30–80 apertures for deliverability consideration, even if we are able to solve such a huge problem directly. We, therefore, adopt a column generation method that iteratively solves a subproblem and a master problem. In each iteration, a subproblem (also called the *pricing problem*) is first solved, which either adds a suitable aperture to a given pool of allowed apertures or concludes that the current solution is clinically acceptable [4,5]. A master problem is then followed to find the optimal intensities associated with those currently generated apertures. The flowchart for implementing our DAO algorithm is shown in Figure 17.3.

17.3.3 The Master Problem

The master problem aims to find the optimal intensities for those selected apertures. Note that the FMO model aims to find the optimal intensity for each beamlet. Mathematically, the optimization model of the master problem is equivalent to the FMO model introduced previously. Therefore, the same gradient projection algorithm for FMO and the same GPU CUDA code are used here to solve the master problem.

17.3.4 The Pricing Problem

In each iteration, the pricing problem identifies one aperture that decreases the objective value most if added to the master problem. Let us denote the

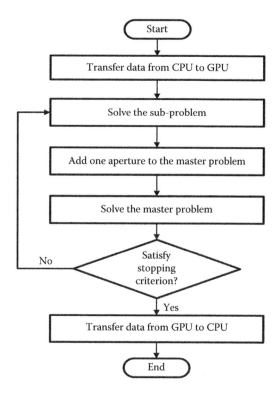

FIGURE 17.3 Flowchart of the GPU-based column generation method for solving the DAO problem. (Reprinted from Men, C. et al., *Phys. Med. Biol.*, 55, 4309, 2010.)

set of beams by B and the set of apertures that can be delivered by an MLC system from beam direction $b \in B$ by K_b. We further denote the dual multiplier associated with constraints $d_j = \sum_{k \in K} D_{kj} y_k$ by π_j for $j \in V$ (for more detailed explanation of the dual multiplier concept, please refer to any optimization text books, e.g., [6]). The pricing problem can then be written in the following form:

$$\min_{k \in K_b} \sum_{i \in A_k} \left(\sum_{j \in V} D_{ij} \pi_j \right). \qquad (17.7)$$

Before solving this pricing problem, we can compute the value placed in parentheses, $w_i \equiv \sum_{j \in V} D_{ij} \pi_j$, for each beamlet $i \in N$ explicitly. Assume that there are n beamlets in each row of an MLC, our problem then becomes that for each row of MLC finding a consecutive set of beamlets for which the sum of their w_i values is minimized. In fact, we can find such a set of beamlets for a given row by searching through the n beamlets from left to right only once. At an intermediate step of this searching process, let the cumulative value of w_i over all beamlets considered so far, the maximum cumulative value found so far, and the best value found so far be denoted by v, $\bar{v}$, and v^*, respectively. In addition, let l and r denote the left and right MLC leaf positions at current searching step and l^* and r^* denote the best left and right MLC leaf positions so far, respectively. We can then solve this problem through a polynomial time algorithm on GPU. The algorithm can be summarized as follows:

Initialization:

$$v = \bar{v} = v^* = 0; \quad l = l^* = 0; \quad r = r^* = 1.$$

Main iterative loop:

1. $v = v + w_r$.

2. If $v > \bar{v}$, then $\bar{v} = v$ and $l = r$.

3. If $v - \bar{v} < v^*$, then $v^* = v - \bar{v}$, $l^* = l$ and $r^* = r + 1$.

4. If $r < n$, then $r = r + 1$, go to Step 1; else stop.

In the CUDA implementation, each thread independently runs the algorithm for an MLC row in parallel to obtain l^*, r^*, and v^* for that row. Then the summation of v^* for all MLC rows within each beam can be calculated. The aperture from the beam that attains the smallest summation value will be the solution to the pricing subproblem, and it is then added to the pool of apertures.

The dose deposition coefficient matrix is a sparse matrix, and we therefore store both D_{ij} and D_{kj} in a CSR format. In each iteration of our column generation method as shown in Figure 17.3, the dose deposition coefficient sparse matrix and the matrix transpose have to be updated (because a new aperture is added to the model). While transposing a sparse matrix might be easy for traditional serial CPU computing, it is quite a challenge for efficient parallel GPU implementation. To improve the performance of our CUDA code, we use a template library *Thrust* from NVIDIA to reorganize (sorting, counting, etc.) the data and then obtain the matrix transpose.

17.3.5 Experimental Results

To test our implementation, we use five clinical cases of prostate cancer (Cases P1–P5) and five clinical cases of head-and-neck cancer (Cases H1–H5). For prostate cancer cases, nine 6 MV coplanar beams are evenly distributed around the patient and the prescription dose to PTV is 73.8 Gy. For head-and-neck cases, five 6 MV coplanar beams are evenly distributed around the patient and the prescription dose to PTV1 is 73.8 Gy, and the prescription dose to PTV2 is 54 Gy. PTV1 consists of the gross tumor volume (GTV) expanded to account for both subclinical disease and daily setup errors and internal organ motion. PTV2 is a larger target that also contains high-risk nodal regions and is again expanded for the same reasons. For all cases, we use a beamlet size of 5×5 mm^2 and a voxel size of $2.5 \times 2.5 \times 2.5$ mm^3 for target and OARs. For unspecified tissue (i.e., tissues outside the target and OARs), we increase the voxel size in each dimension by a factor of 2 to reduce the optimization problem size. The full resolution is used when evaluating the treatment quality (DVHs, dose color wash, isodose curves, etc.). Table 17.2 shows the dimensions of these 10 cases in the DAO models.

We tested our CUDA implementation on an NVIDIA Tesla C1060 GPU card, which is the same as the one we used to solve the FMO models.

TABLE 17.2 Case Dimensions and GPU Running Time on an NVIDIA Tesla C1060 GPU Card for Direct Aperture Plan Optimization Implementations

Case	# Beamlets	# Voxels	# Nonzero D_{ij}'s	Running Time (s)
P1	7196	45,912	2,763,243	1.7
P2	7137	48,642	2,280,076	0.7
P3	5796	28,931	1,765,294	0.8
P4	7422	39,822	2,717,424	2.3
P5	8640	49,210	3,086,884	1.6
H1	5816	33,252	1,576,418	1.0
H2	8645	59,615	3,162,752	2.4
H3	9034	74,438	3,500,188	1.8
H4	6292	31,563	1,596,168	1.8
H5	5952	42,330	2,215,202	2.5

Source: Reprinted from Men, C. et al., *Phys. Med. Biol.*, 55, 4309, 2010.

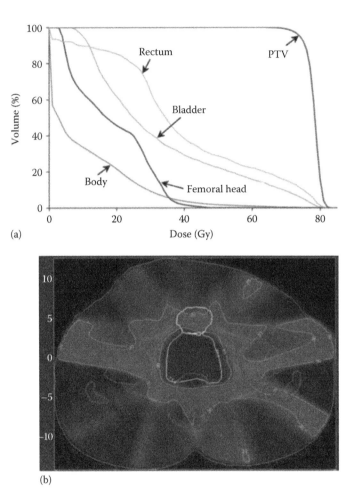

FIGURE 17.4 The treatment plan obtained from our GPU-based DAO implementation for Case P1 (prostate cancer): (a) DVHs; (b) dose color wash/isodose on a representative CT slice. (Reprinted from Men, C. et al., *Phys. Med. Biol.*, 55, 4309, 2010.)

In previous work, we have developed two stopping criteria to determine the number of apertures required [5]. In this work, we are more interested in how fast we can solve our DAO model on GPU. We, therefore, generate a fixed number of apertures, namely, 50, for all test cases. The running time for DAO plan optimization for these cases on a GPU is shown in Table 17.2. The amount of time required ranged in 0.7–2.5 s. In contrast, it takes 2–3 min to solve such problems on an Intel Xeon 2.27 GHz CPU using exactly the same column generation algorithm.

We then analyze the accuracy of the results. The obtained results on GPU are very close to those obtained on CPU with only 10^{-2} to 10^{-3} relative difference in the objective values. The difference can be attributed to single floating point precision on GPU and is negligible in clinical practice. In fact, by checking the DVHs, dose color wash, and isodose curves, no differences could be observed between the CPU and the GPU results. Figures 17.4 and 17.5 show the DVHs and isodose/dose color wash superimposed on representative CT slices of a prostate cancer case (Case P1) and of a head-and-neck cancer case (Case H1), corresponding to the DAO treatment plans, respectively.

17.4 CONCLUSIONS AND DISCUSSIONS

We investigate the performance of our CUDA implementation on different GPUs with similar clock speeds, including NVIDIA's GeForce 9500 GT, GTX 285, Tesla C1060, and Tesla S1070. The GeForce 9500 GT has only four multiprocessors and produced limited speedup (1×–3×). Both GTX 285 and S1070 have 30 multiprocessors (same as C1060) and deliver similar speedup results as described for the C1060. In fact, we find that the speedup factors solely depend on the number of multiprocessors per GPU.

The memory required for the largest dataset tested (Scenario 3) for the FMO model, and any test cases for the DAO model, is less than 1 GB, which could be accommodated by all of the GPUs that are tested.

We also notice that the speedup factor was different for different CUDA kernels. For the FMO model, in Scenario 3, the speedup factor is 46 for the dose calculation kernel, 18 for the objective value calculation kernel, and 12 for the gradient calculation kernel. The dose and objective calculations are parallelized based on voxels while the gradient calculation is parallelized based on beamlets. In general, the number of voxels is much greater than the number of beamlets. Therefore, it is not surprising to see that we can obtain better speedup for the dose and the objective function calculations. The speedup for the objective function calculation is not as good as that for the dose calculation due to the communication cost of the parallel reduction algorithm.

For the DAO model, we also evaluate the efficiency of our algorithm using various beamlet and voxel sizes. We notice that the running time increases/decreases if the number of voxels increases/decreases by changing voxel resolution. However, it is interesting to see that the GPU running time does not change too much for both the prostate and the

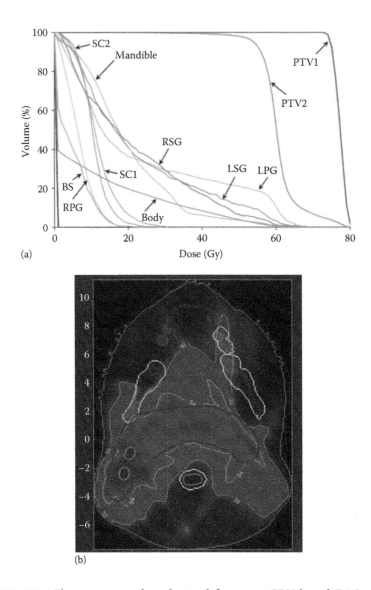

(a)

(b)

FIGURE 17.5 The treatment plan obtained from our GPU-based DAO implementation for Case H1 (head-and-neck cancer): (a) DVHs; (b) dose color wash/
isodose on a representative CT slice. (BS, brain stem; SC1, spinal cord 1; SC2,
spinal cord 2; LSG, left submandibular gland; RSG, right submandibular gland;
LPG, left parotid gland; RPG, right parotid gland). (Reprinted from Men, C.
et al., *Phys. Med. Biol.*, 55, 4309, 2010.)

head-and-neck cancer cases even though more/less beamlets are used. For the FMO model, the GPU running time highly depends on the size of the clinical case, that is, the number of beamlets and voxels. For the DAO model, the number of beamlets does not play a critical role. Remember that using column generation method, we iteratively solve a master problem and a subproblem. Subproblem is to identify an aperture that decreases the objective function most if added to the master problem. We identify this aperture by passing through beamlets once for each beamlet row, and hence, the number of beamlets in each row and the number of rows decide the size of the problem. Since we exploit fine-grained parallel algorithm, we are able to solve the subproblem very fast (<0.1 ms) for all cases and then the effect of solving subproblem can be negligible with respect to the total GPU running time. In fact, we notice that solving the master problem is the most time-consuming part. For the master problem, the decision variables are the intensities for all selected apertures. All cases generated the same number of apertures; therefore, the efficiency of the master problem highly depends on the number of voxels. Thus, even though more/less beamlets were used in the clinical cases, it does not result in longer/shorter GPU running time. Higher beamlet resolution will lead to more accurate treatment plan. Using traditional CPU-based computational tools and/or FMO models, higher beamlet resolution leads to longer computational time, which limits the size of beamlets that can be used. Using our GPU-based DAO treatment plan (re-)optimization algorithm, we can handle a very large number of beamlets without losing efficiency.

In this work, we only consider the GPU implementations to solve the FMO and DAO problems. We have shown the feasibility of using GPU to speed up the optimization process. Other optimization problems for IMRT planning, such as BAO, LSO, and ASO, can also be implemented on GPUs, and the speedup is also expected.

REFERENCES

1. Men C, Gu X, Choi D J, Majumdar A, Zheng Z, Mueller K, and Jiang S B, GPU-based ultrafast IMRT plan optimization, *Phys. Med. Biol.*, 2009, 54, 6565–6573.
2. Men C, Jia X, and Jiang S B, GPU-based ultra-fast direct aperture optimization for online adaptive radiation therapy, *Phys. Med. Biol.*, 2010, 55, 4309–4319.
3. Bell N and Garland M, Efficient sparse matrix-vector multiplication on CUDA, *NVIDIA Technical Report*, 2008 (SantaClara, CA: NVIDIA Corporation).

4. Romeijn H E, Ahuja R K, Dempsey J F, and Kumar A, A column generation approach to radiation therapy treatment planning using aperture modulation, *SIAM J. Optim.*, 2005, 15, 838–862.

5. Men C, Romeijn H E, Tas Z C, and Dempsey J F, An exact approach to direct aperture optimization in IMRT treatment planning, *Phys. Med. Biol.*, 2007, 52, 7333–7352.

6. Bazaraa M, Sherali H, and Shetty C, *Nonlinear Programming: Theory and Algorithms*, 2006 (Hoboken, NJ: Wiley).

CHAPTER 18

Treatment Plan Optimization for Volumetric-Modulated Arc Therapy (VMAT)

Fei Peng, Zhen Tian, H. Edwin Romeijn, and Chunhua Men

CONTENTS

18.1 INTRODUCTION

V OLUMETRIC-MODULATED ARC THERAPY (VMAT) is a new modality of cancer treatment. Its initial form was introduced in 1995 as intensity-modulated arc therapy (IMAT) [16]. As an augmentation of intensity modulated radiation therapy (IMRT), IMAT treatments rotate the gantry around the patient in one or multiple arcs while the radiation source stays on. In the meantime, intensity modulation is achieved by using multileaf collimator (MLC) system to dynamically change the field shape.

Although IMAT is only slowly tested and adopted over the course of more than a decade, with preliminary research showing that either multiple arcs or long treatment time is necessary to deliver high-quality treatment in a short period of time [4,7,8], the advent of dedicated treatment systems that incorporate the ability to dynamically change the dose rate (source output rate) and the gantry rotation speed during the treatment allowed for much better control of the delivery process and the potential for improved treatment quality in drastically shorter delivery time. Termed VMAT since then, this new modality differs from IMRT in a few different ways:

- IMRT delivers radiation from a predetermined set of beam angles; VMAT has the ability to utilize all angles on the treatment arc.

- IMRT treatment machine requires nontrivial setup time between beam angles; many types of VMAT treatments can be completed in one arc and each arc is delivered in one continuous gantry rotation.

- IMRT usually relies on constant dose rate, and the overall radiation output is controlled by varying the time the source is turned on; VMAT can dynamically change the dose rate it is using at different angles.

One major consequence of these fundamental differences is that VMAT treatment can usually be finished in significantly shorter time. This is highly desirable because of the potential for increased equipment turnover and alleviated uncertainty from intrafraction motion during treatments. Moreover, the ability to use a large number of angles to deliver the treatment can potentially produce more conformal dose distributions.

Despite its benefits, VMAT treatments are much more complex to design than IMRT treatments because of their different nature.

In addition, clinics nowadays increasingly demand faster planning time in an effort to achieve adaptive replanning of radiation therapy, making it necessary to not only design optimization algorithms that are capable of producing good VMAT plans but also incorporate suitable implementation techniques that offer low computation time, such as graphics processing unit (GPU) computing. In this chapter, we describe the VMAT optimization problem and focus on GPU-based optimization techniques that have successfully been shown to generate high-quality results fast. Section 18.2 describes the VMAT plan optimization problem and its mathematical formulation; Section 18.3 describes optimization algorithms proposed by Men et al. [10] and Peng et al. [13]. GPU implementation details and experimental results are provided; Section 18.4 concludes this chapter by briefly discussing the applicability of GPU-based parallel optimization to other VMAT optimization methods.

18.2 VMAT OPTIMIZATION MODEL

18.2.1 Arc Discretization and Control Points

The continuous delivery of the treatment in VMAT dictates that the movement of leaves in the MLC, and the aperture formed by those leaves, must comply with the physical restrictions of the treatment machine. This means the gantry speed employed as the gantry rotates between any two angles on the arc needs to allow the apertures formed by the MLC to complete its transition. When both the gantry speed and apertures are decision variables defined over the continuous space of treatment angles, this optimization problem quickly becomes intractable. Therefore, rather than directly considering the continuous treatment arc(s), most approaches to VMAT optimization deal with a discretization of an arc into a set of *control points*, which represent beam directions along the arc, and determine the gantry speed, dose rate as well as the aperture shape at these points. The treatment machine, on the other hand, is then configured so that these parameters smoothly transition between consecutive control points.

Following this convention, the total number of (not necessarily equispaced) control points along one or more treatment arc(s) is denoted by K (in case of multiple arcs, K represents the total number of control points along all arcs, which will be ordered so that the first control point on the second arc succeeds the last control point on the first arc, etc.). A control point k along with an aperture A_k, a dose rate r_k (in MU/s), and a gantry speed s_k (in degrees/s) at this control point represents a "snapshot" of the continuous gantry rotation as the gantry passes through control point k. Depending on the manufacturer of the treatment equipment, there may be different physical restrictions that apply.

18.2.2 Deliverability Constraints

As mentioned in Section 18.2, the gantry speeds and apertures specified at consecutive control points need to be compatible with each other. This compatibility requirement stems from the characteristics of the MLC system used to form the apertures during the treatment. As the gantry travels from control point k to control point $k+1$, the leaves of the MLC system need to shift from positions described by aperture A_k to those by A_{k+1}. Since the speed is limited, the time spent by the gantry moving between the control points needs to be sufficiently large (and hence the gantry speed sufficiently small) to allow the required leaf movement to take place. The maximum gantry speed that will allow aperture A' at control point $k+1$ to be reached from aperture A' at control point k is denoted by $S_{k,k+1}^{U}(A, A')$ for $k = 1,\ldots, K$. For convenience, a dummy control point $K+1$ is added at the end of the arc(s), and $S_{k,k+1}^{U}(A, A') = \infty$ is defined implicitly for all pairs of apertures A and A'.

Several other machine restrictions, when present, also need to be taken into account. First, there usually exist upper and lower bounds on the gantry speed (denoted by S^U and S^L) and dose rate (denoted by R^U and R^L). Moreover, the treatment machine may only be able to sustain change in gantry speed up to a certain level, which may be given per degree or per control point. This upper bound on the change in speed between control points k and $k+1$ is denoted by ΔS_k. Finally, denote by $\mathcal{A}$ the set of all apertures deliverable by the MLC system. In particular, $\mathcal{A}$ includes MLC apertures where the left and right leaves within each leaf pair can take any nonoverlapping continuous position within the range of the leaf row. $\mathcal{A}$ can also capture requirements that allow/prohibit interdigitation of leaves in adjacent rows.

18.2.3 Basic VMAT Optimization Model

For tractability reasons, the dose delivered to each voxel during the treatment is calculated by making the approximation that the aperture, dose rate, and gantry speed are constant throughout the arc spanning from one control point to the next. Let δ_k denote the angular distances between pairs of control points k and $k+1$, let $\mathcal{V}$ denote the set of voxels of interest, and let z_j denote the dose delivered to voxel $j \in \mathcal{V}$; this approximation allows us to express

$$z_j = \sum_{k=1}^{K} D_{kj}(A_k)\delta_k \frac{r_k}{s_k} \quad j \in \mathcal{V}, \qquad (18.1)$$

where $D_{kj}(A)$ represents the dose received by voxel j from aperture A at control point k at unit fluence. For each control point, the radiation beam is divided into a grid of *beamlets*. In a preprocessing step, the dose contribution to each voxel from each of these beamlets at unit intensity is calculated, which allows the coefficient $D_{kj}(A)$ to be calculated efficiently by adding up that of all beamlets inside aperture A, provided that partially covered beamlets are treated properly. Note that (18.1) is essentially a "step-and-shoot" approximation. It is shown that this approximation will be sufficiently accurate if δ_k is sufficiently small [12].

By establishing a quantitative measure of the treatment quality, $\mathcal{F}(\mathbf{z}, \mathbf{r}, \mathbf{s}, \mathbf{A})$, which obtains its minimum at an optimal solution, the basic VMAT optimization formulation is then

Minimize$_{\mathbf{z},\mathbf{r},\mathbf{s},\mathbf{A}}$ $\mathcal{F}(\mathbf{z}, \mathbf{r}, \mathbf{s}, \mathbf{A})$

subject to

$$z_j = \sum_{k=1}^{K} D_{kj}(A_k)\delta_k \frac{r_k}{s_k} \quad j \in \mathcal{V}, \qquad (18.2)$$

$$\mathbf{r}, \mathbf{s}, \mathbf{A} \text{ deliverable} \qquad (18.3)$$

$\mathcal{F}(\mathbf{z}, \mathbf{r}, \mathbf{s}, \mathbf{A})$ is usually a function of the delivered dose $\mathbf{z}$, which depends on $\mathbf{r}$, $\mathbf{s}$, and $\mathbf{A}$, but other types of functions can also be used.

Compared to its IMRT counterparts, the VMAT optimization problem is significantly larger in size both in terms of the number of model components (variables and constraints) and in terms of the amount of data. For example, the number of control points involved is frequently on the order of a couple of hundred or hundreds for a typical VMAT treatment, whereas IMRT rarely considers more than a dozen treatment beams. Moreover, even though the model can be formulated in a way that the objective function is convex [6,10,13], it is still in essence a nonconvex optimization problem because of, for example, the inverse relationship between the dose $\mathbf{z}$ and the gantry speed $\mathbf{s}$ represented by constraint (18.2). Furthermore, the dose delivered to each voxel from an aperture follows a complex physical relationship that depends on the patient's internal structures that lie in between the aperture and the voxel, the position of the aperture in the beam opening, among other things. This contributes to the fact that two apertures that are similar in size or shape can have drastically different dose distributions. Finally, because each individual MLC leaf can be placed anywhere within a continuous range prescribed by the MLC, the number of possible configurations in $\mathcal{A}$ is infinite. Choosing the optimal set of apertures from this set is thus difficult.

These reasons led to the emergence of many different directions for designing heuristic VMAT optimization methods. In the following sections, we describe how GPU enables implementation of a family of fast VMAT optimization algorithms and briefly discuss the applicability of these implementation techniques to other existing VMAT planning algorithms.

18.3 GPU-BASED OPTIMIZATION METHODS

The column generation-based methods for VMAT optimization problems were first proposed in Men et al. [10], and formalized in Peng et al. [13] to handle more realistic constraints. This type of method is carried out in an iterative process, where each iteration augments the solution by determining an aperture at a certain control point that both satisfies the deliverability constraints and leads to the most marginal improvement of the objective function value.

Formally, by first defining the *fluence rate* as the source output per degree of gantry rotation

$$y_k = \frac{r_k}{s_k},$$

and explicitly modeling restrictions included in (18.3), the complete VMAT optimization model reads

$$(\text{FP}) \quad \min_{\mathbf{r},\, \mathbf{s},\, \mathbf{A},\, \mathbf{z},\, \mathbf{y}} \quad \mathcal{F}(\mathbf{z}, \mathbf{r}, \mathbf{s}, \mathbf{A})$$

subject to

$$z_j = \sum_{k=1}^{K} D_{kj}(A_k)\delta_k y_k \quad j \in V, \tag{18.4}$$

$$y_k = \frac{r_k}{s_k} \quad k=1,\ldots,K \tag{18.5}$$

$$|s_k - s_{k+1}| \leq \Delta S_k \quad k=1,\ldots,K-1 \tag{18.6}$$

$$s_k \in [S^L, S^U] \quad k=1,\ldots,K \tag{18.7}$$

$$r_k \in [0, R^U] \quad k=1,\ldots,K \tag{18.8}$$

$$s_k \leq S^U_{k,k+1}(A_k, A_{k+1}) \quad k=1,\ldots,K \tag{18.9}$$

$$A_k \in \mathcal{A} \quad k=1,\ldots,K. \tag{18.10}$$

Here, the deliverability constraint (18.3) is fully specified by constraints (18.6 through 18.10): ΔS_k limits the change in gantry speed between control points k and $k+1$. By defining $S^U_{k,k+1}(A_k, A_{k+1})$ as the maximum gantry speed to allow the MLC transition from A_k to A_{k+1}, constraint (18.9) captures the relationship between the gantry speed and aperture shapes at consecutive control points. Recall that the terminal aperture A_{k+1} in (18.9) is added for convenience only and can be chosen arbitrarily. Note again that, because of the nonconvexity embedded in, for example, constraints (18.4) and (18.5), this problem is not convex and is difficult to solve directly. Model (FP) as well as Section 18.3.1 is based on Section 2 of Peng et al. [13].

18.3.1 Optimization Algorithm

Model (FP) is first simplified by setting all the gantry speeds at their lower bound SL, and scaling the dose rates by a factor of $S^L/s_k \leq 1$, $k=1,\ldots,K$, to compensate. For any feasible solution s, r, y, z, A, doing this allows the same apertures and the same values of y and z to be used (and hence the resulting solution corresponds to the same objective function value). As a result, the gantry speed variables are eliminated from the model.

In addition, the upper bound constraints on the dose rates are replaced by upper bounds on fluence rate, and thus, the dose rate variables are eliminated as well. The problem thus reduces to the following *master problem* (MP):

(MP) minimize $F(\mathbf{z})$

subject to

$$z_j = \sum_{k=1}^{K} D_{kj}(A_k)\delta_k y_k \quad j \in \mathcal{V}, \tag{18.11}$$

$$y_k \in [0, Y^U] \qquad\qquad k = 1,\ldots, K$$

$$s^L \le S_{k,k+1}^U(A_k, A_{k+1}) \quad k = 1,\ldots, K \tag{18.12}$$

$$A_k \in \mathcal{A} \qquad\qquad k = 1,\ldots, K, \tag{18.13}$$

where $Y^U \equiv R^U/S^L$. The nonlinearities embedded in this model, especially in the dose deposition coefficient $D_{kj}(A_k)$, make model MP nonconvex.

Problem (MP) is solved with a column generation-based algorithm. Starting with an empty plan without any apertures, the algorithm adds one aperture in each iteration and progressively fills up the control points. In each iteration, which is characterized by a set $C \subseteq \{1,\ldots, K\}$ of control points and the corresponding collection of apertures $\{\bar{A}_k, k \in C\}$, an aperture is automatically selected out of all unoccupied control points based on marginal improvement in the cost function. This is done by solving the so-called *pricing problem* (PP). The so-called *restricted master problem* (RMP) is then solved to find the best intensity configuration of the existing set of apertures. This process is repeated until termination conditions are reached. The flowchart of the algorithm is shown in Figure 18.1.

18.3.1.1 Solving Problem (MP)

The RMP at each iteration of the algorithm is obtained from MP by fixing the apertures $A_k = \bar{A}_k$ for all control points in the current set $C \subseteq \{1,\ldots, K\}$ and setting the fluence rates $y_k = 0$ for all control points $k \notin C$:

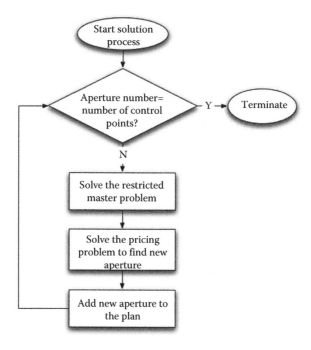

FIGURE 18.1 Flowchart for the column generation-based heuristic for MP. (From Peng, F. et al., *Phys. Med. Biol.*, 57(14), 4569, 2012.)

(RMP$^{(C)}$)   minimize $F(\mathbf{z})$

　　　　subject to

$$z_j = \sum_{k\in C} D_{kj}(\bar{A}_k)\delta_k y_k \quad j \in V$$

$$y_k \in [0, Y^U] \qquad\qquad k \in C.$$

Any feasible solution to RMP$^{(C)}$ should correspond to a deliverable plan that can be fully specified by completing the sequence of apertures at control points $k \notin C$ in a way that is compatible with apertures $\bar{A}_k$, $k \in C$, and setting $y_k = 0$, $k \notin C$. This is the case only if the set of apertures $\bar{A}_k$, $k \in C$ in the restricted problem satisfy

$$S^U_{k,k^+}(\bar{A}_k, \bar{A}_{k^+}) \geq S^L, \quad k \in C, \tag{18.14}$$

where k^+ is the successor of k in C. The PP is designed to ensure that the apertures added at each iteration of the column generation algorithm satisfy (18.14).

Based on first-order optimality conditions, the PP searches for the aperture/control point combination that leads to the largest rate of *improvement* (i.e., decrease) in objective function value per unit fluence rate:

$$(\text{PP})\max_{c \notin C} \max_{A \notin \mathcal{A}_c(s)} \sum_{j \in V} \pi_j(\bar{z}) \, D_{cj}(A_c(s)) \delta_c$$

where

$(\pi_j(\bar{z}): j \in V)^\top = \pi(\bar{z}) \equiv -\nabla F(\bar{z})$ is the gradient of the objective function at the optimal solution $\bar{z}$ to RMP.

$\mathcal{A}_c(s) \subseteq \mathcal{A}$ is a set of deliverable apertures that can feasibly be added to the current treatment plan, parameterized by a lower bound s on the speed between two control points:

$$\mathcal{A}_c(s) = \left\{ A \in \mathcal{A} : S^U_{c^-c}(\bar{A}_{c^-}, A) \geq s, \, S^U_{cc^+}(A, \bar{A}_{c^+}) \geq s \right\} \quad \text{for } s \geq S^L.$$

Choosing a smaller set $\mathcal{A}_C$ (corresponding to a larger s value) results in lesser flexibility in the current selection of apertures. On the other hand, this can potentially allow for more flexibility in later iterations of the algorithm.

The outcome of PP either provides an improving aperture or determines that no one can be found and the algorithm is terminated. It is easy to see that PP decomposes into a collection of independent problems for the different candidate control points c, and, absent of the interdigitation constraints, further decomposes by MLC row. For a particular control point c and row m, the price for the aperture row depends on the leaf pair (ℓ, r). Using precomputed beamlet-based dose deposition coefficients D_{cmnj} for beamlets $n = 1,\ldots, N$ in this row and making the approximation

$$D_{cmj}(\ell, r) = \int_\ell^r \phi_{cmj}(x)\,dx,$$

where $\phi_{cmj}: [0, N] \rightarrow \mathbb{R}^+$ is the following step function:

$$\phi_{cmj}(x) = D_{cmnj}, \quad n - 1 < x \leq n; \, n = 1,\ldots, N,$$

the problem can be solved efficiently by examining $O(N)$ candidate points for ℓ and r.

18.3.1.2 Postprocessing

The column generation algorithm will return a feasible solution $\bar{y}, \bar{z}$ and $\bar{A}_k$, $k=1,\ldots, K$ to the problem (MP). As a final step of the solution procedure, the gantry speed and dose rate vectors **s** and **r** are computed, with the goal of minimizing treatment time by solving the following convex optimization problem:

Minimize
$$\sum_{k=1}^{K} \frac{\delta_k}{s_k}$$

subject to

$$s_k \in \left[S^L, S_k^U \right] \quad k = 1,\ldots, K$$

$$|s_k - s_{k+1}| \le \Delta S_k \quad k = 1,\ldots, K-1,$$

where $S_k^U = \min\{S^U, S_{k,k+1}(\bar{A}_k, \bar{A}_{k+1}), R^U/\bar{y}k\}$ is the maximum gantry speed within the restrictions of fixed $\bar{y}$ and $\bar{A}_k$ for $k=1,\ldots, K$ determined earlier in the solution process. If $\bar{s}_k$ is an optimal solution to this problem, then the corresponding dose rates are $\bar{r}_k = \bar{y}_k \bar{s}_k$, $k=1,\ldots, K$.

Finally, it may be desirable to have similar dose rates at consecutive control points. This is not considered in the optimization process and can be achieved by solving the following convex optimization problem instead:

Minimize
$$\sum_{k=1}^{K} \frac{\delta_k}{s_k} + \gamma \sum_{k=1}^{K} (r_k - r_{k+1})^2$$

subject to

$$\bar{y}_k = \frac{r_k}{s_k} \quad k = 1,\ldots, K$$

constraints (18.6 through 18.8).

Stricter restrictions on dose rate changes can be satisfied by increasing the constant γ on the smoothing term in the objective function. However, it may not be possible to meet arbitrary smoothness goals given fixed $\bar{y}_k$

values. Alternatively, the approach employed by Men et al. [10] may be used, which assumes constant gantry speed during the optimization process and, after the last column generation iteration, ensures smoothness in dose rates by adjusting the fluence rates if necessary.

18.3.2 GPU Implementation

18.3.2.1 Single GPU

Both the MP and the PP lend themselves very well to GPU-based parallel implementation. Each iteration of the optimization algorithm, when decomposed into small computational tasks, comprises the following steps:

1. Solving the master problem with a projected gradient method, which involves iteratively:

   a. Calculating dose z at the current solution. This step is equivalent to multiplying the dose deposition matrix D_{kj} by the vector of aperture intensities y and is implemented as a sparse matrix–vector multiplication [2]

   b. Calculating the objective function value $F(z)$—parallelizable by voxel

   c. Calculating the gradient direction, parallelizable by aperture

   d. Determining the step size to move along the gradient direction with a backtracking line search, which further requires:

      i. Determining an initial step size and a shrinkage factor

      ii. (Tentatively) moving the solution along the gradient direction according to the step size—parallelizable by aperture

      iii. Updating the dose vector z

      iv. Calculating the objective function value at the new dose level

      v. Stop if the termination criteria (e.g., Armijo's rule) are satisfied, otherwise shrink the step size by the shrinkage factor

   e. Updating the solution by moving the current solution in the gradient direction with the step size determined in the step earlier

2. Solving the PP by:

a. Calculating the price $\delta_c \sum_{j \in \mathcal{V}} \pi_j(\overline{z}) D_{cmnj}$ for each beamlet n in row m, control point c. By precalculating the coefficients D_{cmnj}, this step is essentially calculating the gradient of a piecewise quadratic function followed by a sum–product operation for each beamlet, and is parallelizable by beamlet.

b. Determining, for each control point in $\{1, \ldots, K\} \backslash C$, the range of the left and right leaves in each row that satisfies deliverability—parallelizable by rows and by control point.

c. Scanning the $O(N)$ candidates in each beamlet row to find the leaf positions ℓ and r that minimizes $D_{cmj}(\ell, r)$. This step is parallelizable by control point and by row. When interdigitation is not allowed, the rows in a control point cannot be considered independently anymore, and this as well as the previous step must be modified to handle this situation. However, the required operations can still be parallelized by control point (we refer interested readers to Peng et al. [13] for more details).

d. Adding up the best row prices to obtain the aperture price—parallelizable by control point.

e. Selecting solution A_k and updating the dose deposition coefficient matrix D_{kj}.

All major tasks shown earlier can be parallelized by using one GPU thread to carry out one computational component, each of which is independent of one another. Because the number of voxels and beamlets is on the order of tens of thousands, and the number of apertures is on the order of hundreds, the computation can benefit greatly from distributing small chunks of work onto parallel threads rather than running them in serial.

The dose deposition coefficient matrices D_{cmnj} and D_{kj} are usually sparse due to the fact that a beamlet or aperture only penetrates part of the body and thus contributes to limited number of voxels. These matrices are stored in the compressed sparse row (CSR) format in the global memory of the GPU, which is accessible by all threads. The Thrust library* provides GPU-enabled operations such as sort, reduction, and binary search.

---

* NVIDIA Developer. Thrust. 2015. http:// developer.nvidia.com/thrust. Accessed on May 30, 2015.

And the template library CUSPARSE* can be used for sparse matrix-related calculations such as matrix–vector multiplication and matrix transpose.

Once the column generation phase is terminated, the postprocessing problem is a simple convex optimization problem and can be solved very efficiently with many commercial or open-source solvers such as Cplex[†] or cvx.[‡]

18.3.2.2 Multi-GPU

The memory size on the GPU is usually fairly limited, making it very difficult for a single GPU to handle the entire D_{cmnj} matrix, especially for cases with large target size, multiple arcs, and/or small beamlets. One simple way of overcoming the memory limitation is to delete small elements of the matrix, or, alternatively, the matrix can be copied from the CPU to the GPU part by part repeatedly. Another solution may be to move the calculations that need access to D_{cmnj} from the GPU to the CPU. Nonetheless, using any of these methods will inevitably lead to adverse impacts on either the plan quality or speed of the computation.

With the reduced cost of GPU cards, the best solution to this problem is to utilize a platform with multiple GPUs, so that neither the plan quality nor computational efficiency has to be sacrificed. However, since each individual GPU only holds its own memory, the distribution of data and computational tasks need to be designed so that inter-GPU data communication is infrequent, and the additional overhead remains low.

The computational steps outlined in Section 18.3.2 are divided into two groups. One is associated with data that are too large to fit in a single GPU, and thus require multi-GPU implementation; the other contains steps that interact with data enough to be handled on a single GPU. In particular, only the step (ii) that calculates the price of each beamlet in the pricing problem belongs to the first type, where access to matrix D_{cmnj} is needed.

One specific implementation, detailed in Zhen et al. [17], designated one GPU (GPU1) as primary to implement all steps in the latter group, and used all available GPUs in storing data and performing computations for steps in the first group. At the beginning of computation, beamlets are

---

* NVIDIA Developer. cuSPARSE. 2015. developer.nvidia.com/cuSPARSE. Accessed on May 30, 2015.
† IBM. CPLEX Optimizer: High-performance mathematical programming solver for linear programming, mixed integer programming, and quadratic programming, 2015. ibm.com/software/commerce/optimization/cplex-optimizer/. Accessed on May 30, 2015.
‡ CVX Research. MATLAB® Software for Disciplined Convex Programming. 2015. cvxr.com/cvx. Accessed on May 30, 2015.

divided into four blocks and each GPU stores a submatrix corresponding to one block, again in CSR format. The parameters involved in this step are then transferred from CPU to all the GPUs. Each GPU is then only responsible for calculating part of the beamlet price.

With the delegation of beamlet price calculation to multiple GPUs, data transfer between GPUs needs to be designed carefully to ensure both efficiency and accurate alteration between the PP and MP. Note that only the dose vector $\mathbf{z}$ needs to be copied to all GPUs before, and individual price subvector needs to be copied back to GPU1 after the price calculation. Therefore, a broadcast scheme can be performed. In order to copy the dose distribution $\mathbf{z}$ at the current solution obtained by MP on GPU1, to all other GPUs at each iteration for PP, in the case of four total GPUs, GPU1 first copies $\mathbf{z}$ to GPU3, then the same operation is conducted from GPU1 to GPU2 and from GPU3 to GPU4 simultaneously. Since each GPU is responsible for calculating only part of the beamlet price vector, a parallel reduction among GPUs is conducted to get the complete price vector on GPU1. As such, GPU2 first passes its beamlet price subvector to GPU1. The same operation is performed simultaneously from GPU4 to GPU3. After each GPU carries out their share of the price calculation, the individual price subvectors need to be aggregated back to GPU1 for the subsequent calculations. The reduction is conducted by transferring the subvectors to GPU1 via a reversed path to what is employed in the broadcasting stage, first simultaneously from GPU4 to GPU3 and from GPU2 to GPU1, then from GPU3 to GPU1. The broadcast and reduction schemes are shown in Figure 18.2b.

18.3.3 Experimental Results

This section describes several experiments for GPU-based VMAT optimization algorithms. Results for prostate and head-and-neck (H&N) cases have been reported, with very promising run-time statistics compared to the state-of-the-art commercial packages, as well as high plan quality.

Table 18.1 includes dimensions of five prostate and five H&N cases tested by Men et al. [10]. For prostate cases, the prescription dose to planning target volume (PTV) was 73.8 Gy and for the H&N cases, the prescription dose was 73.8 Gy to PTV1 and 54 Gy to PTV2. PTV1 consists of the gross tumor volume (GTV) expanded to account for both subclinical disease and daily setup errors and internal organ motion; PTV2 is a larger target that also contains high-risk nodal regions and is again expanded for the same reasons. For all cases, the beamlet size is 10×10 mm^2 and voxel

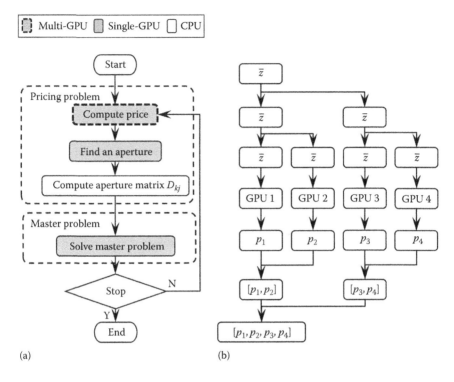

(a)   (b)

FIGURE 18.2 (a) Division of computational tasks between single-GPU and multi-GPU; (b) illustration of broadcast and reduction scheme on multiple GPUs when computing the beamlet price. (From Zhen, T. et al., *Med. Phys.*, 42, 2841, 2015)

TABLE 18.1   Problem Dimensions of Different Cases (after Downsampling)

Case	# Voxels	# Beamlets	# $D_{cmnj} \neq 0$	$\delta_k \ (k=2,\dots,175)$	δ_1, δ_{176}
1	13,047	21,417	89,360,831	$\dfrac{330}{175}$	$\dfrac{1}{2} \cdot \dfrac{330}{175}$
2	10,130	17,700	63,359,780	$\dfrac{330}{175}$	$\dfrac{1}{2} \cdot \dfrac{330}{175}$
3	12,102	29,913	106,889,530	$\dfrac{330}{175}$	$\dfrac{1}{2} \cdot \dfrac{330}{175}$
4	9,602	21,417	79,889,713	$\dfrac{330}{175}$	$\dfrac{1}{2} \cdot \dfrac{330}{175}$
5	13,769	25,488	114,315,187	$\dfrac{340}{175}$	$\dfrac{1}{2} \cdot \dfrac{330}{175}$

Source: Peng, F. et al., *Phys. Med. Biol.*, 57(14), 4569, 2012.

size is $2.5 \times 2.5 \times 2.5$ mm³ for target and organs at risk (OARs). For tissues outside the target and OARs, the voxel size in each dimension is increased by a factor of 2 to reduce the optimization problem size. The full resolution was used when evaluating the treatment quality. The objective function is of the form

$$F(\mathbf{z}) + G(\mathbf{y}),$$

where

$$F(\mathbf{z}) = \sum_{j \in \mathcal{V}} (\alpha_j (\max\{0, T_j - z_j\})^2 + \beta_j (\max\{0, z_j - T_j\})^2) \quad (18.15)$$

is a piecewise quadratic function that penalizes both over- and under-dosing a voxel j from threshold T_j and $G(\mathbf{y})$ is a regularization function that prevents large variations in the dose rate in consecutive control points, as described in Section 18.3.1.

Figure 18.3 shows the resulting dose–volume histograms (DVHs) and dose wash for a prostate and a H&N case, respectively. For clarity of presentation, critical structures that received very low dose, such as brain stem, optic nerve, spinal cord, etc., are not represented in Figure 18.3. Compared to the serial MATLAB® implementation, the GPU computation times, in general an order of magnitude faster, are clearly very beneficial.

In Peng et al. [13], the algorithm was evaluated on five prostate cases, using 1×1 cm² beamlets and $4 \times 4 \times 2.5$ mm³ voxels. The optimization problem used a voxel grid downsampled by a factor of 2 in each direction in critical structures and by a factor of 4 in unspecified tissue. For all cases, the number of control points $K = 176$. The machine parameters are shown in Table 18.2 (where $\Delta S_k = \Delta S$ for all $k = 1, \ldots, K$). The prescription dose to the PTV was set to 79.2 Gy, corresponding to a 44-fraction treatment with 1.8 Gy delivered in each fraction. The objective function used in the optimization model is of the same piecewise quadratic form as $F(\mathbf{z})$ in (18.15) (Table 18.3).

Studying the effect of using different values for s, the parameter that dictates how "greedy" the heuristic is, revealed that the treatment plan quality is not sensitive to the choice of s. However, the overall treatment time is reduced dramatically when a larger s value is used. A benchmark was then established by solving an IMRT treatment plan optimization problem with the same objective function and using the beam angles that

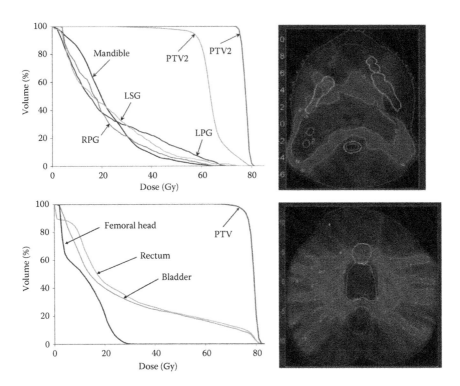

FIGURE 18.3 VMAT plans for a head-and-neck case (top) and a prostate case (bottom) (LSG, left submandibular gland; LPG, left parotid gland; RPG, right parotid gland). (From Men, C. et al., *Med. Phys.*, 37(11), 5787, 2010.)

TABLE 18.2 Machine Parameters

R^L (MU/s)	R^U (MU/s)	S^L (Degree/s)	S^U (Degree/s)	ΔS (Degree/s)	v (cm/s)
0	10	0.83	6.0	0.75	2.25

Source: Peng, F. et al., *Phys. Med. Biol.*, 57(14), 4569, 2012.

correspond to the 177 control points in the VMAT model. Comparison between plans obtained with $s = S^U$ (shown in Table 18.4) and the benchmark showed that the VMAT treatment plans exhibit high quality in general, closely resembling the IMRT treatment plans when evaluated with the same set of clinical criteria. Figure 18.4 shows the DVHs of the 177-beam IMRT and VMAT (with $s = S^U = 6$) treatment plans for Case 2.

Note that the run times are around half a minute for all test cases (Table 18.4). In contrast with the current commercial treatment planning systems, this is a reduction of the planning time by one or two orders of

TABLE 18.3   Case Dimensions and CPU/GPU Running Time on an Intel Xeon 2.27 GHz CPU and an NVIDIA Tesla C1060 GPU

Case	# Beamlets	# Voxels	# Nonzero D_{ij}'s ($\times 10^7$)	CPU Time (s)	GPU Time (s)
P1	40,620	45,912	2.3	340	22
P2	59,400	48,642	3.2	265	18
P3	38,880	28,931	1.8	276	20
P4	43,360	39,822	2.6	410	26
P5	51,840	49,210	3.0	348	23
H1	51,709	33,252	2.5	290	21
H2	78,874	59,615	5.0	468	27
H3	90,978	74,438	5.5	342	25
H4	71,280	31,563	2.6	363	25
H5	53,776	42,330	3.5	512	31

Source: Men, C. et al., *Med. Phys.*, 37(11), 5787, 2010.

TABLE 18.4   Performance of VMAT Treatment Plans with $s = S^U$

Structure	Threshold Dose (Gy)	Volume Criterion (%)	Case				
			1	2	3	4	5
PTV	73.7	≥99	99	99	99	99	99
	79.2	≥95	95	95	95	95	95
	87.1	≤10	1	1	3	4	4
Rectum	75	≤15	5	17	15	10	11
	70	≤25	7	20	19	14	16
	65	≤35	10	23	22	20	20
	40	≤45	26	40	36	35	40
Bladder	65	≤17	7	24	14	45	12
	40	≤35	14	45	24	70	25
Femoral heads (L/R)	50	≤10	0/5	0/4	1/1	4/3	7/0
	45	≤25	0/9	0/9	3/2	10/9	13/3
	40	≤40	7/15	0/16	8/5	14/13	24/9
Runtime (s)			24.8	35.2	29.4	22.8	28.5
Treatment time (s)			128.1	128.4	121.7	121.7	122.5

Source: Peng, F. et al., *Phys. Med. Biol.*, 57(14), 4569, 2012.

magnitude. This type of performance is especially desirable in situations when adaptive or replanning of the treatment is necessary.

18.3.3.1 Multi-GPU

Experiments conducted in Zhen et al. [17] used a H&N patient case with two coplanar arcs, each with 356 equispaced control points. Beamlet size is 1×1 cm^2 and voxels are of size $0.195 \times 0.195 \times 0.25$ cm^3. The dose deposit

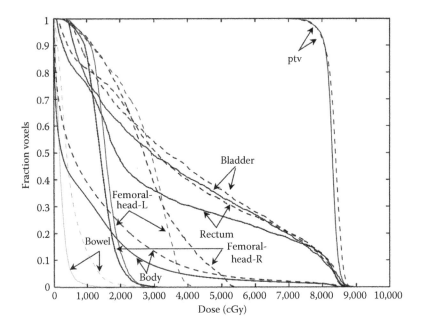

FIGURE 18.4   Treatment plans for Case 2. Solid: 177-beam IMRT; Dashed: VMAT. (From Peng, F. et al., *Phys. Med. Biol.*, 57(14), 4569, 2012.)

coefficient matrix D_{cmnj} for this problem occupies 4.93 GB of memory space. The multi-GPU system was built on a desktop workstation with two Nvidia GeForce GTX590 cards, each containing two identical GPUs (labeled as GPU 1–4). There are 512 thread processors at a clock speed of 1.26 GHz on each GPU and they share the same 1.5 GB GDRR5 global memory at a 164 GB/s memory bandwidth. Data transfer between GPUs is conducted through the motherboard via PCIe-16 bus.

While it is possible to fit this data into a single high-end GPU card, performing the optimization is not possible because of the additional space required in carrying out the algorithm. Three approaches are implemented on a single GPU for comparison purposes. The first method (S1) deletes small elements in D_{cmnj}; the second method (S2) divides the matrix into smaller chunks, and transfers them from the CPU to the GPU one at a time when calculating beamlet prices; the third one (S3) moves the calculation of beamlet prices to the CPU. Note that the multi-GPU implementation (M) and implementations S2 and S3 should result in the same plan, whereas S1 will lead to an inferior plan due to the truncation in the dose deposit coefficient matrix. Shown in Figure 18.5a,

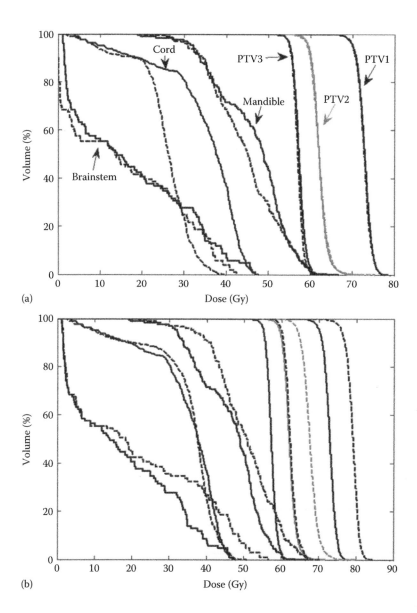

FIGURE 18.5 DVHs of the resulting plans. Solid lines for Plan-M; dashed lines for Plan-S1. Dose for Plan-S1 illustrates only optimized dose in (a) and actual dose in (b) (From Zhen, T. et al., *Med. Phys.,* 42, 2841, 2015)

TABLE 18.5 Time Comparison for Multi-GPU versus Three Single-GPU VMAT Implementation

		Plan-M	Plan-S1	Plan-S2	Plan-S3
D_{cmnj} matrix size (GB)		4.93	1.19	4.93	4.93
GPU cards		4	1	1	1
Time (s)	Total	60.60	50.13	255.32	1802.51
	Read data	16.64	9.9	16.76	16.69
	Data transfer	9.25	7.2	176.83	0.01
	Computation	9.48	14.88	37.96	1762.03

Source: Zhen, T. et al., *Med. Phys.*, 42, 2841, 2015.

Plan-M achieved good quality for this H&N patient case. Although it seems that the OARs are better spared in Plan-S1 compared to Plan-M, after putting back the deleted small elements ignored in (S1), Plan-S1 is clearly a drastic deterioration compared to Plan-M, as can be seen in Figure 18.5b.

The computation times for different implementation strategies are shown in Table 18.5. For multi-GPU implementation, T(data transfer) also includes the time for inter-GPU data transfer, which took only 30 ms overall. These results demonstrated that multi-GPU-based optimization can handle large-scale VMAT cases without sacrificing the plan quality or computational efficiency.

18.4 CONCLUSION

This lineage of highly parallelizable algorithms that were designed to take advantage of the power of the GPU, and their success in enabling fast treatment planning, has shown that VMAT plan optimization can benefit greatly from GPU-based parallelization, achieving up to 70× speedup compared to serial implementations.

It is conceivable that many other algorithms (or significant components) for VMAT treatment plan optimization may also be implemented on the GPU and realize the potential of substantially shortened optimization time. Examples of these algorithms in the literature include the following:

- Methods that rely on an initial IMRT plan [1,3,5,6,9,14,15]. The problem of treatment plan optimization for IMRT has been investigated in Men et al. [11], and results showed that a GPU-based implementation can achieve up to 40× speedup compared to serial implementations of the same algorithm. Moreover, the initial IMRT plan usually leads to a partial plan that needs to be adjusted or further optimized to satisfy VMAT deliverability requirements or to improve treatment quality. This process involves computing the delivered dose and/or objective value multiple times and, as discussed in the earlier sections, can be parallelized well.

- Metaheuristic-based algorithms, such as the one proposed in Otto [12]. The search process in these algorithms usually involves determining one or more new candidate decisions for machine parameters and computing the associated dose vector and objective function value. The latter can often times be done independently for each voxel and is suitable for GPU-based parallelization.

Due to the wide availability and relatively low cost of GPU parallelization, it is expected that the technology will be adapted by and incorporated into commercial planning systems as well as more research prototypes for VMAT treatment plan optimization and contribute to the advancement of planning efficiency and treatment outcomes.

REFERENCES

1. J.L. Bedford. Treatment planning for volumetric modulated arc therapy. *Medical Physics*, 36(11):5128–5138, 2009.
2. N. Bell and M. Garland. Efficient sparse matrix–vector multiplication on cuda. Technical report, Nvidia Technical Report NVR-2008-004, Nvidia Corporation, Santa Clara, CA, 2008.
3. K. Bzdusek, H. Friberger, K. Eriksson, B. Hardemark, D. Robinson, and M. Kaus. Development and evaluation of an efficient approach to volumetric arc therapy planning. *Medical Physics*, 36(6):2328–2339, 2009.
4. C. Cameron. Sweeping-window arc therapy: An implementation of rotational IMRT with automatic beam-weight calculation. *Physics in Medicine and Biology*, 50:4317–4336, 2005.
5. D. Cao, M.K.N. Afghan, J. Ye, F. Chen, and D.M. Shepard. A generalized inverse planning tool for volumetric-modulated arc therapy. *Physics in Medicine and Biology*, 54:6725–6738, 2009.
6. D. Craft, D. McQuaid, J. Wala, W. Chen, and T. Bortfeld. Multicriteria VMAT optimization. Technical report, Massachusetts General Hospital, Boston, MA, 2011.
7. S.M. Crooks, X. Wu, C. Takita, M. Watzich, and L. Xing. Aperture modulated arc therapy. *Physics in Medicine and Biology*, 48:1333–1344, 2003.
8. M.A. Earl, D.M. Shepard, S. Naqvi, X.A. Li, and C.X. Yu. Inverse planning for intensity-modulated arc therapy using direct aperture optimization. *Physics in Medicine and Biology*, 48:1075–1089, 2003.

9. S. Luan, C. Wang, D. Cao, D.Z. Chen, D.M. Shepard, and C.X. Yu. Leaf-sequencing for intensity-modulated arc therapy using graph algorithms. *Medical Physics*, 35(1):61–69, January 2008.

10. C. Men, H. Edwin Romeijn, X. Jia, and S.B. Jiang. Ultrafast treatment plan optimization for volumetric modulated arc therapy (VMAT). *Medical Physics*, 37(11):5787–5791, 2010.

11. C. Men, X. Gu, D. Choi, A. Majumdar, Z. Zheng, K. Mueller, and S.B. Jiang. GPU-based ultrafast IMRT plan optimization. *Physics in Medicine and Biology*, 54(21):6565, 2009.

12. K. Otto. Volumetric modulated arc therapy: IMRT in a single gantry arc. *Medical Physics*, 35(1):310–317, 2008.

13. F. Peng, X. Jia, X. Gu, M.A. Epelman, H. Edwin Romeijn, and S.B. Jiang. A new column-generation-based algorithm for VMAT treatment plan optimization. *Physics in Medicine and Biology*, 57(14):4569, 2012.

14. D.M. Shepard, D. Cao, M.K.N. Afghan, and M.A. Earl. An arc-sequencing algorithm for intensity modulated arc therapy. *Medical Physics*, 34(2):464–470, February 2007.

15. C. Wang, S. Luan, G. Tang, D.Z. Chen, M.A. Earl, and C.X. Yu. Arc-modulated radiation therapy (AMRT): A single-arc form of intensity-modulated arc therapy. *Physics in Medicine and Biology*, 53:6291–6303, 2008.

16. C.X. Yu. Intensity-modulated arc therapy with dynamic multileaf collimation: An alternative to tomotherapy. *Physics in Medicine and Biology*, 40:1435–1449, 1995.

17. T. Zhen, F. Peng, X. Jia, and S.B. Jiang. Multi-GPU implementation of a VMAT treatment plan optimization algorithm. *Medical Physics*, 42, 2841–2852, 2015.

18. NVIDIA Developer. Thrust. 2015. https://developer.nvidia.com/thrust. (Accessed on May 30, 2015).

CHAPTER 19

Non-Voxel-Based Broad Beam Framework

A Summary

Weiguo Lu and Mingli Chen

CONTENTS

19.1 INTRODUCTION

THE ADVANCEMENT OF THE general purpose graphic processing unit (GPU) with data parallel streaming processor evoked many innovative approaches of handling massive computation, such as in the field of medical physics as reviewed in References 1,2. Many algorithms were redesigned to take advantage of GPU's parallel processing power and to accommodate its relatively small memory compared with the central processing unit (CPU). This is no exception in the field of intensity-modulated radiation therapy (IMRT). The nonvoxel-based broad-beam (NVBB) framework [3] was developed to handle IMRT optimization and proved to be effective. This chapter gives a summary review of the NVBB framework.

In IMRT, treatment planning involves large-scale or very large-scale optimization problems. The conventional approach, which uses voxel and beamlet representation in dose calculation, objective function, and derivative calculation, requires precalculation and large memory storage of beamlet matrices. We refer to the voxel and beamlet representation approach as the voxel-based beamlet superposition (VBS) framework. The number of voxels and the number of beamlets involved can be on the order of 10 M and 100 K, respectively, for typical dimensions of $256 \times 256 \times 256$. Thus, if saved in full, a beamlet matrix can be as large as 1T (=10 M × 100 K), which is too large to be handled by the state-of-the-art workstation. The advantage of beamlet representation is in its ease of dose and derivative formulation. However, in addition to large memory and long calculation time, the beamlet representation suffers from inaccuracy due to linear modeling for nonlinear effects and lack of flexibility for changes of treatment parameters in plan optimization. These prompted the research on the NVBB framework [3] for IMRT treatment planning.

A major distinguish of the NVBB framework from the VBS framework is the adoption of the continuous viewpoint. Discrete sampling is inevitable for computer implementation but will be done at a later phase. The initial formulations are in continuous format, including fluence, dose, objective function, and derivatives calculation. This viewpoint avoids the beamlet matrix and, along with other modifications that are fully parallelizable, allows calculation on the fly. Furthermore, the continuous viewpoint increases the flexibility of objective function formulation, which enables direct treatment parameter optimization (DTPO).

The NVBB framework published 4 years ago has been implemented in a commercial treatment planning system (TPS): the Accuray VoLo™

technology. User reports were positive [4,5]. The NVBB implementation demonstrated efficiency without sacrificing plan quality. It also results in huge savings in hardware and maintenance cost by implementing the framework on a single personal computer (PC) as opposed to a computer cluster and significant improvements in planning throughput.

19.2 NVBB FRAMEWORK AND ITS COMPONENTS

Though the physical world is continuous, it must be discretized for the computational purpose. In the NVBB framework for IMRT planning, however, discretization is done at a later phase; initial formulations are in continuous format.

Table 19.1 is an outline of the process of the NVBB framework for IMRT optimization.

It is worth pointing out the philosophies of the steps given in Table 19.1:

1. The treatment parameters and physical constraints are modeled in the fluence map, which is defined as a continuous function. Thus, the derivatives of the fluence map w.r.t. the treatment parameters can be calculated.

2. In Steps 2 and 3, doses are calculated by the NVBB ray tracing to transmit two-dimensional (2D) fluence to 3D dose.

3. Derivatives $d\Im/dp$ of the objective w.r.t. the treatment parameters also involve the NVBB ray tracing.

TABLE 19.1   Outline of the Process of the NVBB Framework

1. Generate an initial guess p0 for the treatment parameter $p = p0$ and calculate the initial fluence map $f0 = f = f(p0)$.
2. Calculate the full dose $D(f)$.
3. Calculate the approximate dose $D_a(f)$.
4. Calculate the difference $\Delta D = D(f) - D_a(f)$.
5. Evaluate the objective $\Im(D)$.
6. If the clinical goal achieved, return.
7. Calculate the derivatives $d\Im/dp$.
8. Update the treatment parameter p based on $\Im$ and $d\Im/dp$.
9. Calculate the fluence map $f = f(p)$.
10. If $norm(f - f0) > \varepsilon \cdot norm(f0)$, let $f0 = f$ and go to Step 2.
11. Calculate the approximate dose $D_a(f)$.
12. Let $D = D_a(f) + norm(f)/norm(f0) \cdot \Delta D$ and go to Step 5.

4. Sampling is done at the time of the NVBB ray tracing and results in samples on the beam's eye view (BEV) grids. The differential volume in the NVBB ray naturally accounts for beam divergence.

5. The chain rule is applied to evaluate $d\mathfrak{J}/dp$ (Step 7).

6. If deviation of the updated fluence f from the initial fluence f0 is within a given tolerance, then dose update only involves approximate dose calculation rather than full dose calculation (Steps 10–12). Otherwise, the process starts over: the updated fluence is assigned to the initial fluence and dose update involves full dose calculation (Step 2). This approach is referred to as the adaptive full dose calculation.

7. The adaptive full dose calculation is efficient in the sense that it utilizes approximate dose to speed up calculation and perform full dose to gain accuracy when needed.

The key components, fluence map calculation, the NVBB ray tracing, adaptive full dose calculation, and NVBB derivative calculation are described in the following.

19.2.1 Fluence Map Calculation

In general, dose calculation is decomposed in two steps: (1) from treatment parameters to fluence and (2) from fluence to dose. The first step, where accurate machine modeling, such as the penumbra, the tongue and groove (T&G) effect, leakage, etc., are taken into account, is highly nonlinear. Fortunately, it involves only a 2D fluence f calculation, and thus, its computational demand is much lower and it is less time consuming than the second step that involves a 3D volume, even though the second step is linear w.r.t. the 2D fluence map f.

For DTPO, the fluence map $f = f(p)$ and its derivatives $\partial f/\partial p$ w.r.t. the treatment parameters that are to be optimized need to be explicitly written down. For example, in the TomoTherapy's dynamic jaw and binary multi-leaf collimator (MLC) modeling, the treatment parameters to be optimized are positions of left and right jaws and individual leaf open time, respectively, and the fluence map needs to be expressed as a function of those parameters [3]. Other physical properties, such as cone effects, leakage, and T&G, also need to be included in the modeling. 2D MLC consisting of individual leaf pairs can be modeled similarly to dynamic jaws.

19.2.2 NVBB Ray Tracing

Ray tracing is widely used in physics to calculate distributive (e.g., dose calculation) or accumulative (e.g., Radon transform) quantity along a particle path. The conventional approach of ray tracing is voxel based, where the rays are lines of zero width, the space is partitioned into voxels, and the path length of the ray within each voxel needs to be calculated. Several technical difficulties arise, such as calculation of the path length within each voxel [6], normalization to account for inhomogeneous ray distribution at each cross section, energy conservation, and beam divergence. On the other hand, the NVBB ray tracing [3], which can be regarded as a dual representation of the conventional voxel-based ray tracing, naturally accounts for those difficulties. In the NVBB ray tracing, a ray is represented as a ray pyramid (Figure 19.1), and ray tracing is performed along

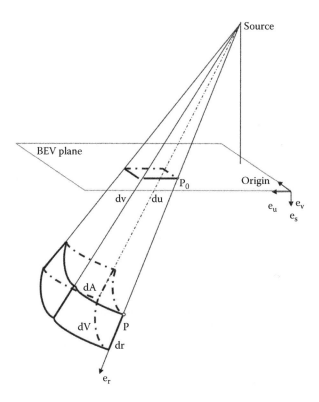

FIGURE 19.1 Differential divergent beam as the NVBB ray in the BEV coordinate system. dV denotes an infinitesimal divergent volume element with the radial component between r and r + dr.

the center axis of a ray pyramid in the BEV coordinate system. The outcome distributions are samples along the center axes, and then interpolation is applied to obtain values on the desired grid points. Total energy is naturally conserved, because energy distribution is based on the partition by the intersection areas of fluence map and ray pyramids. Divergence is corrected by averaging energy in the differential volume dV indicated in Figure 19.1.

19.2.3 Adaptive Full Dose Calculation

Dose calculation is actively utilized during IMRT treatment planning. Physicists and physicians rely heavily on dose calculation to define plans [7]. There are typically two places where dose calculation is needed: during optimization and after optimization. During IMRT optimization, dose is calculated whenever a plan is updated, which is used as the driving force for next iterations to reach a desirable plan. After a plan is optimized, the dose is recalculated to include modeling of all machine constraints for final plan evaluation and for comparison with the measurement. The dose calculated during plan optimization is referred to as the "iteration dose," and the dose of the last optimization iteration is referred to as the "optimization dose." The dose of final evaluation is referred to as the "final dose." Ideally, the optimization dose should match the final dose, and the final dose should match the measurement. On one hand, dose calculation must be accurate enough to make plan evaluation meaningful. On the other hand, dose calculation must be fast enough to complete hundreds of iteration in a reasonable amount of time for typical IMRT optimization.

Trade-offs between accuracy and speed can be seen in various dose calculation algorithms. For example, the Monte Carlo (MC) and full convolution/superposition (C/S) methods are accurate full dose calculations, but they are very time consuming. Some approximate dose engines, such as the finite size pencil beam (FSPB) or the fluence convolution broad beam (FCBB) algorithm [8], are much faster than MC and C/S but have limited accuracy, especially when there exist significant heterogeneities.

In the VBS framework, the trade-off between accuracy and speed is undertaken to some extent via precalculation of the beamlet matrix B, an off-line process that utilizes a slow full dose engine. During optimization iterations, dose is calculated as a simple matrix product $D = Bw$. However, there are drawbacks of the VBS framework that we want to avoid as described in Section 19.1. Instead, we use an "adaptive full dose correction" scheme [9] that combines advantages of the approximate dose engine (speed) and the full dose engine (accuracy).

Full dose calculation may take minutes to hours of CPU time for a complex IMRT plan.

Therefore, full dose calculation is not affordable for every iteration, and an alternative scheme must be employed to reduce calculation time without sacrificing too much accuracy.

To take advantage of the accuracy of full dose calculation (e.g., MC or collapsed convolution superposition [CCCS] methods) and the efficiency of approximate dose calculation (e.g., the FSPB methods), hybrid methods were proposed by various investigators [9–11]. In the NVBB framework, we adopt the "additive correction matrix" method proposed by Siebers et al. [10] with some modifications as described later and detailed in Reference 3.

Note that as outlined in Table 19.1, the approximate dose engine is invoked whenever the fluence changes. Therefore, the approximate dose is calculated once in every iteration. On the other hand, the full dose engine is only called when the updated fluence f deviates from the initial fluence f0 by a preset threshold. Then, f replaces f0, and the process starts over. As optimization converges, the full dose will be calculated less and less frequently. Also note that the derivative of the iteration dose w.r.t. the fluence map is determined by the approximate dose only and independent of the full dose.

Such feature makes the derivative calculation relatively easy, provided that the approximate dose has a simple formulation, which is indeed the case as in FCBB [8]. We chose to use the CCCS method as the full dose engine and the FCBB method as the approximate dose engine.

The full and approximate dose engines have the same first step for fluence calculation but differ in the second step from fluence to dose, which is the more time-consuming part. Next, we show that the full dose calculation, CCCS, accomplished in two steps, Total Energy Released per unit MAss (TERMA) calculation and C/S energy deposition, were redesigned to be fully parallelizable with linear complexity, and both TERMA and the approximate dose calculation, the FCBB dose, are implemented by a simple distributive NVBB ray tracing.

19.2.3.1 Full Dose Calculation

C/S dose calculation [12] consists of two independent parts: TERMA calculation [13–15] and C/S energy deposition [16–18]. TERMA calculation includes the fluence phase space modeling, primary photon ray tracing and modeling of the interaction with material, and TERMA sampling. C/S energy deposition is a means of spreading the released energy to the media by convolving TERMA with a precalculated MC kernel [19].

19.2.3.1.1 TERMA   TERMA calculation is a direct application of the distributive ray tracing. The energy fluence f(u, v) defined on the BEV plane yields energy of f(u, v)·du × dv, the product of fluence, and the infinitesimal area, within the differential divergent beam as illustrated in Figure 19.1, where du × dv represents the intersection area by the beam and the BEV plane. The energy is then released along the divergent beam, and the released energy is regarded as new energy sources for subsequent primary deposition and scattering. The calculation of TERMA only concerns the released energy. Note that TERMA is defined as a density property; therefore, the infinitesimal divergent volume dV in Figure 19.1 naturally accounts for beam divergence. After simplification [3], TERMA can be calculated as a product of fluence, beam-hardening-corrected attenuation $A(r_\rho)$, and a divergence factor a(r): T(u, v, r) = f(u, v)·$A(r_\rho)$·a(r). The beam-hardening-corrected attenuation $A(r_\rho)$ is a function of radiological distance r_ρ, and the radiological distance r_ρ is the integration of the electron density along the beam path from the source to the geometric location (u, v, r) in the BEV coordinates. For TERMA calculation, $A(r_\rho)$ is first constructed as a look-up table. Therefore, its calculation complexity is the same as that of the radiological distance, which has a linear complexity.

19.2.3.1.2 C/S Energy Deposition   In the NVBB work, energy deposition is calculated using the CCCS method, rather than using a full volume convolution. In CCCS, a 3D point kernel is collapsed into multiple line kernels representing the center axes of the cones that partition the 3D volume, where the apex of the cones is at the center of the point kernel. Therefore, the computation complexity is reduced from $O(N^7)$ to $O(N^4M)$, where N is the dimension of each side of the 3D volume and M is the number of cones. Because the energy deposition kernel has a sharp gradient near the center, the cumulative–cumulative kernel (CCK) algorithm [17] is adopted to reduce sampling artifact. We also adopted analytical kernels instead of tabulated kernels with two exponential components (exponential kernels) as proposed in Reference 16. A separate optimization process is used to determine the parameters of the exponential kernels by fitting the tabulated kernels. Based on these, a recursive formula [20] can be constructed. The recursive formula expresses dose as a linear combination of local deposition, where the TERMA point and the target point are in the same voxel, and remote deposition, where the target point receives energy from all preceding TERMA points. Such recursive formula for TERMA calculation exists because of the use of the exponential kernels. The recursive formula has a linear complexity, which further reduces the complexity from $O(N^4M)$ of the CCCS with the tabulated kernels to $O(N^3M)$ of the CCCS with the exponential kernels.

When implemented in GPU, the calculation of C/S energy deposition is confined by memory access, a costly operation. Thus, a coalesced memory access pattern [20] was designed to reduce the number of memory transactions. Other tricks, such as precalculating a look-up table for exponential kernels and grouping ray tracing of two opposing CC directions in a single GPU function, are implemented to increase the calculation speed.

19.2.3.2 Approximate Dose Calculation

As discussed earlier, the model-based dose calculation can be decomposed into two independent steps: (1) from treatment parameters to fluence and (2) from fluence to dose. The first step is machine dependent and requires accurate machine modeling, such as T&G, leakage, latency, etc. It is highly nonlinear and sensitive to any modeling and calculation errors. In fact, any error in the first step will propagate to the second step. Therefore, we need to pay special attention to this step to make the calculated dose match measurement regardless of how machine parameters change. In addition, the fluence map also needs to be calculated using a grid fine enough to reduce the sampling errors. Fortunately, the first step only involves 1D or 2D data, and therefore, it has much lower computation demands than the second step that involves 3D data. Because of its high importance and marginal computation demand, we should always use accurate modeling and fine grids in the first step calculation. The second step, which models energy transportation in the patient body, is patient dependent. It involves 3D data and its computational demand is so high that we need to pay special attention to throughput. Full C/S is an accurate algorithm but too time consuming to be used in every optimization iteration. In the NVBB framework, we introduced an approximate dose engine based on the FCBB algorithm [8], which uses exactly the same "fluence map calculation" as in the full C/S dose engine but uses approximation in the second step.

Given a fluence map, there are three main components that determine dose to the patient: beam divergence, fluence attenuation, and body scatter.

Recall that full C/S dose calculation consists of two steps: TERMA calculation and C/S energy deposition. TERMA calculation models beam divergence and fluence attenuation. C/S energy deposition mainly models body scatter. Beam divergence is patient independent, whereas both fluence attenuation and body scatter are patient (density) dependent. For the photon beam used in radiotherapy, the heterogeneity correction is much more important in the primary beam (fluence attenuation and forward/backward scatter) modeling than that in the secondary beam (lateral scatter) modeling.

Based on the previous analysis, the FCBB algorithm decouples the three components (divergence, primary beam attenuation and scatter, and lateral scatter), applies heterogeneity correction only for the primary beam (analogous to fluence attenuation in TERMA calculation), and ignores the heterogeneity correction for lateral scatter contribution. Detailed descriptions of the FCBB algorithm, its commissioning, and validation tests are given in Reference 8. Here, we give a brief derivation.

The FCBB dose engine is based directly on the fluence map without resorting to FSPBs [21,22], and it is implemented in the BEV coordinate system: $x_{Cartesian} = (u, r)_{BEV}$. Ideally, dose can be expressed as the integration of the product of fluence and dose of infinitesimal beams: $D(x) = \int f(u)B(u, x)du$, where the dose of an infinitesimal beam $B(u, x)$ is approximately the product of three dominant components: central axis component $c(r_\rho)$, divergence factor $a(r)$, and lateral spread function $k(u)$. Here, r_r indicates radiological distance. This approximation was verified by simulations [8]. Another approximation employed was to use the same central axis components for different beams for the same radiological distance. These approximations allow the expression of dose as the product of the central axis component, divergence, and convolution of fluence and the lateral spread function: $D(x) = c(r_\rho)a(r) \int f(u)k(u - u')du'$. The last component effectively accounts for scatter. It can be seen that the expression has the same form as TERMA calculation. Therefore, the same NVBB ray tracing approach can be applied here for approximate dose calculation.

19.2.4 NVBB Derivative Calculation

The optimization objective $\mathfrak{I}$ can be formally expressed as an integral of the composite of dose $D(x)$ and a function $F(D)$ of dose: $\mathfrak{I}(D) = \int F(D(x))dx$, where the integration variable x is the spatial points in the patient or phantom volume. The objective $\mathfrak{I}$ is also called a functional of D. The derivatives $\partial\mathfrak{I}/\partial p$ of the objective with respect to treatment parameters p can be calculated formally by the chain rule: $\partial\mathfrak{J}/\partial p = \int (\partial F/\partial D)(\partial D/\partial p)dx$. The first term $\partial F/\partial D$ can be easily calculated, whereas the second term $\partial D/\partial p = \iint(\partial D/\partial f)(\partial f/\partial p)du$, involving 2D samples u on the fluence plane (BEV plane) and if calculated in brute force, would result in a large complexity of $O(N^5)$ for $\partial\mathfrak{J}/\partial p$. Instead, we calculate directly the derivative of dose w.r.t. treatment parameters and use the approximate dose. Recall that fluence with scatter convolution is a multiplicative factor of the approximate dose, so the derivative of fluence with scatter convolution would still be a multiplicative factor of the derivative of dose: $\partial D/\partial p(x) = c(r_p)$ $a(r)\int(\partial f/\partial p(u)) k(u - u')\,du'$. It turns out that the derivative of the optimization objective can be expressed as the integral of the product of the derivative of fluence with scatter convolution $h(u) = \int(\partial f/\partial p(u))k(u - u')du'$ and accumulation of the central axis component weighted by the derivative of the function of dose, $e(u) = \int(\partial F/\partial D)(u, r)c(r_p(u, r))dr$. That is, $\partial\mathfrak{J}/\partial p = \int e(u)$ $h(u)du$. The integral can be calculated as a 2D summation, which has a complexity of $O(N^2)$. The derivative of fluence with scatter convolution $h(u)$ also has a complexity of $O(N^2)$. The accumulation $e(u)$ of the central axis component weighted by the derivative of the function F can be calculated via accumulative NVBB ray tracing, which has a complexity of $O(N)$. Therefore, the derivative of the objective can be evaluated with a complexity of $O(N^3)$.

19.3 RESULTS

Through recursive dose grid calculation utilizing both the exponential kernels and the CCK algorithm and optimizing the memory access pattern, the CCCS implementation achieved significant speed gain. Table 19.2 shows the breakdown of the speed gain. For example, the speedup factors of exponential kernels relative to tabulated kernels using CPU are shown in the second column, and the speedup factors of GPU relative to CPU using tabulated kernels are shown in the third column.

TABLE 19.2 Speedup due to Exponential Kernels and GPU Implementation, Respectively

Grid Size	Speedup due to Exp Kernel		Speedup due to GPU	
	CPU	GPU	Tab Kernel	Exp Kernel
$64\times64\times64$	9	12	16	21
$128\times128\times128$	16	19	53	64
$256\times256\times256$	30	39	73	97

Source: Adapted from Chen, Q. et al., *Med. Phys.*, 38(3), 2011, 1150.

The NVBB framework was implemented in the C++ programming language on both CPU and GPU architectures, and both CPU and GPU implementations were incorporated into the TomoTherapy TPS. The CPU version runs on a computer cluster, and the GPU version runs on a single workstation with an NVIDIA GeForce GTX 295 graphic card. Extensive verification/validation tests were performed both internally and externally. Table 19.3, adopted from Reference 3, shows the benchmark comparisons between the NVBB framework on a workstation with 2.66 GHz CPU and one NVIDIA GeForce GTX 295 card and the VBS framework on a TomoTherapy 14-node cluster ($14 \times 4 = 56$ 2.66 GHz CPUs). The former is abbreviated as NVBB-GPU and the latter is abbreviated as VBS cluster. The reported times are the preprocessing time, the time for 100 iterations, the time to calculate the full dose, and the time to calculate the final dose. The total time is the sum of the previous four times. For the VBS cluster, a typical TomoTherapy planning took 10–102 min to precalculate beamlet doses. The preprocessing time was reduced to ~10 s for the NVBB-GPU. Excluding the preprocessing time, the NVBB-GPU took only about 30%–90% of iteration time of the VBS cluster for the same number of iterations. As for the full dose and the final dose calculation, the NVBB-GPU had a speedup of about 8–16 times over the VBS cluster. Both VBS cluster and NVBB-GPU use CCCS as the full and final dose engine. The number of collapsed cone directions is 24 (zenith) $\times$ 16 (azimuth) = 384. The VBS cluster uses the tabulated CCK, while the NVBB-GPU uses the exponential CCK.

For the same delivery plans, the differences of final dose between the VBS cluster and the NVBB-GPU were within 1%, 1 mm for all test cases, while the doses of the VBS cluster were well commissioned to match the measurements. For most cases, after the same number of iterations, the plan quality under these two TPSs has no clinically significant differences, except for some cases where the NVBB-GPU showed superior plan quality than the VBS cluster. For those cases, the VBS cluster has less target dose uniformity. Such inferior plan quality was due to the model limitation and errors from beamlet representation and compression of the VBS cluster. The DTPO and nonvoxel, nonbeamlet nature of the NVBB framework greatly reduced those modeling errors and resulted in a better plan.

19.3.1 Commercial Product

As a result of implementing the NVBB framework, a commercial product, VoLo, was released in ASTRO 2012 and user reports were positive:

TABLE 19.3   Benchmark Comparison between the VBS Cluster and NVBB-GPU

Cases	Dose Grid Size	# Beamlets	TPS	Preprocessing	100 Iters	Full Dose	Final Dose	Total Time
Prostate	328×268×35	4,830	VBS cluster	585	300	52.4	73.4	1010.8
			NVBB-GPU	10	250	7.2	7.4	274.6
Lung	128×134×114	11,827	VBS cluster	1000	257	46.9	64	1367.9
			NVBB-GPU	10	205	5	5.6	225.6
Breast	144×128×176	14,631	VBS cluster	1300	420	46.4	66.4	1832.8
			NVBB-GPU	10	180	4.4	4.6	199
H&N	256×256×125	8,121	VBS cluster	1489	491	82.7	170.8	2233.5
			NVBB-GPU	10	403	12.5	12.6	438.1
TBM	144×128×176	117,027	VBS cluster	6785	1745	153.3	210.7	8894
			NVBB-GPU	10	526	13	13.1	562.1

Source: Adapted from Lu, W., *Phys. Med. Biol.*, 55(23), 2010, 7175.
The processing time is shown in seconds.

significantly improved speed without sacrificing plan quality. Because of the improved speed, the planning system can afford calculation with fine resolution calculation grid (1 CT voxel instead of 2 CT voxels), supersampling in gantry angle (3×) [23], additional optimizations for trying out different parameters of objective or criteria, and even exploiting online dose-guided delivery or replanning for adaptive radiotherapy.

19.4 DISCUSSION

The advantages provided by the NVBB framework are summarized as follows:

1. The continuous viewpoint increases the flexibility of objective function formulation, enabling derivative evaluations and direct optimization of various treatment parameters. Fluence that models nonlinear effects, such as the tongue and groove effect, leakage, and different leaf latencies, can be easily incorporated in treatment plan optimization.

2. Without the voxel and beamlet representation, it reduces dose calculation errors due to the voxel size effect that blurs high-dose gradient.

3. The implementation has linear spatial and temporal complexity. It reduces the problem size, lessens memory demand, and increases speed. Thus, it allows dose calculation and IMRT optimization in a much finer grid, better spatial resolution than currently affordable.

4. The full parallelization and low memory nature of this framework are very suitable for GPU implementation, such that calculation can be accomplished on the fly. Thus, a single PC or laptop computer can efficiently perform VLS IMRT optimization and still be faster than the conventional VBS framework run on a computer cluster.

5. The adaptive full dose correction algorithm combines the advantages of full dose (accuracy) and approximate dose (efficiency), making the iteration dose approach the full dose and the optimization dose approach the final dose with a high level of accuracy.

19.4.1 Complexity

The problem size in the conventional VBS framework that representing dose as a matrix product of the beamlet matrix and beamlet weight (fluence) is determined by the product of the number of voxels, $O(N^3)$, and the number of beamlets, $O(N^2)$. The number of beamlets is $O(N^2)$, because there should be one beamlet corresponding to each pixel on the fluence plane. Therefore, the spatial complexity of the VBS framework is $O(N^5)$. The two most time-consuming operations in every optimization iteration in the VBS framework are dose calculation and derivative calculation, which is another matrix multiplication utilizing the transpose of the beamlet matrix. Therefore, the temporal complexity of the VBS framework is $O(N^5)$ as well. Various compression techniques can be used to reduce the problem size to $O(N^5/R)$, where R is the compression ratio. However, the complexity $O(N^5/R)$ still limits the usage of fine resolution, for example, $N = 256$ or 512.

In the NVBB framework, since both dose and derivatives are calculated on the fly through distributive or accumulative ray tracing, no beamlet matrix B is required. Therefore, the spatial complexity is $O(N^3)$ to store the 3D volume such as density, TERMA, dose, etc. In every iteration of the NVBB optimization, there are two NVBB ray tracing operations that are most time consuming:

1. Calculation of approximate dose in the BEV coordinate system via distributive ray tracing, $D(u, r) = c(r_\rho(u, r))a(r)g(u)$

2. Calculation of derivatives via accumulative ray tracing, $e(u) = \int (\partial F/\partial D)(u, r)c(r_\rho(u, r))dr$

Suppose that the fluence map sampling and ray sampling have the same resolution as the 3D volume. Then the number of rays is $O(N^2)$ and the number of samples per ray is $O(N)$. Therefore, the complexity of both ray tracing operations in (1) and (2) is $O(N^3) = O(N) \cdot O(N^2)$. There are several other additional 3D operations, including calculation of $\partial F/\partial D$, converting the approximate dose in the BEV coordinate system to the Cartesian coordinate system, and full dose correction. All of those have complexity proportional to the number of dose samples, that is, $O(N^3)$. Full dose calculations (CCCS) are performed every once in a while when the updated fluence f and initial fluence f0 deviate more than a given threshold. Typically, the CCCS dose calculation happens more frequently at the beginning and less frequently as the solution converges. The TERMA part of CCCS uses the distributive ray tracing, and thus, it also has a complexity of $O(N^3)$. The C/S part typically takes 5–10 times longer than TERMA calculation, rather than 384 (=number of collapsed cone directions) times longer, as memory access for parallel beams is being optimized, and results in about an order of magnitude more expensive than iterations that use approximate dose. However, the frequency of full dose iteration is in principle an order of magnitude less than that of approximate dose and derivative calculation. Therefore, in general, adaptive full dose correction at most doubles the total iteration time compared with using only approximate dose for all iterations. Other operations in the NVBB framework involve only 1D or 2D data with a complexity of $O(N)$ or $O(N^2)$, which can be omitted when compared with $O(N^3)$. Therefore, both the temporal complexity and spatial complexity for the NVBB framework in IMRT optimization are linear w.r.t. dose samples $O(N^3)$, and the linear complexity makes it easy to handle large systems, for example, $N \geq 256$. In addition, the NVBB framework is easily parallelizable and can be efficiently implemented in GPU for the following reasons:

1. The linear spatial complexity $O(N^3)$ allows the problem to fit in the small memory of a GPU, even for large N ($N \geq 256$). For example, for a large dose grid of $256 \times 256 \times 256$, the amount of memory needed for every 3D volume is 64 MB. Suppose we need up to 10 3D volumes (density, TERMA, full dose, derivatives, etc.) for internal storage, then the memory requirement is only 640 MB, which can still fit in a modern GPU card with global memory around 1 GB.

2. The fully parallelizable nature of NVBB ray tracing makes it easy to maintain instruction and data alignment.

3. There is no data write–write conflict in any part of the parallelized code.

4. Trilinear interpolations are inherently implemented via texture structures in modern GPUs with little cost.

19.4.2 Applicability of the NVBB Ray Tracing in CT Reconstruction

The BEV coordinate system and the NVBB ray tracing prove to be very efficient for applications related to divergent beams, including dose calculation and CT reconstruction.

19.5 CONCLUSION

The parallelizable nature, on-the-fly calculation, and low memory demand of the NVBB framework are suitable for GPU implementation.

This framework improves IMRT treatment planning speed without compromising accuracy. It reduces the optimization time manyfold and thus increases clinical throughput.

REFERENCES

1. Jia, X., Ziegenhein, P., and Jiang, S. B. GPU-based high-performance computing for radiation therapy, 2014. *Phys Med Biol* **59**(4), R151–R182.
2. Pratx, G. and Xing, L. GPU computing in medical physics: A review, 2011. *Med Phys* **38**(5), 2685–2697.
3. Lu, W. A non-voxel-based broad-beam (NVBB) framework for IMRT treatment planning, 2010. *Phys Med Biol* **55**(23), 7175–7210.
4. Llagostera, C., Chiavassa, S., and Lisbona, A. Using the new dosimetric calculation system of tomotherapy: Volo®, what impact on the computation time and calculated dose distributions? 2013. *Phys Med* **29**(Suppl. 1), e21.
5. Bolan, C. Technology trends: Expediting the treatment planning process, 2013. *Appl Radiat Oncol* **2**(4), 19–23.
6. Siddon, R. L. Fast calculation of the exact radiological path for a three-dimensional CT array, 1985. *Med Phys* **12**(2), 252–255.
7. Ahnesjo, A. and Aspradakis, M. M. Dose calculations for external photon beams in radiotherapy, 1999. *Phys Med Biol* **44**(11), R99–R155.
8. Lu, W. and Chen, M. Fluence-convolution broad-beam (FCBB) dose calculation, 2010. *Phys Med Biol* **55**(23), 7211–7229.
9. Siebers, J. V., Tong, S., Lauterbach, M., Wu, Q., and Mohan, R. Acceleration of dose calculations for intensity-modulated radiotherapy, 2001. *Med Phys* **28**(6), 903–910.
10. Siebers, J. V., Lauterbach, M., Tong, S., Wu, Q., and Mohan, R. Reducing dose calculation time for accurate iterative imrt planning, 2002. *Med Phys* **29**(2), 231–237.
11. McNutt, T. Dose calculations: Collapsed cone convolution superposition and delta pixel beam, 2002. Pinnacle white paper no 4535 983 02474.
12. Mackie, T. R., Reckwerdt, P., McNutt, T., Gehring, M., and Sanders, C. Photon beam dose computations, 1996. *Proceedings of the 1996 Summer School for the AAPM*, pp. 103–135. Vancouver, British Columbia, Canada.
13. Liu, H. H., Mackie, T. R., and McCullough, E. C. Calculating output factors for photon beam radiotherapy using a convolution/superposition method based on a dual source photon beam model, 1997. *Med Phys* **24**(12), 1975–1985.
14. Mohan, R., Chui, C., and Lidofsky, L. Differential pencil beam dose computation model for photons, 1986. *Med Phys* **13**(1), 64–73.
15. Papanikolaou, N., Mackie, T. R., Meger-Wells, C., Gehring, M., and Reckwerdt, P. Investigation of the convolution method for polyenergetic spectra, 1993. *Med Phys* **20**(5), 1327–1336.
16. Ahnesjo, A. Collapsed cone convolution of radiant energy for photon dose calculation in heterogeneous media, 1989. *Med Phys* **16**(4), 577–592.

17. Lu, W., Olivera, G. H., Chen, M. L., Reckwerdt, P. J., and Mackie, T. R. Accurate convolution/superposition for multi-resolution dose calculation using cumulative tabulated kernels, 2005. *Phys Med Biol* **50**(4), 655–680.

18. Mackie, T. R., Scrimger, J. W., and Battista, J. J. A convolution method of calculating dose for 15-mv x rays, 1985. *Med Phys* **12**(2), 188–196.

19. Mackie, T. R., Bielajew, A. F., Rogers, D. W., and Battista, J. J. Generation of photon energy deposition kernels using the egs monte carlo code, 1988. *Phys Med Biol* **33**(1), 1–20.

20. Chen, Q., Chen, M., and Lu, W. Ultrafast convolution/superposition using tabulated and exponential kernels on GPU, 2011. *Med Phys* **38**(3), 1150–1161.

21. Ahnesjo, A., Saxner, M., and Trepp, A. A pencil beam model for photon dose calculation, 1992. *Med Phys* **19**(2), 263–273.

22. Bourland, J. D. and Chaney, E. L. A finite-size pencil beam model for photon dose calculations in three dimensions, 1992. *Med Phys* **19**(6), 1401–1412.

23. Hardcastle, N., Bayliss, A., Wong, J. H., Rosenfeld, A. B., and Tome, W. A. Improvements in dose calculation accuracy for small off-axis targets in high dose per fraction tomotherapy, 2012. *Med Phys* **39**(8), 4788–4794.

CHAPTER 20

Gamma Index Calculations

Xuejun Gu

CONTENTS

20.1 INTRODUCTION

THE γ-INDEX CONCEPT INTRODUCED by Low et al. [1] has been widely used to compare two dose distributions in cancer radiotherapy. With the recently developed and more sophisticated treatment modalities such as volumetric-modulated arc therapy (VMAT) [2], the comparison of three-dimensional (3D) dose distributions becomes necessary in patient-specific treatment plans quality assurance (QA). However, comparing two 3D dose distributions is still a time-consuming (e.g., many minutes) task, because of the involvement of computational intensive tasks such as interpolation of dose grid and exhaustive search. There is a clinical need to significantly speed up 3D γ-index computations.

In the past few decades, the accuracy and/or efficiency of the original γ-index calculation algorithm has been greatly improved [3–7]. More recently, Wendling et al. [8] speeded up the exhaustive search by presorting involved evaluation dose points with respect to their spatial distances to a reference dose point and performing interpolation on fly in a fixed search radius region. This fixed search region induces an overestimation of γ-index values at cases of large dose difference (DD) in the search region and sharp dose drop just beyond the search region boundary. The algorithm's accuracy relies on a fine dose interpolation. The geometric interpretation of γ-index evaluation technique proposed by Ju et al. [9] implies a linear interpolation by calculating the distance from a reference point to a subdivided simplex formed by evaluation dose points in search regions. Thus, high accuracy and efficiency can be achieved without dose grid interpolation. However, searching the closest distance over all subdivided simplexes is still time consuming. Later, Chen et al. [10] reported a method based on using fast Euclidean distance transform (EDT) of quantized n-dimensional dose distributions. The discretization errors cannot be avoided when quantizing dose distributions. This method also requires more memory space than original search-based algorithm. Thus, a full 3D application of EDT method is limited by its memory requirement. Yuan and Chen [11] applied a k-d tree technique for nearest neighbor searching. The searching time for N^n voxels dose distribution can be reduced to $(N^n)^{1/k}$, where $2 < k < 3$ for 2D and 3D dose distributions. However, the accuracy of this method is limited by the interpolation. In certain cases, the overhead of k-d tree construction time is longer than γ-index calculation time.

In this chapter, we describe a graphics processing unit (GPU)-based fast and accurate γ-index evaluation algorithm, developed by Gu et al. [12]. The GPU is originally designed for graphics rendering. It has recently been introduced into the radiotherapy community to accelerate computational tasks such as cone beam CT (CBCT) reconstruction, rigid and deformable image registration, dose calculation, and treatment plan optimization [13–26]. The GPU is especially well suited for problems that can be expressed as data parallel computations [27]. The γ-index calculation, which evaluates each reference point independently, is suitable for GPU-based parallelization.

The remainder of this chapter is organized as follows. In Section 20.2, we discuss the graphic interpretation of γ-index and the consequent algorithm. In Section 20.3, implementation of the γ-index algorithm and its performance optimization are detailed. In Section 20.4, we present the evaluation of our GPU-based algorithm using eight 3D IMRT dose distribution pairs. The speedup factor achieved by GPU implementation and the effects of DD sorting and the r-index values on the computation time are also studied. The impact of dose distribution resolution and DD and distance-to-agreement (DTA) criteria on computation time are also investigated as well. Conclusion is given in Section 20.5.

20.2 GRAPHIC INTERPRETATIONS OF GAMMA INDEX AND ALGORITHM

20.2.1 Original Gamma Index Algorithm

The γ-index is often used to evaluate dose distribution, which combines two terms: a DD term and a DTA term [1]. Briefly, an evaluated dose distribution $D_e(\mathbf{r}_e)$ is compared to a reference dose distribution $D_r(\mathbf{r}_r)$, where $\mathbf{r}_e$ and $\mathbf{r}_r$ refer to the evaluated and reference dose grid points. The DD is defined as $\delta D(\mathbf{r}_e, \mathbf{r}_r) = D_e(\mathbf{r}_e) - D_r(\mathbf{r}_r)$, and the spatial distance is defined as $\delta r(\mathbf{r}_r, \mathbf{r}_e) = |\mathbf{r}_e - \mathbf{r}_r|$. The generalized Γ function is defined as

$$\Gamma(\mathbf{r}_r, \mathbf{r}_e) = \sqrt{\frac{\delta D^2(\mathbf{r}_e, \mathbf{r}_r)}{\Delta D^2} + \frac{\delta r^2(\mathbf{r}_e, \mathbf{r}_r)}{\Delta d^2}}, \tag{20.1}$$

where ΔD and Δd are the acceptance criteria for the DD and DTA, respectively. The γ-index is defined as the smallest value $\Gamma(\mathbf{r}_r, \mathbf{r}_e)$, which corresponds best to the reference point, defined as

$$\gamma(\mathbf{r}_r) = \min\{\Gamma(\mathbf{r}_r, \mathbf{r}_e)\} \forall \{\mathbf{r}_e\}. \tag{20.2}$$

In a normalized dose-distance space, the γ-index can be defined as a minimum Euclidean distance between an evaluated dose point and a reference dose point. Mathematically, it can be written as

$$\gamma(\mathbf{r}_r) = \min\{\Gamma(\mathbf{r}_r, \mathbf{r}_e)\} \quad \forall \{\mathbf{r}_e\},$$

with

$$\Gamma(\mathbf{r}_r, \mathbf{r}_e) = |\tilde{r}_r - \tilde{r}_e|,$$

$$\tilde{r}_r = \left(\frac{\mathbf{r}_r}{\Delta d}, \frac{D_r(\mathbf{r}_r)}{\Delta D} \right),$$

$$\tilde{r}_e = \left(\frac{\mathbf{r}_e}{\Delta d}, \frac{D_e(\mathbf{r}_e)}{\Delta D} \right). \qquad (20.3)$$

Here, $\tilde{r}_r$ and $\tilde{r}_e$ are redefined reference and evaluation points in dose-distance space. Using these definitions, in common, if $\gamma(\mathbf{r}_r) \leq 1$, then the evaluated dose distribution at that point is acceptable, otherwise it is beyond acceptance criteria.

20.2.2 Geometric Interpretation of Gamma Index

As mentioned earlier, in the dose-distance space the γ-index measures the shortest distance from a normalized reference point $\tilde{r}_r$ to a set of normalized evaluated points $\tilde{r}_e$. Instead of treating the set of normalized evaluated points as discretized points, Ju et al. [9] proposed a geometric method to describe the evaluated dose distribution in a continuous surface that consists of possible interpolated points, referred to as the *dose surface*. With such dose surface definition, the γ-index measurement is to find the shortest distance from a reference point $\tilde{r}_r$ to a dose surface. For 1D dose distribution, the dose surface formed by evaluated dose points in 2D dose-distance space becomes line segments. γ-index is the smallest distance from a reference point to line segments. For 2D dose distribution, the dose surface formed in 3D dose-distance space consists of quadrilateral elements arranged in a rectangular array. γ-index is calculated by a point to a plane surface distance. For 3D dose distribution, the dose surface is a hypersurface lying in the 4D dose-distance space and consisting of cubical elements. γ-index is calculated by a point to a hypersurface distance.

To efficiently find the closest distance of a point to a dose surface, Ju et al. [9] used a graphic method with simplicial meshes, composed of simplexes. Simplex is a generalized notion of a triangle or tetrahedron to arbitrary dimensions. Specifically, in an n-dimensional space, a simplicial mesh consists of k-simplexes S with $0 \le k \le n$ and a k-simplex is the convex hull of its $k+1$ vertices. For example, in a 3D space, we get 0-simplex (point with one vertex), the 1-simplex (a line with two vertices), the 2-simplex (a triangle with three vertices), and the 3-simplex (a tetrahedron with four vertices). With simplex representation of geometry, the distance from a point to a surface can be calculated with a closed form. For a 3D dose distribution, the dose surface will be described in the 4D dose distribution and the dose surface can be represented by tetrahedral mesh (3-simplex). The distance from a point to a simplex is the shortest distance from a point to any point on the boundary or in the interior of the simplex.

With known vertices, any boundary and interior point (v) in a k-simplex can be expressed as $v = \sum_{i=1}^{k+1} \omega_i v_i$, where v_i represents vertex i of the simplex and ω_i is nonnegative weight with $\sum_{i=1}^{k+1} \omega_i = 1$. Thus, finding the shortest distance from point p to the k-simplex (denoted as S) becomes a minimization problem, written as

$$\bar{D}(p,S) = \min \left| p - \sum_{i=1}^{k+1} \omega_i v_i \right|, \qquad (20.4)$$

where $|.|$ represents the Euclidean distance (2-norm distance) between points. Explicitly, the right-hand side of Equation 20.4 can be written as

$$\bar{D}(p,S) = \min \sqrt{\sum_{j=1}^{n} \left(c_j(p) - \sum_{i=1}^{k+1} \omega_i c_j(v_i) \right)^2}, \qquad (20.5)$$

with $c_j(*)$ as the jth coordinate of point*. With $\sum_{i=1}^{k+1} \omega_i = 1$, $\omega_{k+1} = 1 - \sum_{i=1}^{k} \omega_i$, Equation 20.5 can be rewritten as

$$\bar{D}(p,S) = \min \sqrt{\sum_{j=1}^{n} \left((c_j(p) - c_j(v_{k+1})) - \sum_{i=1}^{k} \omega_i (c_j(v_i) - c_j(v_{k+1})) \right)^2}, \qquad (20.6)$$

or in a matrix-norm format:

$$\bar{D}(p,S) = \min|P - WV|;$$

$$\text{with } P = \begin{bmatrix} c_1(p) - c_1(v_{k+1}) \\ \vdots \\ c_n(p) - c_n(v_{k+1}) \end{bmatrix}, \quad W = \begin{bmatrix} \omega_1 \\ \vdots \\ \omega_n \end{bmatrix}, \quad (20.7)$$

$$\text{and } V = \begin{bmatrix} c_1(v_1) - c_1(v_{k+1}) & \cdots & c_1(v_k) - c_1(v_{k+1}) \\ \vdots & \ddots & \vdots \\ c_n(v_1) - c_n(v_{k+1}) & \cdots & c_n(v_k) - c_n(v_{k+1}) \end{bmatrix}.$$

The defined minimization is a quadratic minimization problem, whose minimum can be achieved when

$$W = (V^T V)^{-1} V^T P. \quad (20.8)$$

The solution given by Equation 20.8 calculates the shortest distance from a point to the surface containing the k-simplex. However, the calculated distance may not be an actual distance when such calculation finds a point neither inside nor on the boundary of the simplex. Under such scenario, the shortest distance should be found between the point and the surface.

20.2.3 Modified Gamma Index Algorithm

According to Equation 20.3, γ, the minimum of Γ values, can be obtained by an exhaustive search through all calculated reference points. The exhaustive search can be accelerated by a presorting strategy [8]. Algorithm 20.1 illustrates a CPU implementation of the combined presorting and geometric γ-index algorithm. This modified γ-index algorithm includes a step of presorting the geometric distance $\{L_n\}$ of all the voxels with a search radius $\max(DD(\mathbf{r}_r))\Delta d/\Delta D$.

Algorithm 20.1: A modified γ-index calculation algorithm implemented on CPU

1. Calculate the maximum DD: $\max(DD(\mathbf{r}_r)) = \max(D_r(\mathbf{r}_r) - D_e(\mathbf{r}_r))$, $\forall \{\mathbf{r}_r\}$
2. Calculate the geometric distance set L_n, which defines the maximum search range for each reference point

$$L_n = \sqrt{(i\Delta x)^2 + (j\Delta y)^2 + (k\Delta z)^2}/\Delta d$$

$$\text{with } |i(\text{or } j, k)| \le \frac{\max(DD(\mathbf{r}_r))}{\Delta D} \frac{\Delta d}{\Delta x(\text{or } \Delta y, \Delta z)}$$

$$\text{and } |L_n| < \frac{\max(DD(\mathbf{r}_r))}{\Delta D}$$

where i, j, k are discretized coordinates in x, y, z directions and Δx, Δy, Δz refer to resolutions in x, y, z directions

3. Sort the geometric distance set $\{n, L_n\}$ in ascending order of L_n, with n as the total number of voxel inside the search range

4. For each reference dose point:
   a. Set $\gamma(\mathbf{r}_r) = DD(\mathbf{r}_r)/\Delta D$
   b. For $n = 1$: N (N is the length of $\{L_n\}$)
      For $j = 1$: nS (nS is the number of simplexes in one voxel)
      i. Calculate Euclidean distance $\Gamma(\tilde{r}_r, S_j)$ from reference point $\tilde{r}_r$ to a k-simplex S_j:

$$\Gamma(\tilde{r}_r, S_j) = \begin{cases} \min \left| \tilde{r}_r - \sum_{i=1}^{k+1} \omega_i v_i \right|, & \text{if all } \omega_i > 0 \\ \min_{S_i \in \partial S_i} \Gamma(\tilde{r}_r, S_i), & \text{others} \end{cases}$$

      ii. If $\Gamma(\tilde{r}_r, S_j) < \gamma(\mathbf{r}_r)$: $\gamma(\mathbf{r}_r) = \Gamma(\tilde{r}_r, S_j)$

      End For
      If $\gamma(\mathbf{r}_r) < L_n$, break
      End For
      End For

20.3 GPU IMPLEMENTATIONS AND OPTIMIZATION OF PERFORMANCE

The above CPU γ-index algorithm (Algorithm 20.1) was implemented on GPU using *Compute Unified Device Architecture* (CUDA) programming environment. A key point of the GPU implementation of this algorithm is to ensure all threads in the same batch (strictly speaking *warp* in CUDA terminology) to have similar numbers of arithmetic operations. This is because, if some threads in a warp require much longer execution time, the other threads in this warp will finish first and then wait in idle until the longer execution time threads finish, implying a waste of computational power. Therefore, directly mapping the CPU version of γ-index algorithm (Algorithm 20.1) onto GPU cannot guarantee that all threads in a warp have similar computation workload, and consequentially cannot achieve the maximum speedup.

In Algorithm 20.1, the most time-consuming part is Step 4 where the minimum Γ value is searched around the reference point in a search range of a radius $(DD(\mathbf{r}_r)/\Delta D)\Delta d$. On CPU, this step has to be looped over every reference points in a sequential manner. The GPU implement can parallelize a large number of reference points' computation and execute simultaneously using multiple threads. However, the search range for each reference point is different, which is selected as $\max(DD(\mathbf{r}_r))\Delta d/\Delta D$. Approximately, the computational task on each thread for each reference point is proportional to the dose difference $DD(\mathbf{r}_r)$. In order to have all threads take similar computational tasks in a warp, presorting the reference points with respect to $DD(\mathbf{r}_r)$ (for convenience we call it DD sorting) is conducted and γ-index calculation is performed on GPU according to the presorted voxel order. CUDA Thrust library [28] is utilized for presorting to improve the computational efficiency. For example, the DD-sorting procedure and the presorting geometric distance set $\{n, L_n\}$ are to be parallelized with Thrust sorting functions, which enables sorting a (or multiple) millions-element array(s) within subseconds. Avoiding IF conditions is critical for improving GPU's computational efficiency. To avoid IF statement, which happens in the recursive computation of $\Gamma(\tilde{r}, S_j)$ in Algorithm 20.1 Step 4-b-i, we calculate $\Gamma(\tilde{r}, S_j)$ in all simplexes in the GPU implementation. The completed GPU-based γ-index algorithm is illustrated as follows:

Algorithm 20.2: A modified γ-index calculation algorithm implemented on GPU

1. Transfer dose distribution data from CPU to GPU
2. CUDA Kernel 1: calculate in parallel the dose difference

$$DD(\mathbf{r}_r) = D_r(\mathbf{r}_r) - D_e(\mathbf{r}_r), \ \forall\{\mathbf{r}_r\}$$

3. Sort in parallel {Voxel Index, $DD(\mathbf{r}_r)$} array pair in ascending order of $DD(\mathbf{r}_r)$ using Thrust parallel sorting function and obtain $\max(DD(\mathbf{r}_r))$

4. CUDA Kernel 2: calculate in parallel the geometric distance set $\{L_n\}$
5. Sort in parallel the geometric distance set $\{n, L_n\}$ in ascending order of L_n using Thrust parallel sorting function
6. CUDA Kernel 3: calculate in parallel the γ-index values using the algorithm illustrated in Step 4 of Algorithm 20.1
7. Sort {Voxel Index, γ} back to the original voxel index order
8. Transfer the γ-index data from GPU to CPU

20.4 EXPERIMENTAL RESULTS

20.4.1 Evaluation Data Set and Time Measurement Criteria

The GPU implementation was tested on eight IMRT dose distribution pairs (four lung cases [L1–L4] and four head and neck cases [H1–H4]). The reference data were generated using a Monte Carlo dose engine MCSIM [29], and the evaluation data were created with an in-house developed pencil beam algorithm [20]. All dose distributions were originally calculated with a voxel size of 4.0 mm × 4.0 mm × 2.5 mm and normalized to the prescription dose and interpolated to various resolution levels for comparison studies. CPU computation was conducted on a four-core Intel Xeon 2.27 GHz processor. GPU computation was performed on one single NVIDIA Tesla C1060 card, which has 240 processor cores (1.3 GHz) and 4 GB device memory.

The total computational time T^C for CPU implementation based on Algorithm 20.1 was divided into two parts $T^C = T_p^C + T_\gamma^C$, where T_p^C is the data processing time (Steps 1, 2, and 3 of Algorithm 20.1) and T_γ^C is the γ-index calculation time (Step 4 of Algorithm 20.1). For the GPU implementation based on Algorithm 20.2, the total computation time T^G was split into three parts, that is, $T^G = T_t^G + T_p^G + T_\gamma^G$, where T_t^G is the data transferring time between CPU and GPU (Steps 1 and 8 of Algorithm 20.2), T_p^G is the data processing time (Steps 2–5 and Step 7 of Algorithm 20.2), and T_γ^G is the γ-index calculation time (Step 6 of Algorithm 20.2).

20.4.2 Speedup of GPU versus CPU

In the evaluation study, the resolution of dose distributions is $256 \times 256 \times 144$ (or 160, 206) and 3%/3 mm criterion is used. Computation time for the CPU implementation (Algorithm 20.1) and the GPU implementation (Algorithm 20.2) is listed in Table 20.1, as well as speedup factors with and without CPU–GPU data transferring time, that is, T^C/T^G and $T^C/(T^G - T_t^G)$. Table 20.1 shows that the speedup factors with and without data transferring are quite similar (within 3% for all cases), indicating

TABLE 20.1 Calculation Time of γ-Index for CPU and GPU Implementations for Eight IMRT Dose Distribution Pairs

Case	Voxel Number	CPU (s)			GPU (s)				Speedup Factor	
		T_p^C	T_γ^C	T^C	T_t^G	T_p^G	T_γ^G	T^G	$T^C/(T^G-T_t^G)$	T^C/T^G
L1	256×256×206	0.33	64.93	65.26	0.07	0.18	1.18	1.43	47.99	45.64
L2	256×256×160	0.24	65.64	65.89	0.06	0.15	1.13	1.34	51.47	49.16
L3	256×256×160	0.28	101.46	101.74	0.06	0.14	1.68	1.88	55.90	54.12
L4	256×256×160	0.25	30.10	30.35	0.06	0.14	0.43	0.63	53.25	48.17
H1	256×256×144	0.49	47.73	47.95	0.05	0.11	0.91	1.07	47.27	45.07
H2	256×256×144	0.45	242.23	242.68	0.05	0.12	3.13	3.30	74.67	73.54
H3	256×256×144	0.24	116.14	116.38	0.05	0.12	2.10	2.27	52.42	51.27
H4	256×256×144	0.22	107.61	107.86	0.05	0.12	1.75	1.92	57.66	56.16

that the data transferring time in GPU calculation is not significant compared to γ-index computation time T_γ^G. Also, the data processing time in both CPU and GPU implementations is relatively insignificant compared to the γ-index calculation time (Step 4 of Algorithm 20.1 and Step 6 of Algorithm 20.2). Overall, the GPU implementation can achieve about 45×–75× speedup compared to its CPU implementation as illustrated in Table 20.1.

20.4.3 Effect of DD Sorting on Computation Time T_γ^G

Introducing of DD sorting (Step 3 in Algorithm 20.2) offers better synchronization of CUDA threads computational tasks and consequently reduces computation time. Table 20.2 illustrates the effect of DD sorting on T_γ^G and the speedup achieved by DD sorting is around 2.7–5.5 times.

20.4.4 Effect of γ-Index Values on Computation Time T_γ^C and T_γ^G

As illustrated in Table 20.1, both T_γ^C and T_γ^G change significantly from case to case. Cases L3 and L4 have the same number of voxels, but their γ-index calculation time differs by approximately three times. As is known, the computation time t for each reference point is proportional to the number of searched voxels N_s, that is, $t \propto N_s$. The relationship of N_s with the search length L can be expressed as $N_s \propto L^n$, where n is the dimension of dose distributions and $n = 3$ for all the testing cases in this chapter. Moreover, according to Algorithm 20.1 Step 4-b-ii, the search length L is proportional to the γ value at each reference point, that is, $L \propto \gamma$. Thus, the computation time t for each reference point is proportional to γn: $t \propto \gamma^n$. Figure 20.1a illustrates T_γ^C and T_γ^G versus the summation of γ^3 over all voxels $(\sum \gamma^3)$ for

TABLE 20.2 Speedup Achieved in GPU Computation by Sorting Voxels Based on the Dose Difference Values

Case	T_γ^G (Non-DD Sorting) (s)	T_γ^G (DD Sorting) (s)	Speedup Factor Achieved by DD Sorting
L1	4.37	1.18	3.70
L2	4.62	1.13	4.09
L3	6.71	1.68	3.99
L4	1.18	0.43	2.74
H1	4.99	0.91	5.48
H2	14.31	3.13	4.57
H3	9.01	2.10	4.29
H4	12.54	1.75	7.17

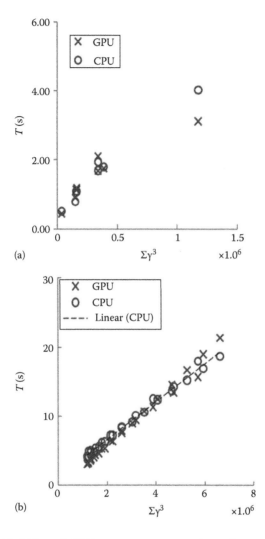

FIGURE 20.1   (a) GPU and CPU computation time for eight testing cases versus the summation of γ^3 values over all evaluated points. (b) GPU and CPU computation time for case H2 with various dose shifts on the evaluation dose distribution versus the summation of γ^3 values over all evaluated points. For convenient purpose, we scaled down CPU computation time by a factor of 60.0 to illustrate them in the same vertical axis of GPU computation time. (Reproduced from Gu, X. et al., *Phys. Med. Biol.*, 56(5), 1431, 2011.)

each of eight testing cases, respectively. We see that both T_γ^C and T_γ^G monotonically increase with the value of $\sum \gamma^3$. To verify the relationship, we shift the evaluation dose distribution of patient data (H2) (normalized to the prescription dose) by −10%, −9%, ... , up to 10%, at a step size of 1%, inside the region of 10% isodose line. Figure 20.1b illustrates the γ-index calculation time T_γ^C and T_γ^G with respect to $\sum \gamma^3$, where T_γ^C versus $\sum \gamma^3$ can be fitted with a straight line (dashed line in Figure 20.1b), indicating that T_γ^C strictly follows the rule $T_\gamma^C \propto \sum \gamma^n$. However, the date points for T_γ^G are much more scattered. This is because the GPU computation time is not only the function of $\sum \gamma^3$ but also the function of γ^n distribution that determines the variation of threads computation time in a warp.

20.4.5 Effect of Dose Distribution Resolution on the Computation Time T_γ^C and T_γ^G

The effect of the dose distribution resolution on computation time T_γ^C and T_γ^G was tested on case H2. The original dose distributions were interpolated to various resolution levels, including $128 \times 128 \times 72$, $128 \times 128 \times 144$, $256 \times 256 \times 72$, $256 \times 256 \times 144$, and $512 \times 512 \times 72$. Figure 20.2 illustrates T_γ^C and T_γ^G change with respect to the resolution changes. As indicated by the power trend lines (dashed lines in Figure 20.2), T_γ^C increases approximately as $N^{1.90}$ while T_γ^G increases approximately as $N^{1.75}$, when the resolution of dose distribution increases N times. As illustrated in Algorithm 20.1, the CPU-based γ-index calculation is completed with two loops: one outer loop over all the reference dose points and one inner loop of an exhaustive search in a limited region around each reference dose point. The computation time of the outer loop is increased linearly with respect to the increase in the resolution of dose distributions, while, the inner loop computation time is proportional to the number of voxels involved, which is increased linearly in a fixed region as the resolution increases. Overall, it leads to a quadratic increase in computational time ($\propto N^{1.90}$) for a linear change of resolution. For the GPU algorithm 20.2, the computation time increases as $(N)^{1.75}$ in this testing case. This slight difference might be caused by the memory accessing time variation.

20.4.6 Effect of DD and DTA Criteria on Computation Time T_γ^C and T_γ^G

This study was performed on Case H2 by varying DD and DTA criteria. The dose resolution is fixed on $256 \times 256 \times 144$. Table 20.3 shows

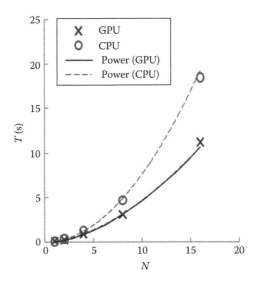

FIGURE 20.2 CPU and GPU computation time as functions of dose distribution resolution. Again, for the convenient purpose, we scaled down CPU computation time by a factor of 60.0 to illustrate CPU computation time in the same axis of GPU computation time. (Reproduced from Gu, X. et al., *Phys. Med. Biol.*, 56(5), 1431, 2011.)

TABLE 20.3 CPU and GPU Computation Time Varies with DD and DTA Criteria Values

		Computational Time (s)		
DD Criteria (%)	DTA Criteria (mm)	T_γ^G	T_γ^C	Speedup Factor T_γ^C/T_γ^G
1	1	3.13	240.37	76.80
1	2	11.33	863.33	76.20
1	3	23.91	1545.09	64.62
2	1	0.97	74.27	76.57
2	2	3.13	265.82	84.93
2	3	6.53	486.68	74.53
3	1	0.55	40.09	72.89
3	2	1.55	119.32	76.98
3	3	3.13	245.34	78.38

the computation time obtained from varying criteria. There are three interesting phenomena: (1) when fixing the DTA criterion value and increasing the DD criterion value, the computation time decreases; (2) when fixing the DD criterion value and increasing the DTA criterion value, the computation time increases; and (3) when increasing both

DD criterion value and DTA criterion value proportionally, for example, from 1%, 1 mm to 2%, 2 mm, or 3%, 3 mm, the computation time does not change. Aforementioned, the γ-index calculation time for each reference dose point t is proportional to γn. For phenomenon (1), when increasing the DD criterion value with the fixed DTA criterion value, the γ-index value decreases. Consequently, computation time decreases due to decreased searching steps. For phenomenon (2), when fixing the DD criterion value, but increasing the DTA criterion value by k times, the γ-index values decrease by k' times, with $k' < k$, and consequently decrease computation time by $(k')^n$ times. However, when the DTA criterion value increases by k times, the resolution in the normalized dose-distance space also increases by kn times, which consequently increases the computation time by kn times. The net change of the computation time should be $(k/k')^n$. Since $k' < k$, the overall computation time then increases. For phenomenon (3), when increasing both DTA and DD criteria values simultaneously by k times, the γ-index values decrease by $k' = k$ times. The increase rate of computation time is $(k/k)^n = 1$. The computation time under this situation does not change. From Table 20.3, we can see that the change in DD and DTA criteria values does not affect the speedup factor achieved with GPU implementation.

20.5 CONCLUSION AND DISCUSSIONS

In this chapter, we described the implementation of a modified γ-index algorithm on GPU. The implementation was evaluated on eight pairs of IMRT dose distributions. Overall, when using one single Tesla C1060 GPU card for GPU computation and one-core Intel Xeon 2.27 GHz processor, GPU implementation can achieve 45×–75× speedup compared to CPU implementation. The γ-index calculation can be finished within a few seconds. The presorting procedure based on the DD speeds up the GPU calculation by about 2.7–5.5 times. The CPU computation time is proportional to the summation of γ^n over all voxels, where n is the dimensions of dose distributions. The GPU computation time is approximately proportional to the summation of γ^n over all voxels, but affected by the variation of γ^n among different voxels. Also, it is interesting to see that increasing the resolution of dose distribution leads to a quadratic increase in computation time on CPU, while less-than-quadratic increase on GPU. Both CPU and GPU computation time decrease when increasing the DD criterion value and fixing the DTA criterion value, increase when increasing the DTA criterion value and fixing the DD criterion value, and do not vary when both DD and DTA criterion values change proportionally.

REFERENCES

1. Low, D.A. et al., A technique for the quantitative evaluation of dose distributions. *Medical Physics*, 1998. **25**(5): 656–661.
2. Teke, T. et al., Monte Carlo based, patient-specific RapidArc QA using Linac log files. *Medical Physics*, 2010. **37**(1): 116–123.
3. Depuydt, T., A. Van Esch, and D.P. Huyskens, A quantitative evaluation of IMRT dose distributions: Refinement and clinical assessment of the gamma evaluation. *Radiotherapy and Oncology*, 2002. **62**(3): 309–319.
4. Bakai, A., M. Alber, and F. Nusslin, A revision of the gamma-evaluation concept for the comparison of dose distributions. *Physics in Medicine and Biology*, 2003. **48**(21): 3543–3553.
5. Stock, M., B. Kroupa, and D. Georg, Interpretation and evaluation of the gamma index and the gamma index angle for the verification of IMRT hybrid plans. *Physics in Medicine and Biology*, 2005. **50**(3): 399–411.
6. Jiang, S.B. et al., On dose distribution comparison. *Physics in Medicine and Biology*, 2006. **51**(4): 759–776.
7. Spezi, E. and D.G. Lewis, Gamma histograms for radiotherapy plan evaluation. *Radiotherapy and Oncology*, 2006. **79**(2): 224–230.
8. Wendling, M. et al., A fast algorithm for gamma evaluation in 3D. *Medical Physics*, 2007. **34**(5): 1647–1654.
9. Ju, T. et al., Geometric interpretation of the gamma dose distribution comparison technique: Interpolation-free calculation. *Medical Physics*, 2008. **35**(3): 879–887.
10. Chen, M.L. et al., Efficient gamma index calculation using fast Euclidean distance transform. *Physics in Medicine and Biology*, 2009. **54**(7): 2037–2047.
11. Yuan, J.K. and W.M. Chen, A gamma dose distribution evaluation technique using the k-d tree for nearest neighbor searching. *Medical Physics*, 2010. **37**(9): 4868–4873.
12. Gu, X., X. Jia, and S.B. Jiang, GPU-based fast gamma index calculation. *Physics in Medicine and Biology*, 2011. **56**(5): 1431–1441.
13. Lu, W. and M. Chen, Fluence-convolution broad-beam (FCBB) dose calculation. *Physics in Medicine and Biology*, 2010. **55**(23): 7211.
14. Sharp, G.C. et al., GPU-based streaming architectures for fast cone-beam CT image reconstruction and demons deformable registration. *Physics in Medicine and Biology*, 2007. **52**: 5771–5783.
15. Zhao, Y. et al., Application of an M-line-based backprojected filtration algorithm to triple-cone-beam helical CT. *Physics in Medicine and Biology*, 2010. **55**(23): 7317.
16. Lu, W., A non-voxel-based broad-beam (NVBB) framework for IMRT treatment planning. *Physics in Medicine and Biology*, 2010. **55**(23): 7175.

17. Samant, S.S. et al., High performance computing for deformable image registration: Towards a new paradigm in adaptive radiotherapy. *Medical Physics*, 2008. **35**(8): 3546–3553.

18. Jacques, R. et al., Towards real-time radiation therapy: GPU accelerated superposition/convolution. *Computer Methods and Programs Biomedicine*, 2010. **98**(3): 285–292.

19. Hissoiny, S., B. Ozell, and P. Despres, Fast convolution-superposition dose calculation on graphics hardware. *Medical Physics*, 2009. **36**(6): 1998–2005.

20. Gu, X.J. et al., GPU-based ultra-fast dose calculation using a finite size pencil beam model. *Physics in Medicine and Biology*, 2009. **54**(20): 6287–6297.

21. Men, C.H. et al., GPU-based ultrafast IMRT plan optimization. *Physics in Medicine and Biology*, 2009. **54**(21): 6565–6573.

22. Jia, X. et al., GPU-based fast cone beam CT reconstruction from undersampled and noisy projection data via total variation. *Medical Physics*, 2010. **37**(4): 1757–1760.

23. Gu, X.J. et al., Implementation and evaluation of various demons deformable image registration algorithms on a GPU. *Physics in Medicine and Biology*, 2010. **55**(1): 207–219.

24. Jia, X. et al., Development of a GPU-based Monte Carlo dose calculation code for coupled electron-photon transport. *Physics in Medicine and Biology*, 2010. **55**(11): 3077–3086.

25. Men, C.H., X. Jia, and S.B. Jiang, GPU-based ultra-fast direct aperture optimization for online adaptive radiation therapy. *Physics in Medicine and Biology*, 2010. **55**(15): 4309–4319.

26. Men, C.H. et al., Ultrafast treatment plan optimization for volumetric modulated arc therapy (VMAT). *Medical Physics*, 2010. **37**(11): 5787–5791.

27. NVIDIA, *NVIDIA CUDA Compute Unified Device Architecture, Programming Guide*, version 3.2. NVIDIA, Santa Clara, CA, 2010.

28. Hoberock, J. and N. Bell, *Thrust: A Parallel Template Library*. NVIDIA, Santa Clara, CA, 2010.

29. Ma, C.-M. et al., A Monte Carlo dose calculation tool for radiotherapy treatment planning. *Physics in Medicine and Biology*, 2002. **47**(10): 1671.

CHAPTER 21

SCORE System for Online Adaptive Radiotherapy

Zhen Tian, Quentin Gautier, Xuejun Gu,
Chunhua Men, Fei Peng, Masoud Zarepisheh,
Yan Jiang Graves, Andres Uribe-Sanchez,
Xun Jia, and Steve B. Jiang

CONTENTS

21.1 INTRODUCTION

IN THE CURRENT RADIATION therapy practice, treatment plans are generated based on a snapshot of the patient's anatomy captured during treatment simulation and then delivered fractionally over a number of weeks. However, the patient's anatomy may vary significantly from fraction to fraction. These variations occur when a patient loses weight, when the tumor regresses during therapy [1–3], when organs such as bladder and rectum have different filling status [4], etc. With the advent of onboard cone-beam computed tomography (CBCT), interfraction geometry variations can be readily measured before each treatment fraction and the static radiation therapy process can potentially be transformed into a dynamic one, that is, online adaptive radiation therapy (ART) [5,6].

Online ART is a novel treatment paradigm, where treatment simulation and planning are seamlessly integrated into the treatment delivery process. It allows real-time treatment adaptations based on up-to-date patient anatomy and is promising in maximally compensating for variations during treatment delivery. In the past decade, a lot of efforts have been made to develop online replanning techniques [7–10]. Letourneau et al. [11] have examined the feasibility of online planning for palliative treatment of spinal bone metastasis. The mean planning time for a two-field treatment is about 10 min. For much more complicated intensity modulated radiation therapy (IMRT) cases, the replanning process may take tens of minutes or even hours. Seeking for fast replanning, researchers examined alternative approaches to morph preexisting multileaf collimator (MLC) apertures or leaf sequences based on the deformation field between planning CT and CBCT [5,12–14]. The feasibility and accuracy of this type of approaches to generate an optimal plan for the new geometry are inherently limited by the fact that it is very difficult, if not impossible, to fully compensate three-dimensional (3D) anatomy deformation by morphing 2D apertures. Another type of approaches is to design multiple candidate treatment plans that attempt to anticipate possible changes in patient geometry [6,14]. However, it is not trivial to accurately predict the anatomical variation. To obtain an optimal plan in real time without scarifying its quality, online replanning has been ported to CPU-clustered supercomputers [15]. However, it turns out that these computers are probably not suitable for routine clinical use due to the high cost in facility deployment, inconvenience for real-time access, and difficulties in maintenance.

Recently, the rapid development of general-purpose computer graphics processing unit (GPU) offers an innovative computational platform, which is powerful, convenient, and affordable for high-performance computing in a clinical environment. Many researchers have adopted GPUs to accelerate heavy computational tasks in radiotherapy [16–25]. A series of GPU tools have been developed to facilitate fast deformable image registration (DIR) [26], dose calculation [27–30], and plan reoptimization [31–34], which implies a great potential to achieve the real-time online replanning. By integrating these GPU tools into a graphical user interface (GUI), a research platform called Supercomputing Online Re-planning Environment (SCORE) has been developed with the main purpose of facilitating clinical research in the online ART regime.

21.2 OVERALL WORKFLOW OF SCORE SYSTEM

The SCORE system consists of (1) GUI for internal workflow control and external communication with commercial treatment planning systems (TPS) and clinical imaging workstations and (2) a series of GPU-based computational tools for fast treatment replanning, which includes three major functional modules: DIR-based autocontouring, plan reoptimization, and dose calculation. These three modules are relatively independent components in the system and have their own visualization windows where user can modify controlling parameters and inspect the results. This relatively independent design allows users to easily add their own modules to the whole system for testing.

The overall workflow of the SCORE system is shown in Figure 21.1. The system first loads an original IMRT or volumetric modulated arc therapy (VMAT) treatment plan (including planning CT images, structure set [RS], RT dose [RD], and RT plan [RP] files) from a commercial TPS in DICOM radiation treatment (RT) format. The patient's current CT/CBCT images in DICOM format are also loaded into the system. Then SCORE system performs some image preprocessing on the planning CT images and treatment CT/CBCT images, such as couch removal to avoid adverse impact of different couches on image registration, image downsampling for subsequent DIR due to GPU's very limited memory, image truncation/stitching to make the planning CT and treatment CT/CBCT to have the same size, and rigid registration to align these two sets of images. After that, a DIR-based autocontouring module is launched in which a DIR is performed to obtain a deformation vector field (DVF), based on which contours on the planning CT are deformed and propagated onto the current CT/CBCT images. SCORE system allows user interactions such that the transferred contours could be reviewed and revised manually (if necessary) in a visualization window. Once the contours are defined, in the subsequent plan reoptimization module, a dose deposit coefficient (DDC) matrix is precalculated based on the current patient geometry and the contours. Then the original IMRT plan is reoptimized using fluence map optimization (FMO) approach or direct aperture optimization (DAO) approach. VMAT plan could also be reoptimized in the SCORE system using column generation

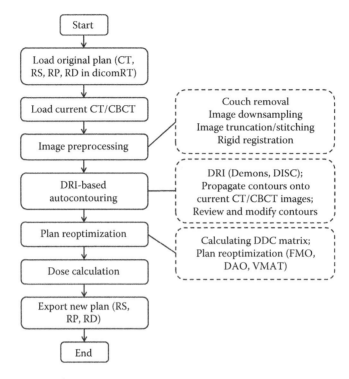

FIGURE 21.1   The flowchart of SCORE system.

approach. With the reoptimized new plan, an accurate final dose calculation is performed to obtain a 3D dose distribution of the new plan for plan evaluation, using either finite size pencil beam (FSPB) dose calculation method or Monte Carlo (MC) simulation method. The related results such as propagated contours, plan specifications of the obtained new plan, as well as its dose distribution will be written into RS, RP, and RD files, respectively, and exported from the SCORE system to commercial TPS for further plan evaluation and/or treatment delivery.

21.3 GPU MODULES IN SCORE SYSTEM

21.3.1 GUI of SCORE System

The GUI of SCORE is developed using Qt, a cross-platform application framework that is widely used for developing application software with a GUI. The major advantage of using Qt is its cross-platform flexibility, which allows users to deploy SCORE across many desktops and embedding operating systems with only a few modifications. Meanwhile, Qt provides a rich set of application building blocks. For example, the embedded

C++ class library provides all the functionality to build advanced applications; the powerful GUI library allows us to construct interactive program, such as contour modification using Qt paint device; DICOM RT reading and writing are also supported by Qt, which provides us a set of toolbox to communicate with a commercial TPS; loading and exporting treatment plans are also realized with Qt functions.

21.3.2 DIR-Based Autocontouring

Rapid segmentation of target and organs at risk (OARs) is required for online ART. Traditional manual contouring is impractical due to its laboriousness. Automatic delineation utilizing DIR is a feasible solution, which transfers contours from planning CT images onto current CT/CBCT images through voxel mapping. In the last few decades, an abundance of image registration algorithms have been developed for autocontour transferring, including thin-plane and B-spline parameterized registration algorithms [35,36], fluid flow and Demons nonparameterized registration algorithms [36–38], and model-based and numerical method-facilitated registration algorithms [39,40]. Among these methods, Demons algorithms have been employed and implemented in SCORE as a CT-to-CT autocontour propagation tool to achieve high efficiency required for online ART [26]. A CT-to-CBCT DIR algorithm with intensity correction, which is called DISC and has been introduced in Chapter 11 in details, is also incorporated in SCORE to propagate contours onto CBCT images [41].

The autocontouring module loads the preprocessed and rigid-registered planning CT and current CT/CBCT as inputs for DIR. Another input for this module is a structure file containing a 3D data array with each element corresponding to a voxel of the patient's body. Within this element that is a 32-bit float number, each bit corresponds to a structure index with value "1" if the voxel belongs to this structure or "0" otherwise. This file is generated during image preprocessing step based on the original contours defined on the planning CT. After DIR is performed, a 3D DVF is obtained as an intermediate result of this module, with its each element representing a 3D displacement of a voxel on the planning CT to match the current images. Then the structure indices for a voxel on the current images could be determined according to its corresponding location on the planning CT. This step is easily parallelizable on GPU with each thread responsible for one voxel. Note that the corresponding location for a given voxel on the current image on the planning CT is not necessary at a voxel center, thus nearest interpolation of the structure indices is needed, which is performed through fast GPU hardware interpolation using its texture memory. The generated new structure file corresponding to the current patient geometry is then written into another file as one output of this step and will be used for subsequent plan reoptimization step, as well as plan evaluation step. Furthermore, the structure array defined on the current image is converted to contours for display, which are also written into RS file for later plan export. The deformed planning CT is another output of this module and will be used for the following calculation of the DDC matrix and final dose calculation.

21.3.3 Plan Reoptimization
21.3.3.1 DDC Matrix Calculation

A DDC matrix contains information about the dose of each voxel contributed by each beamlet at unit intensity. To improve the efficiency of online ART, DDC matrix is precalculated in order to perform fast dose calculation via simple matrix–vector multiplication. This matrix is also needed to evaluate the gradient of an optimization objective function. In SCORE system, the calculation of the DDC matrix is considered as a submodule of plan reoptimization. The finite-size-pencil-beam (FSPB) dose calculation algorithm introduced in Chapter 14 is adopted here to calculate the matrix. Note that for the FSPB code used for dose calculation, its output is usually an accumulated dose of each voxel contributed by all the beamlets. Therefore, for DDC matrix calculation, the FSPB code needs to be repeatedly launched to compute dose from each beamlet.

One challenging issue is the GPU's limited memory size, particularly for cases with large planning target volume (PTV), multiple OARs or VMAT cases. The size of the DDC matrix equals the number of voxels multiplied by the number of beamlets. This matrix cannot be stored on GPU in full matrix form or even sparse matrix form for some cases. To solve this memory issue, several strategies were employed in SCORE. First, only the beamlets contributing to PTV(s) are involved in the optimization step, and hence, only those parts in the DDC matrix for these beamlets are calculated. Second, to reduce the size of the DDC matrix, only the voxels within the patient's body are taken into the calculation, and nonuniform downsampling is performed to impose a fine voxel grid on regions of interests (e.g., PTV and OAR) but a coarse voxel grid on other regions. Third, a truncation threshold is adopted. In fact, if a voxel is too far away from a beamlet, the dose contributed by this beamlet would be small enough to be ignored. Thus, a cone region is assigned for each beamlet, centered by its central axis with the cone angle as a user-defined threshold to control the truncation level. Only the voxels inside the cone are involved for DDC matrix calculation. Fourth, the calculation is performed sequentially for each beam. This is because we would like to store the DDC matrix in a sparse matrix format. However, before performing the computations, it is not possible to know a priori those nonzero matrix elements. Hence, we actually allocate a relatively large space on GPU and compute the DDC matrix in its full matrix format. After that, the full matrix is converted into a sparse matrix format where only nonzero matrix elements and their locations are stored. Because of the relatively large memory space required, we have to process DDC matrix beam by beam to fit the calculations on GPU.

The major steps of this GPU submodule for DDC matrix calculation are listed as follows:

1. Projecting PTV(s) onto the isocenter plane for each beam and only the beamlets inside or close to the PTV projection area are included into DDC matrix. The identification of relevant beamlets is parallelized with each GPU thread responsible for projecting one beamlet on the isocenter plane and checking if it intersects with PTV(s) or not.

2. Performing nonuniform downsampling within the patient's body to decrease the voxels needed for DDC matrix, which is realized on CPU.

3. Precalculating the number of voxels inside a cone for each beamlet given the truncation threshold in order to estimate the memory needed for DDC submatrix of one beam, where each thread is responsible for one beamlet.

4. The calculating DDC submatrix for the beam, where an estimated amount of memory is first allocated on GPU. A GPU thread is responsible for the calculation of one beamlet. It loops over all the voxels inside the cone of this beamlet to calculate the beamlet's dose contribution to them.

5. After the calculation for the beam, the obtained DDC submatrix is transferred onto CPU and at the same time, nonzero elements are identified. The matrix elements and their column and row indices are stored in three separate files, yielding the sparse matrix representation of the submatrix in a coordinate list format (COO).

6. Repeat steps (3)–(5) for all the beams.

21.3.3.2 Plan Reoptimization

After the DDC matrix is obtained, the plan reoptimization module is launched in SCORE. An initial plan is usually developed using a commercial TPS based on the planning CT before the first treatment fraction. This initial plan satisfies physician's clinical requirements by incorporating dose–volume or biological constraints into the optimization model. For online ART, two scenarios may occur in SCORE: (1) If no significant changes are observed in the patient current geometry, an initial plan–guided reoptimization (IPGRO) strategy would be adopted, which allows us to incorporate the physician's preference in the initial plan (e.g., the locations of hot/cold spots) into the new plan, and thus to minimize the necessity for the physician's reapproval of the new plan; (2) If significant changes exist in the current geometry, a prescription dose-guided reoptimization (PDGRO) model will be used, which is designed based on cumulative dose distribution and dose volume histogram (DVH) constraints and allows us to generate a high-quality treatment plan while keeping the new plan not too far away from the initial plan. In SCORE, three different optimization methods have been implemented on the GPU platform for this treatment plan reoptimization step, namely, FMO [31] and DAO [32] for IMRT plans, and VMAT optimization [33]. Their GPU implementations have been introduced in details in Chapters 17 and 18, respectively.

The inputs for this module are the precalculated DDC matrix, target dose for each voxel (prescription dose for PTV, threshold dose for OARs), and overdosing and underdosing weighting factors for PTVs and OARs. Note that some plan specifications of the original plan, such as isocenter location, number of beams for IMRT, number of arcs and control points for VMAT, and gantry angle, couch angle, and collimator angles for each beam or control point, are directly taken from the original plan. For IPGRO strategy, the corresponding dose points of the original plan are loaded onto GPU's constant memory or texture memory (depending on the data size) and used as the target dose in plan reoptimization, with the overdosing and underdosing weighting factors set to be 1. For PDGRO strategy, the target dose and weighting factors are manually input through the GUI of the SCORE system and stored on GPU's constant memory.

The outputs of this plan reoptimization module are different depending on which method is used. The FMO method outputs an optimized fluence map, namely, an optimized intensity for each beamlet. This fluence map is then converted to MLC leaf sequences in a sliding window mode for treatment delivery. The DAO method outputs the shapes of the chosen MLC apertures for each beam through a set of beamlet index of the left and right boundaries of the apertures and an intensity vector of these apertures. These outputs will then be converted to MLC leaf sequences in a step-and-shoot mode. Similarly, the VMAT method outputs the chosen apertures and their intensities for each control point. Apart from those, dose rates and gantry speeds at each control point are also output, since they are parameters optimized for the VMAT problem. These resulting plan specifications are written into RP files for later plan export.

To achieve an efficient GPU parallelization for this plan reoptimization module, we need to maximize working load on each thread and minimize memory access time. As such, we parallelize the dose calculation on voxels. The objective function value is a summation over all voxels, which is achieved through a parallel reduction scheme. The gradient calculation involved in plan reoptimization is parallelized on beamlets for FMO and on apertures for DAO and VMAT in their master problems. Moreover, an efficient sparse matrix–vector multiplication method developed by Bell and Garland [42] is adopted to multiply the DDC matrix with the beamlet or aperture intensity vector for fast dose calculation using the shared memory during optimization. Note that the DDC matrix is required to be converted from COO format to compressed sparse row (CSR) format for this matrix–vector multiplication for coalesced memory access. A template library called Thrust from NVIDIA [43] is utilized to efficiently perform this conversion, as well as matrix transpose needed in DAO and VMAT.

21.3.4 Dose Calculation

Real-time dose calculation is an indispensible part of online ART, which can hardly be achieved using the current CPU-based algorithms. Therefore, both the GPU-based FSPB algorithm [27] and GPU-based MC dose calculation engine gDPM [28,30], introduced in Chapters 14 and 15, respectively, have been integrated into the SCORE system for final dose calculation after plan reoptimization. The deformed planning CT generated by DIR is one of the inputs for this dose calculation module, with CT values being converted to electron densities and/or material types. Texture memory is used again to store this density and/or material type data due to its faster memory caching and hardware interpolation. Fluence map of the optimized new plan is another input for dose calculation and is stored on texture memory as well. For FSPB method, the intrinsic parameters of the pencil beam model are loaded onto GPU's constant memory at the initialization stage of the module. For MC method, the required cross sections for physics interactions are loaded and stored on texture memory to utilize the fast hardware interpolation. When the dose calculation is completed, the resulting 3D dose distribution on the CT grid will be resampled onto the dose grid and is written into RD file for later plan export.

21.4 EXPERIMENTAL RESULTS

The usage of the SCORE system is illustrated here using a prostate cancer patient case. The initial treatment plan in DICOM RT format from a commercial TPS as well as a current CT image in DICOM format is imported into SCORE system, as shown in Figure 21.2. A set of preprocessing procedures on the current CT image are performed, such as rigid registration with the planning CT, structure identification, etc. DIR for autocontouring can be done in ~7 s for CT image of size ~256 × 256 × 100 on an NVIDIA Tesla C1060 GPU card [26]. After DIR, it can be observed that the PTV contour matches the new geometry very well, as shown in Figure 21.3. Figure 21.3 also shows the GUI that allows users to visually inspect and, if needed, manually modify the propagated contours on the current images. With the obtained DVF, the dose distribution of the initial plan is deformed onto the new geometry and used as target dose for the IPGRO model. Then one of the optimization methods is executed to

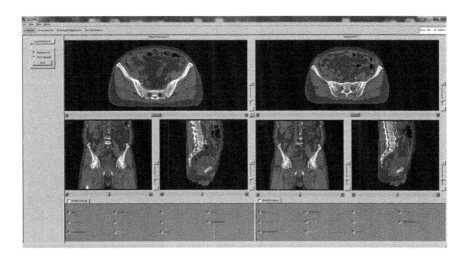

FIGURE 21.2 User interface of SCORE system showing the loading procedure of the initial treatment plan and the current CT image.

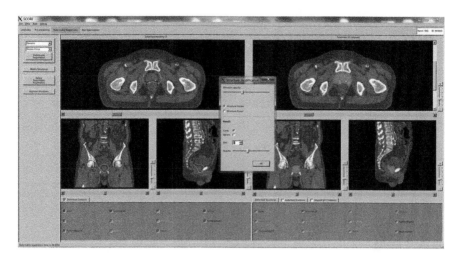

FIGURE 21.3   User interface of SCORE showing the DIR results and manually modification tool of the propagated contours.

generate an optimal plan for the new patient geometry. On an NVIDIA Tesla C1060 GPU card, for a typical nine-field prostate IMRT case with 5×5 mm² beamlet size and $2.5 \times 2.5 \times 2.5$ mm³ voxel size, it takes ~2.8 s for FMO to generate an optimal IMRT plan and 0.7–3.8 s for DAO. Due to the much larger scale and complexity of the optimization problem, it takes about 30–50 s to obtain high-quality VMAT plans for typical prostate and head-and-neck cases. The final dose distribution and DVH curves are calculated using either FSPB or gDPM. It takes less than 1 s to calculate the dose distribution for a typical IMRT patient case using FSPB and 30–40 s using gDPM with less than 1% statistical uncertainty. Figure 21.4 illustrates the dose color wash superimposed on a current CT slice and the DVH curves for both new and initial plans. The new plan can be output in DICOM RT format to a commercial TPS.

21.5 DISCUSSION AND CONCLUSIONS

In this chapter, we have briefly presented our development of a SCORE system for radiotherapy online replanning. The system is built on top of several GPU-based computational tools introduced in previous chapters. On the one hand, each module in the SCORE system is a relatively independent component. They have well-defined inputs and outputs, and GPU is employed to perform computations efficiently in these challenging problems. On the other hand, these modules are seamlessly integrated in the

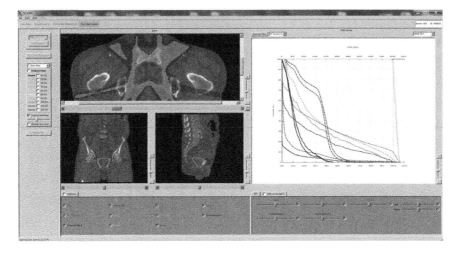

FIGURE 21.4   User interface of SCORE showing the final dose distribution of the new plan with dose color wash and the DVH curves (dot line—initial plan; solid line—new plan).

SCORE system. The modules are performed sequentially. Each of them relies on its previous module to provide some input data and generates output data to be used in the next step. In addition to those modules, a GUI and several other necessary components are developed to form a complete SCORE system. We have demonstrated the functionalities of this system through an example case of prostate cancer replanning, in which a new plan can be finished in a few minutes including loading data and exporting the new plans.

Despite the success, there are still a number of challenges lying ahead before the SCORE system could be used in a clinically realistic environment in an accurate, efficient, robust, and streamlined fashion. Further development of the computational toolbox will be focused on improving algorithm's accuracy, efficiency, and robustness. Computational infrastructure needs to be designed and optimized to facilitate a smooth clinical workflow. Feasible and practicable online quality assurance tools also need to be developed to secure the treatment plan deliverability and safety.

REFERENCES

1. Barker, J.L. et al., Quantification of volumetric and geometric changes occurring during fractionated radiotherapy for head-and-neck cancer using an integrated CT/linear accelerator system. *Int J Radiat Oncol Biol Phys*, 2004. **59**(4): 960–970.

2. Han, C.H. et al., Actual dose variation of parotid glands and spinal cord for nasopharyngeal cancer patients during radiotherapy. *Int J Radiat Oncol Biol Phys*, 2008. **70**: 1256–1262.

3. Tyagi, N.L.J., Yashar, C.M., Vo, D., Jiang, S.B., Mundt, A.J., Mell, L.K., Daily online cone beam computed tomography to assess interfractional motion in patients with intact cervical cancer. *Int J Radiat Oncol Biol Phys*, 2011. **80**: 273–280.

4. Yan, D. et al., Computed tomography guided management of interfractional patient variation. *Semin Radiat Oncol*, 2005. **15**(3): 168–179.

5. Court, L.E. et al., An automatic CT-guided adaptive radiation therapy technique by online modification of multileaf collimator leaf positions for prostate cancer. *Int J Radiat Oncol Biol Phys*, 2005. **62**(1): 154–163.

6. Burridge, N. et al., Online adaptive radiotherapy of the bladder: Small bowel irradiated-volume reduction. *Int J Radiat Oncol Biol Phys*, 2006. **66**(3): 892–897.

7. Wu, C. et al., Re-optimization in adaptive radiotherapy. *Phys Med Biol*, 2002. **47**(17): 3181–3195.

8. Birkner, M. et al., Adapting inverse planning to patient and organ geometrical variation: Algorithm and implementation. *Med Phys*, 2003. **30**(10): 2822–2831.

9. Zhang, T. et al., Automatic delineation of on-line head-and-neck computed tomography images: Toward on-line adaptive radiotherapy. *Int J Radiat Oncol Biol Phys*, 2007. **68**(2): 522–530.

10. Wu, Q.J. et al., On-line re-optimization of prostate IMRT plans for adaptive radiation therapy. *Phys Med Biol*, 2008. **53**(3): 673–691.

11. Letourneau, D. et al., Online planning and delivery technique for radiotherapy of spinal metastases using cone-beam CT: Image quality and system performance. *Int J Radiat Oncol Biol Phys*, 2007. **67**(4): 1229–1237.

12. Mohan, R. et al., Use of deformed intensity distributions for on-line modification of image-guided IMRT to account for interfractional anatomic changes. *Int J Radiat Oncol Biol Phys*, 2005. **61**(4): 1258–1266.

13. Feng, Y. et al., Direct aperture deformation: An interfraction image guidance strategy. *Med Phys*, 2006. **33**(12): 4490–4498.

14. Letourneau, D. et al., Semiautomatic vertebrae visualization, detection, and identification for online palliative radiotherapy of bone metastases of the spine. *Med Phys*, 2008. **35**(1): 367–376.

15. Jiang, S.B. et al., Towards on-line adaptive radiotherapy for cervical cancer, in *Mathematical Methods in Biomedical Imaging and Intensity-Modulated Radiation Therapy (IMRT)*, M.J.Y. Censor and A.K. Louis, Eds., Edizioni Della Normale, Roma, Italy. 2008, pp. 173–184.

16. Sharp, G.C. et al., GPU-based streaming architectures for fast cone-beam CT image reconstruction and demons deformable registration. *Phys Med Biol*, 2007. **52**(19): 5771–5783.

17. Samant, S.S. et al., High performance computing for deformable image registration: Towards a new paradigm in adaptive radiotherapy. *Med Phys*, 2008. **35**(8): 3546–3553.

18. Noe, K.O. et al., Acceleration and validation of optical flow based deformable registration for image-guided radiotherapy. *Acta Oncol*, 2008. **47**(7): 1286–1293.

19. Hissoiny, S., Ozell, B., Despres, P., Fast convolution-superposition dose calculation on graphics hardware. *Med Phys*, 2009. **36**(6): 1998–2005.

20. Jacques, R. et al., Towards real-time radiation therapy: GPU accelerated superposition/convolution. *Comput Methods Programs Biomed*, 2010. **98**(3): 285–292.

21. Jia, X. et al., GPU-based fast cone beam CT reconstruction from undersampled and noisy projection data via total variation. *Med Phys*, 2010. **37**(4): 1757–1760.

22. Jia, X. et al., GPU-based fast low-dose cone beam CT reconstruction via total variation. *J Xray Sci Technol*, 2011. **19**(2): 139–154.

23. Jia, X. et al., GPU-based iterative cone-beam CT reconstruction using tight frame regularization. *Phys Med Biol*, 2011. **56**(13): 3787–3807.

24. Tian, Z. et al., Low-dose CT reconstruction via edge-preserving total variation regularization. *Phys Med Biol*, 2011. **56**(18): 5949–5967.

25. Gu, X., Jia, X., Jiang, S.B., GPU-based fast gamma index calculation. *Phys Med Biol*, 2011. **56**(5): 1431–1441.

26. Gu, X.J. et al., Implementation and evaluation of various demons deformable image registration algorithms on a GPU. *Phys Med Biol*, 2010. **55**(1): 207–219.

27. Gu, X. et al., GPU-based ultra-fast dose calculation using a finite size pencil beam model. *Phys Med Biol*, 2009. **54**(20): 6287–6297.

28. Jia, X. et al., Development of a GPU-based Monte Carlo dose calculation code for coupled electron-photon transport. *Phys Med Biol*, 2010. **55**: 3077.

29. Gu, X.J. et al., A GPU-based finite-size pencil beam algorithm with 3D-density correction for radiotherapy dose calculation. *Phys Med Biol*, 2011. **56**(11): 3337–3350.

30. Jia, X. et al., GPU-based fast Monte Carlo simulation for radiotherapy dose calculation. *Phys Med Biol*, 2011. **56**: 7017–7031.

31. Men, C. et al., GPU-based ultrafast IMRT plan optimization. *Phys Med Biol*, 2009. **54**(21): 6565–6573.

32. Men, C.H., Jia, X., Jiang, S.B., GPU-based ultra-fast direct aperture optimization for online adaptive radiation therapy. *Phys Med Biol*, 2010. **55**(15): 4309–4319.

33. Men, C.H. et al., Ultrafast treatment plan optimization for volumetric modulated arc therapy (VMAT). *Med Phys*, 2010. **37**(11): 5787–5791.

34. Fei, P. et al., A new column-generation-based algorithm for VMAT treatment plan optimization. *Phys Med Biol*, 2012. **57**(14): 4569–4588.

35. Rietzel, E., Chen, G.T.Y., Deformable registration of 4D computed tomography data. *Med Phys*, 2006. **33**(11): 4423–4430.

36. Wu, Z.J. et al., Evaluation of deformable registration of patient lung 4DCT with subanatomical region segmentations. *Med Phys*, 2008. **35**(2): 775–781.

37. Wang, H. et al., Implementation and validation of a three-dimensional deformable registration algorithm for targeted prostate cancer radiotherapy. *Int J Radiat Oncol Biol Phys*, 2005. **61**(3): 725–735.
38. Wang, H. et al., Validation of an accelerated 'demons' algorithm for deformable image registration in radiation therapy. *Phys Med Biol*, 2005. **50**(12): 2887–2905.
39. Brock, K.K. et al., Accuracy of finite element model-based multi-organ deformable image registration. *Med Phys*, 2005. **32**(6): 1647–1659.
40. Lu, W.G. et al., Fast free-form deformable registration via calculus of variations. *Phys Med Biol*, 2004. **49**(14): 3067–3087.
41. Zhen, X. et al., CT to cone-beam CT deformable registration with simultaneous intensity correction. *Phys Med Biol*, 2012. **57**: 1–20.
42. Bell, N., Garland, M., Efficient sparse matrix-vector multiplication on CUDA. NVIDIA Technical Report NVR-2008-004, 2008.
43. Hoberock, J., Bell, N., Thrust: A Parallel Template Library, 2010. https://code.google.com/p/thrust/.

CHAPTER 22

TARGET: A GPU-Based Patient-Specific Quality Assurance System for Radiation Therapy

Yan Jiang Graves, Michael M. Folkerts,
Zhen Tian, Quentin Gautier, Xuejun Gu,
Xun Jia, and Steve B. Jiang

CONTENTS

22.1 INTRODUCTION

MODERN INTENSITY-MODULATED RADIATION THERAPY (IMRT) and volumetric-modulated arc therapy (VMAT) technologies deliver more conformal dose to targets while sparing normal healthy tissues than conventional radiotherapy. To achieve these goals, the treatment planning and delivery process for IMRT and VMAT have become much more complicated and less intuitive to users. For instance, inverse treatment planning is utilized where values of thousands of variables, if not more, are adjusted by a computer program via specialized optimization algorithms to achieve the desired dosimetric goal. During delivery, many components in a linear accelerator (linac) are precisely controlled to deliver the optimized plan to the patient. This process is more error prone than conventional radiotherapy and consequences caused by the errors are probably more severe. Therefore, a quality assurance (QA) procedure is needed before the first treatment fraction to check for potential errors in the patient-specific plan, both in the treatment planning stage and in the plan delivery stage.

Current patient-specific plan QA procedures can be generally categorized into two-dimensional (2D) dosimetry and 3D dosimetry approaches. The current common clinical practice is the 2D dosimetry approach which compares measured planar dose distributions, using radiographic film, portal imagers [1], or matrices of detectors [2,3], with the calculated dose distributions from the treatment planning system at corresponding locations. However, only dose in a single plane is tested in this approach. In addition, most of these practices require phantom and/or detector setup, which is tedious and labor-insensitive. In contrast, the 3D approach attempts to verify the dose distribution in the entire 3D dose space, which is more ideal considering the complexity of IMRT and VMAT. For instance, advanced 3D dosimetric phantoms or detectors [4–7] have been utilized. Yet, those phantoms and detectors also require a careful setup procedure and more importantly, they do not directly verify the dose in the geometry of a specific patient. In particular, tissue heterogeneity is typically ignored. Another 3D approach is to computationally reconstruct 3D dose in the patient geometry based on measured delivery information via electronic portal imaging device (EPID) or the machine log files. In these methods, the accuracy of using EPID to reconstruct plan dose relies on EPID calibrations and geometric accuracy, whereas the accuracy of the recorded information in machine delivery log files has been demonstrated by comparing with film measurement, 2D detector array and portal imagers [8–10]. Hence, linac log file-based dose reconstruction for IMRT and VMAT QA purpose has become an active research topic with great potential for clinical applications [11–13].

With the measured linac information collected from machine log files, the subsequent dose reconstruction process needs a dose calculation engine. Monte Carlo (MC) methods are greatly preferred due to their high accuracy [11–13]. Nonetheless, due to the statistical nature of this method, the computational time is typically very long, limiting the adoption of these novel QA tools in routine clinical practice. With the recent advancement of graphics processor units (GPUs) in general purpose scientific computations, large speedup factors for MC dose calculations have been achieved [14–16]. This holds the potential to substantially improve the efficiency of the patient-specific QA tools and therefore bring them into clinical practice. In this chapter, we present our development of an IMRT and VMAT plan QA system, Treatment Assurance for Radiotherapy with GPU-Enhanced Tools (TARGET). This system performs secondary dose calculations via MC simulations [14,16] to verify the plan dose accuracy of treatment planning. It also reconstructs delivered dose using machine log files to check beam deliverability, as well as to validate the delivered dose accuracy. A GPU-based γ-index tool is used to compare the dose distributions [17]. To reduce the effect of statistical uncertainty of MC-calculated doses on γ-index test results [18], we have also developed a GPU-based denoising method for MC-based doses.

The rest of this chapter is organized as follows. Section 22.2 introduces the overall workflow of the system; all the GPU modules are presented in Section 22.3. Section 22.4 presents experimental tests and results, and finally Section 22.5 concludes this chapter.

22.2 OVERALL SYSTEM STRUCTURE

Figure 22.1 presents the graphical user interface (GUI) of our GPU-based patient-specific plan QA system, TARGET. The interface is developed using HTML5, Python, and Django. It provides a user-friendly way to upload zipped DICOM data and machine log files and set parameters for the MC dose calculation and dose comparison criteria. After the files are uploaded, the MC dose calculation is executed on the remote GPU server and the resulting dose distributions are displayed in the web GUI for the user to review. A 3D γ-index map and DVH curves are also displayed to the user. Finally, a report that summarizes the QA results can be downloaded in PDF format. Such a system is supported by all modern web browsers and it can run on any major platform, even on a mobile device.

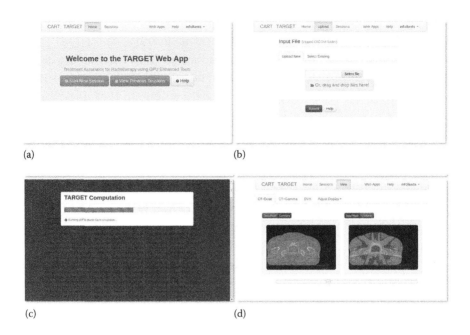

(a) (b)

(c) (d)

FIGURE 22.1 User interfaces of the TARGET system. (a) Main user interfaces of TARGET. (b) Interface for uploading data and setting calculation parameters. (c) A page showing the dose calculation progress. (d) Comparison of dose distributions.

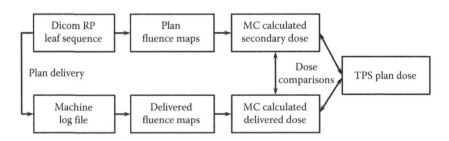

FIGURE 22.2 Workflow for the GPU-based MC-based patient-specific plan QA system.

The overall workflow of the system is shown in Figure 22.2. After a treatment plan is obtained from the treatment planning system (TPS), the motion sequence of the multileaf collimator (MLC) and all beam parameters are extracted. We then generate plan fluence maps. GPU-based dose planning method' (gDPM) [14–16] is then utiliz to perform dose calculations based on the beam configurations and the patient CT images. In this process, a GPU-based denoising algorithm is applied to the resulting dose to reduce statistical uncertainty in the MC dose and to avoid biased γ-index analysis due to the noise. After that, we transfer the treatment plan to the linac and deliver the plan on it without any phantom setup. During the delivery, the machine information is recorded into its log files. When the plan delivery is finished, we can access the log files to extract the actual delivered fluence maps and then perform a delivered dose calculation.

There are three dose distributions available at this point, namely, the TPS calculated plan dose (PD), the secondary dose (SD), and the delivered dose (DD). Dose comparisons with γ-index evaluation tests are conducted among them to check for all possible error sources. (1) SD vs. PD can be used to validate the TPS plan dose calculation accuracy. (2) Since DD and SD are obtained by the same dose calculation algorithm, the difference between them can catch machine deliverability errors. (3) The comparison between DD and PD shows the cumulative errors from both the plan dose calculation and the beam delivery.

22.3 GPU MODULES IN THE TARGET SYSTEM

22.3.1 Fluence Map Generation

In an IMRT or a VMAT plan, an MLC leaf sequence is specifically designed to shape the radiation beam. At each beam angle, a 2D function characterizing the photon fluence on a plane perpendicular to the beam direction, a photon fluence map, is determined. In this section, we present a GPU module that generates a fluence map based on leaf sequences. We use a Varian Truebeam (Pola Alto, CA) linac as an example, which consists of an MLC with 60 leaf pairs.

In this module, we first divide the 2D open field range into a grid of beamlets with a resolution of 0.5×0.2 cm^2. An MLC leaf motion sequence consists of a set of control points, each of which records the leaf positions and fractional MU delivered at that position. Suppose there are K control points for a given beam; the fluence intensity I of a beamlet x_n is the summation of the contributions from all the control points,

$$I(x_n) = \sum_{i=0}^{K} f_i \alpha_i(x_n) \tag{22.1}$$

where

f_i is the fractional MU of the ith control point

$\alpha_i(x_n)$ is the transmission factor at x_n corresponding to the MLC configuration at this moment

The transmission factors are 100% within the MLC leaf opening and are calculated according to the formula in Boyer and Li [19] in the area behind the leafs to consider the round-end effect.

During plan delivery, the Varian TrueBeam system logs segmental machine delivery information every 10 ms from the MLC controller, including MLC leaf positions, beam angle, collimator angle, cumulative dose fraction, jaw positions, beam hold status, etc. Calculating fluence maps from a log file is similar to that for a DICOM file. The fluence intensity I of a beamlet x_n is the summation of the contributions of all recorded segments in the log file,

$$I(x_n) = \sum_{i=0}^{S} f_i \alpha_i(x_n) B_i \qquad (22.2)$$

where
   S is the total number of segments
   f_i is the fractional MU of the ith segment
   $\alpha_i(x_n)$ is the transmission factor of the ith segments at the beamlet x_n

There is an additional factor B_i, which is one or zero depending on the beam status (on or off) for the ith segment.

This fluence map generation code is easily parallelizable on a GPU platform. Each GPU thread calculates the value of the fluence intensity at a given beamlet.

22.3.2 Monte Carlo Dose Engine

In this patient-specific plan QA system, we use gDPM as a dose calculation engine [14–16]. In gDPM, each GPU thread simulates the transport of one particle history. By carefully designed simulation scheme to separate photon transport and electron transport, the so-called GPU thread divergence problem is partially relieved, which would otherwise severely degrade computational efficiency. Besides this simulation scheme, hardware-supported linear interpolation and a high-performance random number generator are also utilized to further speed up the MC calculations. With these novel techniques, gDPM can compute a realistic IMRT or a VMAT plan calculation within 1 min, achieving 69×–87× speedup when compared to single-core CPU-based dose computations.

22.3.3 GPU-Based MC Denoising Algorithm

The noise signal $\widehat{D}(v)$ at a voxel computed by an MC-based dose engine follows from a Poisson distribution. It is important to estimate the true dose value $D(v)$ to avoid any biased conclusions when validating the calculated doses [18]. In our system, this is achieved by solving an optimization problem

$$
\begin{aligned}
D(v) &= \mathrm{argmin}_D E[D] \\
&= \mathrm{argmin}_D \int dv (D - \widehat{D} \log D) + \frac{\beta}{2} \int dv |\nabla D|^2, \quad \text{s.t. } D(v) \geq 0.
\end{aligned}
\tag{22.3}
$$

$E[D]$ is the objective function, in which the first term is a data-fidelity term considering the Poisson noise [20] and the second term is a penalty term to ensure the smoothness of the denoised dose $D(v)$. β is a parameter to control the relative importance of the two terms. Since $E[D]$ is a convex objective function with a nonnegative constraint, we utilized a projection-based gradient descent algorithm to solve the problem. Specifically, at each iteration step k, a new solution is computed as

$$
D_{k+1} = P(D_k - \lambda_k \nabla E(D_k)^\mathrm{T}),
\tag{22.4}
$$

where P denotes the projection of the updated dose onto the feasible set of nonnegative solutions

$$
P(D_k - \lambda_k \nabla E(D_k)^\mathrm{T}) = \max(0, \ D_k - \lambda_k \nabla E(D_k)^\mathrm{T}).
\tag{22.5}
$$

λ_k is the step size. In our algorithm, we combine the Barzilar–Borwein method [21] with the Armijo's line search rule [22]. The line search starts with an initial step size λ_0. After updating the solution with this step size using Equation 22.4, if the new objective function is less than a threshold T_k, we will accept this step size and update the solution. Otherwise, the step size is decreased by a factor >1. This process is repeated until the threshold objective function value is met. The Armijo's rule gives an expression for the threshold as follows with $0 < \varepsilon < 1$:

$$
T_k = E(D_k) - \frac{\varepsilon}{\lambda}(D_{k+1} - D_k)(D_{k+1} - D_k)^\mathrm{T}.
\tag{22.6}
$$

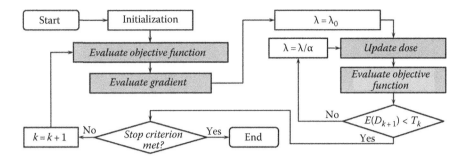

FIGURE 22.3   Flow chart of our GPU-based denoising algorithm. Gray boxes denote CUDA kernels and white boxes are C code on CPU.

The iteration process is terminated when the relative decrease of the objective function between two successive steps is less than a user-defined value $\delta = 10^{-4}$.

Figure 22.3 is the flow chart of our GPU implementation of this dose denoising algorithm. The major computations shown in the gray boxes are parallelized on GPU, for example, objective function evaluation, gradient calculation, and dose update. These steps are easily parallelizable, with each GPU thread being responsible for the computations at a given voxel. The only issue is that the objective function evaluation is a summation over all the voxels. In this case, we use a GPU-friendly parallel sum-reduction scheme available in the thrust library [23]. In addition, simple arithmetic operations shown in the white boxes are quickly computed on the CPU.

22.3.4 GPU-Based Gamma-Index Evaluation

The γ-index algorithm is a computationally intensive task, especially when dealing with 3D γ-index calculations, due to the interpolation of dose grid sample points and an exhaustive search of the closest Euclidean distance in dose-distance space. Gu et al. [17] implemented the γ-index algorithm, gGamma, on GPU using the CUDA programming environment using the geometric method [24]. As opposed to processing data in a serial manner, the search for closest Euclidean distance in dose-distance space can be parallelized for a large number of reference points and executed simultaneously using multiple threads. A radial presorting technique is also employed to sort voxels according to their dose difference to make all threads simultaneously executed by the GPU have similar computational loads. This strategy minimizes the divergence between GPU threads, considerably increasing computational efficiency. With these techniques, the GPU-based γ-index calculation step can accomplish a 3D dose comparison within a few seconds, yielding 45×–75× speedup compared to single threaded CPU implementations.

22.4 EXPERIMENTAL RESULTS

22.4.1 Dose Denoising

Along the chain of the QA process, the MC dose calculation and γ-index calculation results have been reported elsewhere in this book. Here, we present only the results of the dose denosing component.

To verify the accuracy of our denoising algorithm, we conduct numerical experiments using two realistic clinical cases: a 7-beam IMRT prostate plan and a 2-arc VMAT head-and-neck (HN) plan. For each patient case, we first simulate the MC dose distribution using gDPM with a sufficiently large number of particle histories, such that the uncertainty, defined as average uncertainty value normalized to the maximum dose D_{max} within the 50% isodose line, is less than 0.15%. This dose distribution is considered as the ground truth dose distribution, MC_g. The noisy MC dose, MC_n, is obtained by the same MC code, such that the uncertainty level is ~2.0%. After the denoising algorithm is applied to the MC_n to obtain the denoised result, MC_s, it is compared to the ground truth. Three test criteria are used to evaluate the accuracy of the denoisiong algorithm: maximum dose difference, root-mean-square difference (RMSD), and visual inspection of isolines and dose profiles. The maximum dose difference ΔD_{max} and the RMSD are reported in units of D_{max} of MC_g. The spatial resolution of both the cases is $1.953 \times 1.953 \times 2.5$ mm^3.

Table 22.1 summarizes the results for maximum dose difference ΔD_{max} and RMSD values. For both the prostate and the HN cancer patient cases, a significant decrease in the maximum dose difference and the RMSD is observed. Figure 22.4 shows the isodose lines and dose profiles for the ground truth, the smoothed and the noisy dose distributions in a

TABLE 22.1 Quantitative Measures of the Dose Denoising Results

Patient Case	$\Delta D_{max}(\%D_{max})$		RMSD $(\%D_{max})$	
	MC_g vs. MC_n	MC_g vs. MC_s	MC_g vs. MC_n	MC_g vs. MC_s
HN	5.04	0.37	0.37	0.17
Prostate	6.20	1.05	0.46	0.26

transverse slice. The improvements provided by the denoising approach are obvious. In particular, the isodose lines and the dose profiles of the denoised dose are almost the same as those of the ground truth dose.

22.4.2 Overall System Performance

We demonstrate the clinical utility of this GPU-based MC QA tool using a typical IMRT prostate cancer case and a VMAT brain cancer case. The plan dose and the MC-based SD and DD are displayed on an axial CT

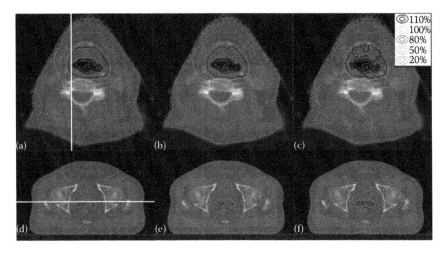

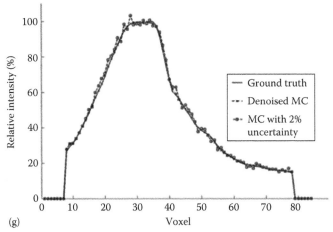

FIGURE 22.4 **(See color insert.)** Isodose lines for a HN case (a through c) and a prostate case (d through f). The three columns are ground truth doses, denoised doses, and noisy doses. (g) Dose profiles along the lines indicated in (a).

(Continued)

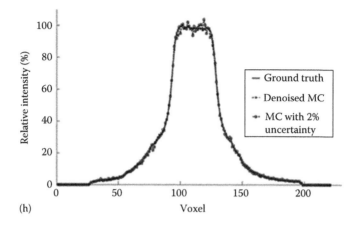

(h)

FIGURE 22.4 (*Continued*) **(See color insert.)** (h) Dose profiles along the lines indicated in (d).

slices in Figures 22.5 and 22.6 together with the γ-index maps for the three pairs of comparisons. Table 22.2 summarizes the γ-index results in a region within the 10% isodose line for the three comparisons. Very good agreement between SD vs PD, SD vs DD, and DD vs PD for both cases were observed. Finally, a QA report generated by the TARGET system for the prostate cancer case is presented in Figure 22.7. Note that both of the two plans have already gone through a standard QA process in our clinic and have been delivered to patients.

22.5 CONCLUSION

In this chapter, we have presented a fast and accurate patient-specific plan QA system called TARGET. This system seamlessly integrates a number of GPU-enabled computational components to achieve the goal of patient and plan-specific quality assurance for IMRT and VMAT plans. Specifically, the plan fluence map is derived from MLC leaf sequence stored in a DICOM file, while the actual delivered fluence map is obtained from machine delivery information recorded in linac log files. A GPU-based MC dose engine is employed to compute dose distributions, which are then postprocessed using a MC denoising algorithm, achieving an independent plan dose calculation, secondary dose (SD), or delivery dose reconstruction, delivered dose (DD). The TPS plan dose calculation accuracy can be verified by comparing SD and the plan dose, whereas the machine deliverability and accuracy is checked by DD–SD comparison.

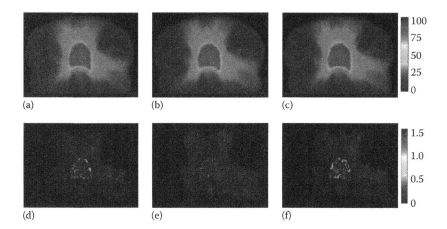

FIGURE 22.5   Dose distributions for an IMRT prostate cancer case. (a) Secondary dose (SD). (b) Delivered dose (DD). (c) Plan dose (PD). (d through f) γ-index map for SD vs. PD, SD vs. DD, and DD vs. PD, respectively.

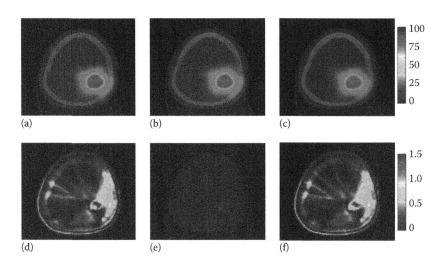

FIGURE 22.6   Dose distributions for a VMAT brain cancer case. Subfigure (a through f) configuration is the same as in Figure 22.5.

TABLE 22.2   γ-Index Result in Low Dose Region for Three Pairs of Dose Comparisons

	SD vs PD		SD vs DD		DD vs PD	
	Mean γ	Passing Rate (%)	Mean γ	Passing Rate (%)	Mean γ	Passing Rate (%)
Brain	0.34	99.0	0.05	100.0	0.34	98.9
Prostate	0.18	99.3	0.09	99.9	0.24	99.5

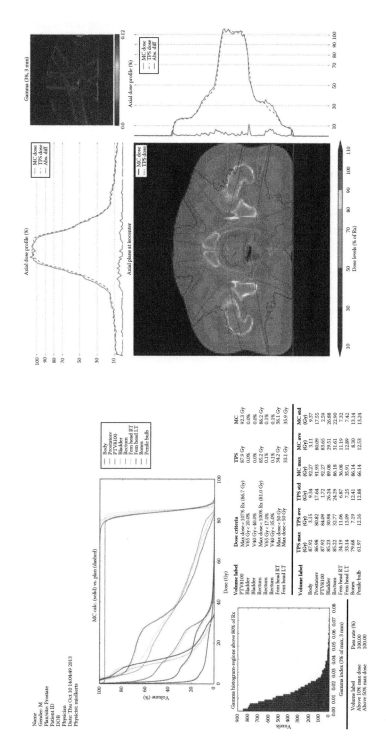

FIGURE 22.7 Two sample pages from the QA report for the prostate IMRT case generated by the TARGET system.

Finally, the comparison between DD and the plan dose identifies the cumulative inaccuracy from both the plan and the delivery processes. To facilitate clinical utilization of the patient-specific plan QA system, a web interface was developed.

Behind the scene of this TARGET system is a number of GPU-based computation modules. These powerful tools are essential for the success of our system. With their help, it is possible to substantially shorten the time of many computationally intensive steps to a clinically acceptable level, making our system routinely usable. We believe a system as such would greatly improve the accuracy, efficiency, and convenience of the patient-specific QA process, which will play an important role in modern radiotherapy by improving treatment safety.

REFERENCES

1. Nicolini, G. et al., The GLAaS algorithm for portal dosimetry and quality assurance of RapidArc, an intensity modulated rotational therapy. *Radiation Oncology*, 2008. **3**: 24.
2. Korreman, S., J. Medin, and F. Kjaer-Kristoffersen, Dosimetric verification of RapidArc treatment delivery. *Acta Oncologica*, 2009. **48**(2): 185–191.
3. Li, J.G., G. Yan, and C. Liu, Comparison of two commercial detector arrays for IMRT quality assurance. *Journal of Applied Clinical Medical Physics*, 2009. **10**(2): 62–74.
4. Richardson, S.L. et al., IMRT delivery verification using a spiral phantom. *Medical Physics*, 2003. **30**(9): 2553–2558.
5. Islam, K.T.S. et al., Initial evaluation of commercial optical CT-based 3D gel dosimeter. *Medical Physics*, 2003. **30**(8): 2159–2168.
6. Feygelman, V. et al., Evaluation of a biplanar diode array dosimeter for quality assurance of step-and-shoot IMRT. *Journal of Applied Clinical Medical Physics*, 2009. **10**(4): 64–78.
7. Letourneau, D. et al., Novel dosimetric phantom for quality assurance of volumetric modulated arc therapy. *Medical Physics*, 2009. **36**(5): 1813–1821.
8. Li, J.G. et al., Validation of dynamic MLC-controller log files using a two-dimensional diode array. *Medical Physics*, 2003. **30**(5): 799–805.
9. Zygmanski, P. et al., Dependence of fluence errors in dynamic IMRT on leaf-positional errors varying with time and leaf number. *Medical Physics*, 2003. **30**(10): 2736–2749.
10. Zeidan, O.A. et al., Verification of step-and-shoot IMRT delivery using a fast video-based electronic portal imaging device. *Medical Physics*, 2004. **31**(3): 463–476.
11. Luo, W. et al., Monte Carlo based IMRT dose verification using MLC log files and R/V outputs. *Medical Physics*, 2006. **33**(7): 2557–2564.
12. Schreibmann, E. et al., Patient-specific quality assurance method for VMAT treatment delivery. *Medical Physics*, 2009. **36**(10): 4530–4535.

13. Teke, T. et al., Monte Carlo based, patient-specific RapidArc QA using Linac log files. *Medical Physics*, 2010. **37**(1): 116–123.
14. Jia, X. et al., Development of a GPU-based Monte Carlo dose calculation code for coupled electron-photon transport. *Physics in Medicine and Biology*, 2010. **55**(11): 3077–3086.
15. Hissoiny, S. et al., GPUMCD: A new GPU-oriented Monte Carlo dose calculation platform. *Medical Physics*, 2011. **38**(2): 754–764.
16. Jia, X. et al., GPU-based fast Monte Carlo simulation for radiotherapy dose calculation. *Physics in Medicine and Biology*, 2011. **56**(22): 7017–7031.
17. Gu, X.J., X. Jia, and S.B. Jiang, GPU-based fast gamma index calculation. *Physics in Medicine and Biology*, 2011. **56**(5): 1431–1441.
18. Graves, Y.J., X. Jia, and S.B. Jiang, Effect of statistical fluctuation in Monte Carlo based photon beam dose calculation on gamma index evaluation. *Physics in Medicine and Biology*, 2013. **58**(6): 1839–1853.
19. Boyer, A.L. and S. Li, Geometric analysis of light-field position of a multileaf collimator with curved ends. *Medical Physics*, 1997. **24**(5): 757–762.
20. Le, T., R. Chartrand, and T.J. Asaki, A variational approach to reconstructing images corrupted by poisson noise. *Journal of Mathematical Imaging and Vision*, 2007. **27**(3): 257–263.
21. Barzilai, J. and J.M. Borwein, 2-point step size gradient methods. *IMA Journal of Numerical Analysis*, 1988. **8**(1): 141–148.
22. Armijo, L., Minimizatio nof functions having Lipschitz continuous first partial derivatives. *Pacific Journal of Mathematics*, 1966. **16**(1): 1.
23. Hoberock, J. and N. Bell. An introduction to thrust. 2008. Available from: https://code.google.com/p/thrust/.
24. Ju, T. et al., Geometric interpretation of the gamma dose distribution comparison technique: Interpolation-free calculation. *Medical Physics*, 2008. **35**(3): 879–887.

Index

Printed and bound by CPI Group (UK) Ltd, Croydon, CR0 4YY

22/10/2024

01777613-0014